Articular Cartilage Injury of the Knee

Basic Science to Surgical Repair

Thieme

Articular Cartilage Injury of the Knee

Basic Science to Surgical Repair

James P. Stannard, MD
J. Vernon Luck Distinguished Professor and Chairman
Department of Orthopaedic Surgery
University of Missouri Hospital
Columbia, Missouri

James L. Cook, DVM, PhD
William and Kathryn Allen Distinguished Professor in Orthopedic Surgery
Comparative Orthopaedic Laboratory
University of Missouri
Columbia, Missouri

Jack Farr, MD
Director, OrthoIndy Sports Medicine Fellowship
Director, OrthoIndy Cartilage Restoration Center of Indiana
Indiana Orthopaedic Hospital
Voluntary Professor of Orthopedic Surgery, IUMC
Indianapolis, Indiana

Thieme
New York • Stuttgart

Thieme Medical Publishers, Inc.
333 Seventh Ave.
New York, NY 10001

Executive Editor: Kay Conerly
Managing Editor: Judith Tomat
Senior Vice President, Editorial and Electronic Product Development: Cornelia Schulze
Production Editor: Torsten Scheihagen
International Production Director: Andreas Schabert
Vice President, Finance and Accounts: Sarah Vanderbilt
President: Brian D. Scanlan
Compositor: Thomson Digital
Printer: Everbest Printing Co.

Library of Congress Cataloging-in-Publication Data

Stannard, James P.
Articular cartilage injury of the knee : basic science to surgical repair / James P. Stannard, James L. Cook, Jack Farr.
p. ; cm.
Includes bibliographical references and index.
ISBN 978-1-60406-858-0 (hardcover) – ISBN 978-1-60406-859-7 (e-ISBN)
I. Cook, James L. (James Lee), 1965- II. Farr, Jack. III. Title.
[DNLM: 1. Knee Injuries—surgery. 2. Cartilage, Articular—injuries. 3. Cartilage, Articular—surgery. 4. Knee Joint—surgery. 5. Tissue Engineering. WE 870]
RD560
617.5'8044–dc23
2013002506

The chapters in this book have been published previously in *The Journal of Knee Surgery*.

Printed in China

5 4 3 2 1

ISBN 978-1-60406-858-0

Also available as an e-book:
eISBN 978-1-60406-859-7

Dedicated to the pioneers of cartilage surgery and the scientists and surgeons who continue to “put the science behind the surgery” in finding better solutions for our patients. May the strong body of work that is reviewed in this book provide the foundation and inspiration for the future.

Contents

Preface

Articular cartilage health is critical to human movement. Because articular cartilage injury and degeneration are commonplace, effective repair and regeneration strategies have become an area of intense research with rapid advancement. Unfortunately, the basic science and clinical data and articles in this area are spread across many disciplines, sources, and journals. This book is our attempt to bring the world's experts together to provide a contemporary update on the topic of articular cartilage repair of the knee in a single resource.

Our first goal is to marry basic science with a comprehensive patient-based approach to diagnosis and treatment. It is impossible to appropriately choose current options or apply exciting new developments without a thorough understanding of the basic science of articular cartilage in health and disease. As such, we have engaged basic science experts who also have strong clinical backgrounds to bridge the divide that can limit effective clinical application of current and emerging treatment options.

The initial chapters of Articular Cartilage Injury of the Knee concentrate on staging and comorbidities, diagnostic imaging now and in the future, and the development and validation of biomarkers for the early diagnosis, staging, clinical decision making, and prognostication of patients with articular cartilage pathology. The subsequent chapters address the basic science and clinical aspects of marrow stimulation, autologous chondrocyte implantation (ACI) and new developments with ACI, particulated allograft cartilage therapy, and nontraditional modifications of articular cartilage.

The next section of the book concentrates on the topic of transplantation of bone cartilage composites with chapters on the bone cartilage interface, autografts, and fresh allografts. This section also includes a chapter on a novel osteochondral allograft preservation system that allows for superior maintenance of chondrocyte viability such that the window of time for implantation of grafts is more than doubled, and individual graft viability can be assessed prior to clinical use.

The final chapters concentrate on scaffolds for cartilage repair from the basic science level to clinical application. We also feature a chapter on the challenges and possibilities of achieving a biologic knee replacement for human patients. A chapter addresses biomechanical outcomes in cartilage replacement therapies while another concentrates on the clinical rehabilitation of these patients. Finally, we provide a chapter on assessment of outcomes after cartilage repair in the knee.

Treatment of articular cartilage pathology of the knee is challenging. Our goal with this text is to provide a comprehensive and up-to-date reference for surgeons and researchers working in this exciting and dynamic field. We hope you find it as useful and interesting as we have while editing the text.

James P. Stannard, MD
James L. Cook, DVM, PhD
Jack Farr, MD

About the cover images
Top: Intraoperative photo of medial femoral condylar articular cartilage lesion. *Bottom:* Photo of lesion treated with investigational cartilage restoration technique. (Courtesy Jack Farr, MD)

List of Contributors

Marco Acuna, MD
Department of Clinical Services
National Institute of Rehabilitation
Mexico City, Mexico

Michael B. Albro, PhD
Department of Mechanical Engineering
Columbia University
New York, New York

Karl F. Almqvist, MD, PhD
Department of Orthopaedic Surgery and Traumatology
Ghent University Hospital
Ghent, Belgium

Gerard A. Ateshian, PhD
Department of Mechanical Engineering
Columbia University
New York, New York

Liming Bian, PhD
Department of Mechanical and Automation Engineering
Biomedical Engineering Programme
The Chinese University of Hong Kong
Hong Kong

William Bugbee, MD
Division of Orthopaedic Surgery
Scripps Clinic
La Jolla, California

Marco Cavallo, MD
II Clinic of Orthopaedics and Traumatology
Rizzoli Orthopaedic Institute
Bologna, Italy

Nadeen O. Chahine, PhD
Department of Bioengineering
The Feinstein Institute for Medical Research
Manhasset, New York

Gaoping Chen, MD
Department of Chemical Engineering
École Polytechnique
Montreal, Quebec, Canada

Brian J. Cole, MD, MBA
Division of Sports Medicine
Department of Orthopedics
Rush University Medical Center
Chicago, Illinois

Cristi Reeves Cook, DVM, MS
Comparative Orthopaedic Laboratory
University of Missouri
Columbia, Missouri

James L. Cook, DVM, PhD
William and Kathryn Allen Distinguished Professor in Orthopedic Surgery
Comparative Orthopaedic Laboratory
University of Missouri
Columbia, Missouri

Aad A. M. Dhollander, MD, PT, PhD
Department of Orthopaedic Surgery and Traumatology
Laboratory of Connective Tissue Biology
Department of Rheumatology
Ghent University Hospital
Ghent, Belgium

Alessandro Di Martino, MD
Biomechanics Laboratory-III Orthopaedic Clinic
Rizzoli Orthopaedic Institute
Bologna, Italy

Richard Evans, PhD
Waukesha, Wisconsin

Jack Farr, MD
Director, OrthoIndy Sports Medicine Fellowship
Director, OrthoIndy Cartilage Restoration Center of Indiana
Indiana Orthopaedic Hospital
Voluntary Professor of Orthopedic Surgery, IUMC
Indianapolis, Indiana

Giuseppe Filardo, MD
Biomechanics Laboratory-III Orthopaedic Clinic
Rizzoli Orthopaedic Institute
Bologna, Italy

Lisa A. Fortier, DVM, PhD
Department of Clinical Science
Cornell University
Ithaca, New York

Bridget C. Garner, DVM, PhD
Department of Veterinary Pathology
University of Georgia
Athens, Georgia

Joeseph T. Garrity, MS
Comparative Orthopaedic Laboratory
University of Missouri Columbia
Columbia, Missouri

Neil Ghodadra, MD
Division of Sports Medicine
Department of Orthopedics
Southern California Orthopedic Institute
Van Nuys, California

Sandro Giannini, MD
II Clinic of Orthopaedics and Traumatology
Rizzoli Orthopaedic Institute
Bologna, Italy

Andreas H. Gomoll, MD
Harvard Medical School
Cartilage Repair Center
Brigham and Women's Hospital
Boston, Massachusetts

Victor R. Guevara Sánchez, MD
Ortopediay Tramatologia
Instituto Nacional de Rehabilitacion
Mexico City, Mexico

Jessica Guzmán-Morales
Department of Chemical Engineering
École Polytechnique
Montreal, Quebec, Canada

Joshua D. Harris, MD
Department of Orthopaedics
The Ohio State University
Columbus, Ohio

Catherine Hayter, MBBS
Department of Radiology and Imaging
New York, New York

Timothy E. Hewett, PhD, FACSM
Sports Health and Performance Institute
Columbus, Ohio

Caroline D. Hoemann, PhD
Department of Chemical Engineering
Institute of Biomedical Engineering
Groupe de Recherche en Sciences et Technologies Biomedicales
École Polytechnique
Montreal, Quebec, Canada

Clark T. Hung, MSE, PhD
Department of Biomedical Engineering
Columbia University
New York, New York

Vasili Karas, BS
Division of Sports Medicine
Department of Orthopedics
Rush University Medical Center
Chicago, Illinois

Elizaveta Kon, MD
Biomechanics Laboratory-III Orthopaedic Clinic
Rizzoli Orthopaedic Institute
Bologna, Italy

Ellen Kroin, BS
Division of Sports Medicine
Department of Orthopedics
Rush University Medical Center
Chicago, Illinois

Keiichi Kuroki, DVM, PhD
Comparative Orthopaedic Laboratory
University of Missouri
Columbia, Missouri

Charles-Hubert Lafantaisie-Favreau, MSc
Institute of Biomedical Engineering
École Polytechnique
Montreal, Quebec, Canada

Viorica Lascau-Coman, MSc
Department of Chemical Engineering
Groupe de Recherche en Sciences et Technologies Biomedicales
École Polytechnique
Montreal, Quebec, Canada

Christian Lattermann, MD
Department of Orthopaedic Surgery and Sports Medicine
University of Kentucky
Lexington, Kentucky

Eric G. Lima, PhD
Department of Mechanical Engineering
Columbia University
New York, New York

Matthew R. Luckett, MD
Department of Orthopaedic Surgery and Sports Medicine
University of Kentucky
Lexington, Kentucky

Maurilio Marcacci, MD
Biomechanics Laboratory-III Orthopaedic Clinic
Rizzoli Orthopaedic Institute
Bologna, Italy

C. Wayne McIlwraith, DVM, PhD
Orthopaedic Research Center
Department of Clinical Sciences
Colorado State University
Fort Collins, Colorado

Brett McCoy, MD
Department of Orthopaedic Surgery
The Cleveland Clinic Foundation
Garfield Heights, Ohio

Anthony Miniaci, MD
Department of Orthopaedic Surgery
The Cleveland Clinic Foundation
Garfield Heights, Ohio

Kai Mithoefer, MD
Harvard Vanguard Orthopedics and Sports Medicine
Boston, Massachusetts

Grace D. O'Connell, PhD
Department of Biomedical Engineering
Columbia University
New York, New York

Debabrata Patra, PhD
Department of Orthopaedic Surgery
Washington University School of Medicine at Barnes-Jewish Hospital
St. Louis, Missouri

Hollis Potter, MD
Department of Magnetic Resonance Imaging
Hospital for Special Surgery
New York, New York

Carmen E. Quatman, MD
Sports Health and Performance Institute
Department of Orthopaedics
The Ohio State University
Columbus, Ohio

Linda J. Sandell, PhD
Department of Cell Biology and Physiology
Washington University School of Medicine at Barnes-Jewish Hospital
St. Louis, Missouri

Robert Schaal, PT, BSPT
Department of Rehabilitation
Missouri Orthopedic Institute
Columbia, Missouri

Seth Sherman, MD
Division of Sports Medicine
Department of Orthopedics
Rush University Medical Center
Chicago, Illinois

James P. Stannard, MD
J. Vernon Luck Distinguished Professor and Chairman
Department of Orthopaedic Surgery
University of Missouri Hospital
Columbia, Missouri

Aaron M. Stoker, MS, PhD
Comparative Orthopaedic Laboratory
Department of Veterinary Medicine and Surgery
University of Missouri–Columbia
Columbia, Missouri

Jennifer Yasu Stone, PT, DPT, OCS
Department of Rehabilitation
Missouri Orthopedic Institute
Columbia, Missouri

Gust Verbruggen, MD, PhD
Laboratory of Connective Tissue Biology
Department of Rheumatology
Ghent University Hospital
Ghent, Belgium

René Verdonk, MD, PhD
Department of Orthopaedic Surgery and Traumatology
Ghent University Hospital
Ghent, Belgium

Peter C. M. Verdonk, MD, PhD
Department of Orthopaedic Surgery
Stedelijk Ziekkenhuis Roeselare
Roeselare, Belgium
Department of Orthopaedic Surgery and Traumatology
Ghent University Hospital
Ghent, Belgium

I

Diagnosis and Treatment Planning

1

Staging and Comorbidities

Christian Lattermann and Matthew R. Luckett

Articular cartilage injuries are common.[1–3] The spectrum of these injuries ranges from small, superficial defects (focal chondral defects) to complete degenerative delamination of entire condyles with or without involvement of the subchondral bone and adjacent structures (osteoarthritis). In an ideal world, focal chondral defects exist in isolation, have clearly defined borders, are solitary defects, and are located in ideally accessible anatomic locations in young patients that are physically active. These types of lesions are the standard that is currently being used to enroll patients into randomized clinical trials investigating the efficiency of articular cartilage procedures. Whereas these studies are important and necessary to compare different techniques, the reality is that most patients (95%) that are presenting with clinically symptomatic cartilage lesions do not fit these clear-cut criteria.[4] This presents a dilemma to the surgeon as the cartilage lesions most commonly treated are usually less clear cut and often involve "best clinical judgment" to perform an adequate assessment. This assessment process, or "staging," is necessary to guide both patient and physician toward a clinically feasible and satisfying solution for the knee cartilage injury patient. The staging process requires knowledge about frequency and prevalence of cartilage defects, their clinical symptoms, arthroscopic grading, and sizing as well as assessment of the joint environment. Furthermore, specific comorbidities have to be taken into account prior to performing cartilage repair procedures as many of them require additional staged or concomitant surgical procedures. In this chapter, we will sequentially discuss the most pertinent factors that influence the decision-making process in patients with symptomatic cartilage lesions of the knee.

◆ Frequency and Prevalence of Cartilage Injuries

Damage to articular cartilage is common and can result from acute traumatic injuries, early posttraumatic degenerative changes, developmental factors affecting the subchondral bone such as osteochondritis dissecans (OCD) lesions, or acquired metabolic factors such as avascular necrosis (AVN).[1–3]

Articular cartilage lesions are frequently encountered in routine knee arthroscopies. Curl et al reported articular cartilage lesions in as many as 63% of over 35,000 knee arthroscopies in the United States.[3] This high incidence was corroborated by Hjelle et al in Norway and Widuchowski et al in Poland, who reported an incidence of 61 and 60%, respectively.[1,2] The average age of patients reported in these studies is high, and thus the percentage of treatable lesions in younger patients is likely much lower. In fact, upon further subanalysis of Curl's data, 60% of the reported lesions were grade III lesions and thus were potentially treatable lesions. Only 1,750 patients out of 31,516 were under the age of 40 and had Outerbridge grade III lesions. Based upon this study, one can estimate that ~ 5% of patients under 40 undergoing knee arthroscopies may present with a chondral lesion that would be considered optimal for current therapies. While these studies provide some data regarding prevalence of these types of lesions among patients, no information is available regarding how many of the lesions are clinically symptomatic. Interestingly, the mere presence of a lesion does not seem to lead to an increase in the osteoarthritic rate over time in large cross-sectional studies, as the long-term natural history study conducted by Widuchowski et al in 2010 suggests.[5] Shelbourne et al found that 123 out of 2,700 patients with anterior cruciate ligament (ACL) injuries and cartilage lesions at the time of surgery showed lower subjective Noyes scores 8 years after ACL reconstruction compared with the patients who did not have cartilage lesions at the time of surgery.[6] Another study suggests that the presence of cartilage lesions can lead to rapid progression of radiographic osteoarthritis (OA), as documented by Messner and Maletius.[7] These findings underline the importance of identifying the patient who has a clinically symptomatic cartilage lesion that may benefit from early treatment.

◆ Lesion Location and Size

The location of cartilage lesions is spread between the three compartments of the knee. Lesions are most commonly found in the weight-bearing femoral condyle (43 to 58%). Patellar lesions are frequently encountered and account for 11 to 36% of all lesions. Trochlear lesions overall are less frequent (6 to 16%).[1–3]

When analyzed for the lesion size, Hjelle et al were able to show that the majority of lesions (88%) were below 4 cm.[1,2]

Widuchowski et al found that 60% of knees (average 39 years old) contained chondral/osteochondral lesions, 68% of which were focal chondral lesions, 3% being OCD lesions and 29% being osteoarthritic lesions.[8]

◆ History and Physical Examination

The clinical evaluation of patients with symptomatic cartilage lesions in the knee is difficult and follows the recommendations of a thorough history and physical exam of the knee joint. No true evidence-based approach is available to guide the clinician, but several factors that may be important should be pointed out.

Upon initial evaluation, it is important to discover the history of symptoms that may be related to a cartilage lesion. Duration of symptoms has been associated with clinical outcome in patients undergoing microfracture. Mithoefer et al could show that patients with symptoms longer than 1 year had lower overall subjective outcome results than patients with more acute cartilage injuries.[9] There is a correlation of worse overall clinical outcomes after cartilage procedures in patients who receive workmen's compensation.[10,11] History of smoking and family history of OA are often considered negative predictive factors for cartilage repair procedures; however, no clear evidence exists to actually link those two isolated factors to clinical outcomes.

History should include the documentation of the body mass index (BMI). Whereas a BMI up to 35 does not seem to affect the overall outcomes in patients undergoing cell-based cartilage procedures,[12,13] a higher BMI clearly affects the results of patients undergoing microfracture treatment.[14] Similar consideration needs to be given to the age factor. Several studies have shown that higher age influences clinical outcome negatively in patients undergoing microfracture procedures.[14,15] The data for cell-based procedures are somewhat conflicting. A clear correlation between age and clinical outcome has not been shown. Basic science studies, however, suggest that chondrocytes from older donors (> 40 years of age) have a lower proteoglycan and collagen production and thus may respond more slowly and less vigorously to the challenging intra-articular environment after implantation.[16] A little-researched topic that is of importance is the willingness to comply with postoperative treatment protocols and rehabilitation procedures. Current protocols are not based upon evidence but rather on anecdotal experience or small case series by individual surgeons and rehabilitation specialists.[17–19] Nevertheless, it is felt that adherence to these basic protocols is important. A history of noncompliance may therefore be a warning sign to the cartilage surgeon potentially indicating the patient's lack of understanding or a significant difference in the goals that the treatment is aiming to achieve.

Pain

Pain assessment is an important part of the preoperative exam. Localized pain may be able to pinpoint a specific area of articular cartilage damage or it may indicate injury to associated structures such as the meniscus. The shorter the history of pain, the more reliably it can be considered to indicate the affected area.

No reliable data exist about the correlation of pain with a symptomatic cartilage lesion. However, the more chronic in nature the pain is, the less likely it is that a cartilage procedure alone is going to address the problem.

Most commonly utilized are visual analog scales (VAS) or a Likert scale for pain.

In absence of any clear evidence-based guidelines regarding pain, there are some pearls of wisdom that may help the less-experienced cartilage surgeon. The ideal patient should not report maximal pain other than perhaps with heavy exertion. Likewise, patients with minimal or no pain are less likely to benefit from cartilage surgery. Typically, the patient reporting pain in the midrange is considered an acceptable patient for treatment. It is also important to assess pain with and without medication (particularly narcotic pain medication) in this context.

Physical Examination

The physical exam should evaluate the overall dynamic and static alignment, antalgic gait, range of motion, muscle envelope, as well as ligamentous stability of the tibiofemoral and patellofemoral joint.

A crude visual gait analysis in the office usually allows for detection of an antalgic gait, quadriceps avoidance gait, or a dynamic varus or valgus thrust. Any of these findings, if present, can point the examiner toward further underlying pathologies that may have a significant impact on the chosen treatment options. A varus thrust, for example, may point out an insufficiency of the lateral ligamentous structures (posterolateral corner, lateral collateral ligament [LCL]) and a triple varus. A quadriceps avoidance gait may indicate chronic anterior instability.

Knee joint effusions are generally felt to be a significant sign for symptomatic cartilage injuries. It is important to understand, however, that intra-articular effusions can exist without pain and therefore can be present longer than the actual onset of pain.

Range of motion assessment should be a routine part of the physical examination and has to be assessed in comparison with the uninjured side. Although small deficits in knee flexion can be observed with knee joint effusions, they are not normal in patients who have no effusion. An extension deficit is an important finding as these are very difficult to correct and may indicate progression to OA already beyond the scope of cartilage repair. Significant loss of motion is considered a relative contraindication for cartilage repair procedures.

Mechanical symptoms, locking during the range of motion exam, or acute inability to flex or extend the knee joint may indicate an unstable meniscus or articular cartilage fragment or a loose body.

A clinical sign that utilizes this concept is the Wilson sign. This test was originally performed to diagnose OCD lesions in the medial femoral condyle. The knee is flexed to 90 degrees. The tibia is forced into internal rotation. Under gradual extension and external rotation of the tibia, the patient may report pain when the lesion rotates into the area of the soft spot of the medial femoral condyle.[20] This test can be modified by pushing the thumb slightly into the soft spot. Another helpful test is the direct palpation of the medial and lateral patella facette. If palpation is reproducing the patient's pain, this can be a sign for a clinically symptomatic lesion in this area and will need to be correlated with the imaging results. Cartilage lesions do not typically hurt directly at the joint line. Direct palpation at the joint line is more likely associated with meniscal pathology.

Ligamentous stability is a prerequisite for cartilage procedures. It is therefore necessary to perform a full ligament examination of the knee joint. This usually includes varus and valgus stress at 0 and 30 to test the collateral ligaments; the Lachman test; the pivot shift exam, which evaluates ACL competency; the posterior drawer test at 90 degrees of knee flexion; and the posterior sag sign, which evaluates posterior cruciate ligament (PCL) sufficiency. In case of a potential posterolateral corner injury, the dial test and the flexion rotation drawer can be performed. Often forgotten is the stability exam of the patella. The medial and lateral patella glide and tilt as well as the competency of the medial patellofemoral ligament (MPFL) and the lateral retinaculum should be assessed. The patellar apprehension test is helpful to rule out the previous patellar sub/dislocation. Q-angle and patellar stability throughout the flexion should be carefully evaluated.

◆ The "Character" of the Lesion

To assess the actual severity of a lesion, arthroscopic evaluation is imperative. The grading of the severity can be done using several different classification systems. The International Cartilage Research Society (ICRS) has developed a universally accepted and comprehensive grading system that should be utilized to allow for the generalization of arthroscopic findings (**Fig. 1.1**).

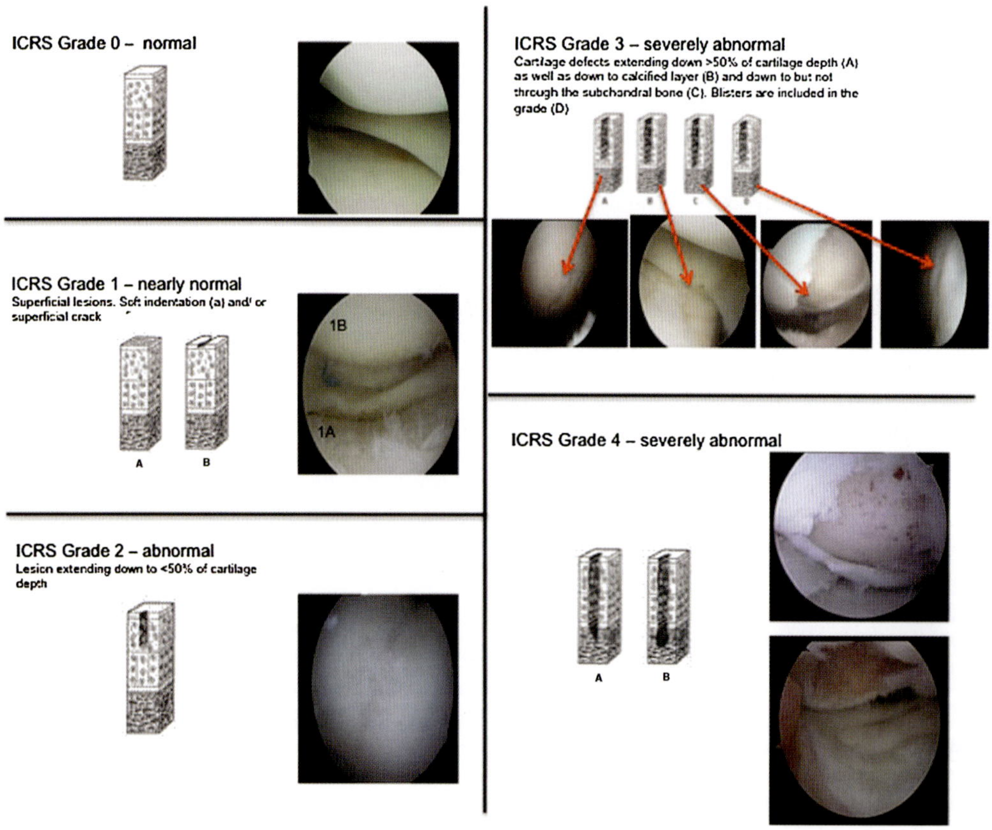

Fig. 1.1 ICRS grading scheme for cartilage defects. ICRS, International Cartilage Research Society.

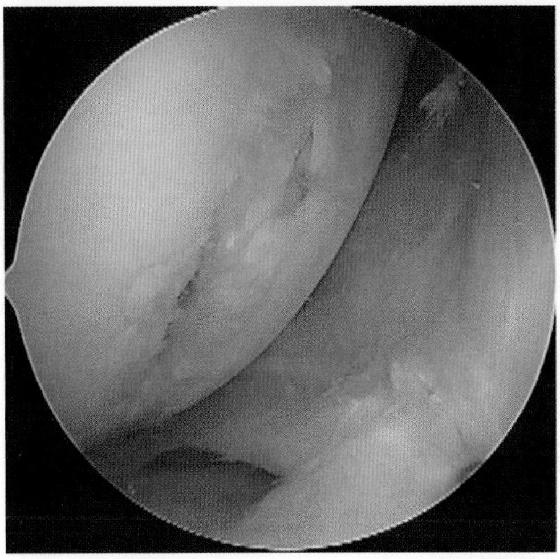

Fig. 1.2 This is a Grade 4b lesion in a medial femoral condyle after direct trauma. This patient was involved in a motor vehichle accident 3 months prior to this image and had a penetrating trauma to the knee.

To restore articular cartilage, it is important to understand the reason for the initial failure of the cartilage surface to maintain its integrity. In a few cases, this can be associated with an acute injury (**Fig. 1.2**). In many cases, however, the underlying reason is more subtle. Even more importantly, it is imperative to assess the true extent of the chondral lesion. Diagnostic imaging has made incredible advances over the last decade and is invaluable to characterize the lesion and its surroundings better. Although it is not the focus of this chapter (see chapter 2), it needs to be understood that imaging provides information about the articular cartilage as well as the subchondral bone, the synovial envelope, and the ligamentous structures of the knee joint. All of those need to be assessed to create an overall picture or "character" of a knee joint. An invaluable tool to help synthesize all of the abovementioned aspects of information about the patient's knee is the arthroscopic evaluation of the knee. For some procedures that allow for immediate point-of-care intervention, such as the microfracture or the cartilage autologous implant system (CAIS—investigational and not currently available in the United States), this evaluation will be followed by an immediate final treatment approach. In other cases, it will be necessary to perform either a biopsy with or without a minor procedure such as a chondroplasty, a partial meniscectomy, or a removal of a loose body. The arthroscopy offers the unique opportunity to assess and verify the location, grading, and actual size of the lesion. Additionally, it allows for assessment of the entirety of a compartment, including the status of the articular cartilage surfaces of the tibial and femoral condyles surrounding a full-thickness lesion, as well as the status of the meniscus, which often has been treated in a prior procedure (the majority of patients undergoing cartilage repair procedures have had more than one previous surgical procedure[10,12]). Particularly, globalized findings such as compartment-wide grade 1 or 2 changes (ICRS) can elude radiographic assessment but may indicate a more generalized chondropenia in the affected compartment. Development of osteophytes along the medial or lateral condylar rim is another sign for more generalized changes in the knee that can easily be missed in X-ray and MRI examination but may be a factor to be taken into account for the assessment of the future success of a cartilage procedure. This arthroscopic evaluation may also help to advise the patient regarding the return to higher-level activities postsurgically. **Figure 1.3** is an example of an isolated focal chondral defect in an otherwise pristine knee joint (**Fig. 1.3a**). This is contrasted with an example of an isolated lesion in a knee joint displaying grade 2 changes throughout the entire compartment (**Fig. 1.3b**) indicating beginning chondropenia.

As a final pearl regarding the arthroscopic examination, it should be noted that a video documentation of the lesion and the involved compartment says more than an isolated picture. In today's world, video documentation is easy, and it facilitates communication with colleagues and greatly improves the surgeon's recall of the character of a lesion in case of a likely time delay between the initial arthroscopy and the final restorative procedure. Another excellent alternative to improve communication is to combine images of defects taken with an intra-articular ruler and combine this with a map indicating the size and location of the lesions.

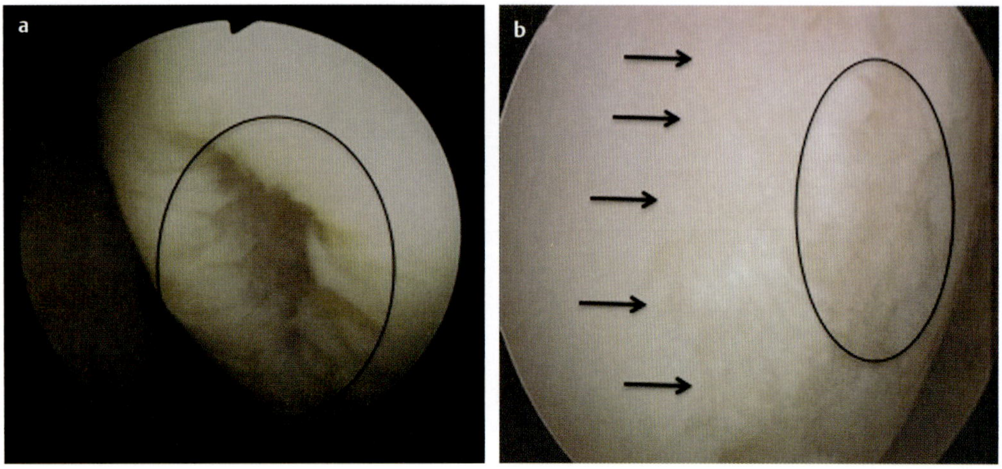

Fig. 1.3 (**a**) An isolated Grade 3b defect in an otherwise pristine-appearing knee joint. This patient went on to receive a microfracture and did well. (**b**) A similar-size Grade 3b lesion (indicated with the circle) surrounded by areas of Grade 2 lesions. This patient failed an initial microfracture and went on to receive an autologous chondrocytes implantation involving the majority of her condyle (2.2 × 4.8 cm). Even though this is obvious on the video of this lesion, it is difficult to document this significant difference in the character of this lesion in pictures.

◆ Comorbidities

Prior to considering a cartilage repair procedure, it is essential to perform a thorough analysis of comorbidities that potentially influence the success of the procedure or may even be contraindications.

Absolute contraindications for cartilage repair procedures are the documented presence of inflammatory arthritis (i.e., psoriatic, gouty, and rheumatoid) or established compartmental OA with radiographic changes indicating joint space collapse (Kellgren Lawrence III–IV) or malignancy in the involved limb. Uncorrected axial malalignment is an absolute contraindication for tibiofemoral cartilage repair procedures, as is chronic uncorrected ligamentous instability. The same holds true for the patellofemoral joint. Malalignment or instability in the patellofemoral joint is considered a contraindication if it remains uncorrected; however, most cartilage surgeons will address obvious patellar malalignment and instability in face of a cartilage repair in the patellofemoral joint.[21] Significant loss of range of motion or arthrofibrosis is also considered to be an absolute contraindication.

Consensus exists that in the younger patient the potential for a successful outcome is higher. For this reason, most surgeons will consider the age of 50 a cutoff point for cell-based procedures or allografts; however, some autologous procedures may be performed in patients up to the age of 60.[22–24] It needs to be understood that the biological age of the patient plays a larger role than the chronologic age. This may account for the relatively soft recommendation of the age cutoff for these procedures.[25–27] Malalignment, meniscus deficiency, or ligamentous instability, even though they represent contraindications to a cartilage repair procedure, can be overcome by either a staged or a simultaneous operation to correct the condition.

◆ Axial Malalignment

Varus or valgus malalignment of the knee is the major contributing factor to compartment overload and thus has to be addressed when a cartilage repair procedure is considered to address a cartilage defect in the overloaded compartment.[25–27] When addressing a cartilage defect surgically, the goal is to restore the normal load distribution that allows the repair cartilage to adjust to physiologic rather than nonphysiologic loads. The goal for axial alignment correction in cartilage repair procedures is therefore not an overcorrection, as popularized by Coventry[28] and others, but rather to correct back to neutral alignment. It is imperative that the origin of the malalignment be identified. Generally, varus alignment originates in the proximal tibia and valgus alignment in the distal femur. However, in some cases this may be different. It is therefore prudent to do a full axial alignment measurement of the tibia and femur rather than just the overall mechanical axis evaluation on the long leg alignment full cassette X-ray. With today's hardware options, low-profile plates can be utilized to perform well-controlled open wedge high tibial or distal femoral osteotomies to address varus or valgus alignment up to 10 degrees. Malalignment correction above 10 degrees may require additional bone grafting or alternate techniques.

◆ Patellofemoral Malalignment

Cartilage injuries in the patellofemoral joint are amongst the most difficult to treat. Technically these lesions are easily accessible, but the analysis of concomitant pathologies is difficult. This fact explains the initial disappointing results that Brittberg et al reported. They saw five out of seven patients undergoing autologous chondrocyte implantation (ACI) of the patellofemoral joint fail.[29] The authors recognized the importance of patellofemoral alignment and tracking at a later time point and advocated the combination of the ACI procedure with concomitant, or staged, unloading and normalization of the patellar tracking in the PF joint. As of 2011, cell-based cartilage procedures in patients with PF malalignment were routinely combined with an anteromedialization (AMZ) of the tibial tubercle.[30]

Since, the clinical experience has been promising. Brittberg et al reported 11 of 17 patients with good and excellent results at 2 years and slightly better results (13/19) at 9 years, indicating a long initial postoperative recovery time with improvement over 1 year, postoperatively.[27] In Minas and Bryant's study of 45 patients, the authors performed an AMZ in over 60% and reported 71% good and excellent results.[31] Henderson and Lavigne reported their results in a group of patients that was divided into ACI (patients with normal PF alignment) and ACI with AMZ (patients with clinically present PF malalignment).[32] Interestingly, the group that did not receive the AMZ because they did not have patellofemoral malalignment did worse than the group with patellofemoral malalignment requiring an AMZ. This study suggested that there is either an additional effect of the anterior unloading of the patellofemoral joint or perhaps some subtle patellofemoral malalignment that was not detected as this study was published prior to the establishment of the tibial tubercle–trochlear groove (TT-TG) measurements that are used today to determine patellofemoral alignment.[33] The potential to unload the patellofemoral joint by doing an anteriorization of the tibial tubercle by less than 1 cm has been shown by Rue et al, who concluded that the patellofemoral contact pressures measured by Tekscan can be reduced by 20%.[34] Overall, cartilage procedures in the patellofemoral joint can be considered a valuable treatment option as long as an adequate evaluation and concomitant treatment of an underlying PF malalignment are performed.

◆ Meniscal Deficiency

The menisci are critical for load sharing and shock absorption. They act as a transmission within the knee linking the femoral condyle with the tibia. They also contribute to joint lubrication and knee stability. Particularly, the medial meniscus has been shown to be the most important secondary stabilizer against anterior translation of the knee.[35] This critical role is commonly impaired as meniscus injuries are the most common knee injury requiring arthroscopic surgery in the United States. Biochemical, biomechanical, as well as clinical, radiographic, and patient-related outcomes data have clearly established a direct relation of loss of meniscus tissue to impairment of all these parameters.[36] Impressive data have been published by Baratz et al, who showed an increase in contact pressures of 75% and an overall increase of 235% in peak-contact pressures after subtotal meniscectomy.[37,38] Lee et al showed that the periphery of the meniscus is more important for the overall pressure distribution in the compartment than the central portion.[39] These data are encouraging and may indicate that patients after partial meniscectomy still have a nearly normal pressure distribution in the joint. An isolated partial meniscectomy therefore may not pose a significant short-term risk for a cartilage repair procedure. However, long-term data exist linking partial meniscectomies to the development of OA over a 15-year time span. These data are even more compelling in conjunction with a ligamentous instability.[40]

Patients who have undergone a subtotal or complete meniscectomy or have suffered a nonrepairable radial tear have pathologic pressure distribution that is detrimental to the weight-bearing articular cartilage and any repair tissue. In these cases, a meniscus transplant may need to be considered. Although the indications for meniscal transplant are still evolving, they are generally considered in patients who are young, have unicompartmental pain, a history of previous meniscectomy, normal ligamentous stability, and normal or correctable alignment. Gomoll et al have published their series of seven patients undergoing cartilage restoration, high tibial osteotomy, and meniscus transplantation. They reported encouraging results in this small series with significant improvement of the International Knee Documentation Committee subjective score, Knee Injury and Osteoarthritis Outcome Score, and Lysholm score after 24 months (average) follow-up.[41] As these patients are a very challenging group, they can achieve significant improvement if all three major factors (axial alignment, focal chondral defect, and meniscal deficiency) are addressed adequately.

◆ Ligamentous Instability

A knee ligament insufficiency such as an ACL insufficiency has been clearly linked to an increased risk of OA over time.[40] Articular cartilage lesions in ACL-injured patients are not uncommon. Not all of these lesions are acute and clinically symptomatic[42]; however, ACL instability will over time contribute to a significant increase in the size of the cartilage lesion, as Murrell could show.[43] He evaluated patients for 2 months and 2 years after ACL tear prior to stabilization and found a six times larger loss of cartilage in patients with longer-standing ACL insufficiency. In patients who had a combination of ACL injury and meniscal tear, this rate increased to 18 times after 2 years. It has been shown that knee ligament stability is important to preserve meniscal integrity. Particularly, the interaction of the medial meniscus and the ACL is important as the lack of the medial meniscus may lead to early failure of the ACL graft due to the meniscus's function as a secondary restraint to anterior tibial translation.[35] In patients with chronic ACL instability and pain, it is important to evaluate the primary factor—pain or instability. Patients who have only instability-related pain episodes may be served well with a correction of the instability alone. Patients who have pain only may benefit from an osteotomy. Lattermann and Jakob showed in a retrospective study that ACL-insufficient patients with varus alignment who predominantly have pain but no instability may significantly improve after high tibial osteotomy and may not require any other procedure. In these cases, a staged approach may be beneficial.[42]

◆ Conclusion

The careful evaluation of patients undergoing cartilage repair procedures is of foremost importance because these patients generally require very individualized care. Thorough examination and judgment of comorbidities and their impact on the cartilage procedure are imperative. Unfortunately, there are no evidence-based guidelines or clear-cut recommendations for the majority of the patients that are encountered in the practice setting. However, with careful clinical decision making, evaluation of malalignment and other comorbidities, and careful staging of the lesion during arthroscopy the cartilage surgeon can make good choices that will lead to good clinical outcomes as reported in the literature. It is important to communicate the complexity of the decision-making process to the patient and make the patient aware that the proposed treatment is not a "routine" straightforward, standardized procedure.

References

1. Hjelle K, Solheim E, Strand T, Muri R, Brittberg M. Articular cartilage defects in 1,000 knee arthroscopies. Arthroscopy 2002;18(7):730–734
2. Widuchowski W, Lukasik P, Kwiatkowski G, et al. Isolated full thickness chondral injuries. Prevalance and outcome of treatment. A retrospective study of 5233 knee arthroscopies. Acta Chir Orthop Traumatol Cech 2008;75(5):382–386
3. Curl WW, Krome J, Gordon ES, Rushing J, Smith BP, Poehling GG. Cartilage injuries: a review of 31,516 knee arthroscopies. Arthroscopy 1997; 13(4):456–460
4. Engen CNE, Aroen A. Knee cartilage defect patients enrolled in randomized controlled trials are not representative of patients in orthopaedic practice. Cartilage 2010;1(4):312–319
5. Widuchowski W, Widuchowski J, Faltus R, et al. Long-term clinical and radiological assessment of untreated severe cartilage damage in the knee: a natural history study. Scand J Med Sci Sports 2011;21(1):106–110
6. Shelbourne KD, Jari S, Gray T. Outcome of untreated traumatic articular cartilage defects of the knee: a natural history study. J Bone Joint Surg Am 2003;85-A(Suppl 2):8–16
7. Messner K, Maletius W. The long-term prognosis for severe damage to weight-bearing cartilage in the knee: a 14-year clinical and radiographic follow-up in 28 young athletes. Acta Orthop Scand 1996;67(2):165–168
8. Widuchowski W, Widuchowski J, Trzaska T. Articular cartilage defects: study of 25,124 knee arthroscopies. Knee 2007;14(3):177–182
9. Mithoefer K, Williams RJ III, Warren RF, Wickiewicz TL, Marx RG. High-impact athletics after knee articular cartilage repair: a prospective evaluation of the microfracture technique. Am J Sports Med 2006;34(9):1413–1418
10. McNickle AG, Provencher MT, Cole BJ. Overview of existing cartilage repair technology. Sports Med Arthrosc 2008;16(4):196–201
11. Minas T, Gomoll AH, Solhpour S, Rosenberger R, Probst C, Bryant T. Autologous chondrocyte implantation for joint preservation in patients with early osteoarthritis. Clin Orthop Relat Res 2010;468(1):147–157
12. Zaslav K, Cole B, Brewster R, et al; STAR Study Principal Investigators. A prospective study of autologous chondrocyte implantation in patients with failed prior treatment for articular cartilage defect of the knee: results of the Study of the Treatment of Articular Repair (STAR) clinical trial. Am J Sports Med 2009;37(1):42–55
13. Rue JP, Yanke AB, Busam ML, McNickle AG, Cole BJ. Prospective evaluation of concurrent meniscus transplantation and articular cartilage repair: minimum 2-year follow-up. Am J Sports Med 2008;36(9):1770–1778
14. Mithoefer K, Williams RJ III, Warren RF, et al. The microfracture technique for the treatment of articular cartilage lesions in the knee. A prospective cohort study. J Bone Joint Surg Am 2005;87(9): 1911–1920
15. Mithoefer K, McAdams T, Williams RJ, Kreuz PC, Mandelbaum BR. Clinical efficacy of the microfracture technique for articular cartilage repair in the knee: an evidence-based systematic analysis. Am J Sports Med 2009;37(10):2053–2063
16. Adkisson HD IV, Martin JA, Amendola RL, et al. The potential of human allogeneic juvenile chondrocytes for restoration of articular cartilage. Am J Sports Med 2010;38(7):1324–1333
17. Hambly K, Bobic V, Wondrasch B, Van Assche D, Marlovits S. Autologous chondrocyte implantation postoperative care and rehabilitation: science and practice. Am J Sports Med 2006;34(6):1020–1038
18. Ebert JR, Robertson WB, Lloyd DG, Zheng MH, Wood DJ, Ackland T. Traditional vs accelerated approaches to post-operative rehabilitation following matrix-induced autologous chondrocyte implantation (MACI): comparison of clinical, biomechanical and radiographic outcomes. Osteoarthritis Cartilage 2008;16(10):1131–1140
19. Hirschmüller A, Baur H, Braun S, Kreuz PC, Südkamp NP, Niemeyer P. Rehabilitation after autologous chondrocyte implantation for isolated cartilage defects of the knee. Am J Sports Med 2011; Epub ahead of print
20. Wilson JN. A diagnostic sign in osteochondritis dissecans of the knee. J Bone Joint Surg Am 1967; 49(3):477–480
21. Farr J. Autologous chondrocyte implantation improves patellofemoral cartilage treatment outcomes. Clin Orthop Relat Res 2007;463:187–194
22. Hangody L, Kish G, Kárpáti Z, Udvarhelyi I, Szigeti I, Bély M. Mosaicplasty for the treatment of articular cartilage defects: application in clinical practice. Orthopedics 1998;21(7):751–756
23. Steadman JR, Ramappa AJ, Maxwell RB, Briggs KK. An arthroscopic treatment regimen for osteoarthritis of the knee. Arthroscopy 2007;23(9):948–955
24. Braun S, Steadman JR, Rodkey WG, Briggs KK. [Microfracture and specific rehabilitation for treating osteoarthritis of the knee. Indications, surgical technique, and rehabilitation protocol.] Z Rheumatol 2009;68(10):811–818
25. Ghodadra NB, Verma NN. Biologic joint reconstruction: alternatives to arthroplasty. In: Cole BGA ed. 1st ed. Thorofare, NJ: Slack Inc; 2009:13–22
26. Lindahl A, Brittberg M, Peterson L. Cartilage repair with chondrocytes: clinical and cellular aspects. Novartis Found Symp 2003;249:175–186, discussion 186–189, 234–238, 239–241
27. Brittberg M, Tallheden T, Sjögren-Jansson B, Lindahl A, Peterson L. Autologous chondrocytes used for articular cartilage repair: an update. Clin Orthop Relat Res 2001;(391, Suppl):S337–S348
28. Coventry MB. Proximal tibial varus osteotomy for osteoarthritis of the lateral compartment of the knee. J Bone Joint Surg Am 1987;69(1):32–38
29. Brittberg M, Lindahl A, Nilsson A, Ohlsson C, Isaksson O, Peterson L. Treatment of deep cartilage defects in the knee with autologous chondrocyte transplantation. N Engl J Med 1994;331(14):889–895
30. Gomoll AH, Minas T, Farr J, Cole BJ. Treatment of chondral defects in the patellofemoral joint. J Knee Surg 2006;19(4):285–295
31. Minas T, Bryant T. The role of autologous chondrocyte implantation in the patellofemoral joint. Clin Orthop Relat Res 2005;(436):30–39

32. Henderson IJ, Lavigne P. Periosteal autologous chondrocyte implantation for patellar chondral defect in patients with normal and abnormal patellar tracking. Knee 2006;13(4):274–279

33. Schoettle PB, Zanetti M, Seifert B, Pfirrmann CW, Fucentese SF, Romero J. The tibial tuberosity-trochlear groove distance; a comparative study between CT and MRI scanning. Knee 2006;13(1): 26–31

34. Rue JP, Colton A, Zare SM, et al. Trochlear contact pressures after straight anteriorization of the tibial tuberosity. Am J Sports Med 2008;36(10):1953–1959

35. Allen CR, Wong EK, Livesay GA, Sakane M, Fu FH, Woo SL. Importance of the medial meniscus in the anterior cruciate ligament-deficient knee. J Orthop Res 2000;18(1):109–115

36. Englund M, Guermazi A, Lohmander LS. The meniscus in knee osteoarthritis. Rheum Dis Clin North Am 2009;35(3):579–590

37. Baratz ME, Rehak DC, Fu FH, Rudert MJ. Peripheral tears of the meniscus. The effect of open versus arthroscopic repair on intraarticular contact stresses in the human knee. Am J Sports Med 1988;16(1):1–6

38. Baratz ME, Fu FH, Mengato R. Meniscal tears: the effect of meniscectomy and of repair on intraarticular contact areas and stress in the human knee. A preliminary report. Am J Sports Med 1986; 14(4):270–275

39. Lee SJ, Aadalen KJ, Malaviya P, et al. Tibiofemoral contact mechanics after serial medial meniscectomies in the human cadaveric knee. Am J Sports Med 2006;34(8):1334–1344

40. Lohmander LS, Englund PM, Dahl LL, Roos EM. The long- term consequence of anterior cruciate ligament and meniscus injuries: osteoarthritis. Am J Sports Med 2007;35(10):1756–1769

41. Gomoll AH, Kang RW, Chen AL, Cole BJ. Triad of cartilage restoration for unicompartmental arthritis treatment in young patients: meniscus allograft transplantation, cartilage repair and osteotomy. J Knee Surg 2009;22(2):137–141

42. Lattermann C, Jakob RP. High tibial osteotomy alone or combined with ligament reconstruction in anterior cruciate ligament-deficient knees. Knee Surg Sports Traumatol Arthrosc 1996;4(1):32–38

43. Jones HP, Appleyard RC, Mahajan S, Murrell GA. Meniscal and chondral loss in the anterior cruciate ligament injured knee. Sports Med 2003;33(14): 1075–1089

2

Magnetic Resonance Imaging of Cartilage Repair Techniques

Catherine Hayter and Hollis Potter

The field of cartilage repair is expanding rapidly and encompasses a wide range of techniques, including microfracture, first- and second-generation autologous chondrocyte implantation, autologous osteochondral transplantation, and allograft transplantation. Although assessment of patient outcome is clinically relevant, objective evaluation of repair allows insight into the natural history of cartilage repair and may allow detection of early signs associated with a poorer prognosis. Objective assessment has traditionally been performed with second-look arthroscopy and histologic evaluation of biopsy specimens.[1] Magnetic resonance (MR) imaging, however, offers a noninvasive method to assess cartilage repair. The information gained from MR imaging is therefore complementary to more subjective clinical outcome instruments that evaluate pain and function, and plays a valuable role in patient follow-up after cartilage repair.[2]

This chapter discusses the MR imaging techniques available for the assessment of articular cartilage, including advanced imaging techniques that allow assessment of cartilage biochemistry. The MR imaging appearance and assessment of microfracture, autologous chondrocyte implantation, and osteochondral autograft and allograft transplantation are reviewed.

◆ Basic Structure of Articular Cartilage

An understanding of the structure of articular cartilage is crucial to understanding the MR imaging appearance of normal and abnormal cartilage, as well as the imaging appearance following cartilage repair. Articular cartilage is composed of chondrocytes that are embedded in an organized extracellular matrix composed primarily of water (65 to 80%), collagen, and proteolgycan.[3] The material properties of articular cartilage are imparted mainly by the collagen and proteolgycan components of the extracellular matrix.

Compressive strength is imparted by the proteolgycan molecules, which are composed of negatively charged glycosaminoglycans radiating from a protein core.[4] These monomers bind to hyaluronic acid to form large aggregates, which resist compression because of their hydrophilic nature. The structural framework and tensile strength of articular cartilage is imparted by collagen. Type II collagen fibers make up 95% of the collagen in articular cartilage. These fibers have a long length-to-thickness ratio, thereby providing tensile stiffness and strength.[5]

Articular cartilage can be divided into four distinct zones on histology, which can be depicted with cartilage-sensitive MR pulse sequences. The superficial zone and lamina splendens consist of highly organized collagen fibers oriented parallel to the cartilage surface, providing high tensile strength. The transitional zone has lower collagen content and consists of randomly oriented collagen fibers; this zone has a higher compressive strength than the superficial zone. The radial zone consists of highly organized collagen fibers oriented perpendicular to the cartilage surface; this zone has the highest proteolgycan content and the lowest water content. The deepest zone is the calcified cartilage layer. The tidemark represents the boundary between the uncalcified and calcified cartilage.[6] At clinically relevant field strengths using traditional MR sequences, the tidemark is indistinguishable from the subchondral bone plate.

The signal characteristics of articular cartilage on MR imaging reflect both the mobility of water and the degree of organization of the tissue. Therefore, in the radial zone, where the collagen is highly ordered, lower signal intensity is generated when compared with the transitional zone where the collagen is more randomly oriented. The lamina splendens is also highly ordered and when visualized on MR images appears of lower signal intensity. It is important to recognize this normal "gray-scale stratification" when performing MR imaging of cartilage, as loss of the normal gray-scale stratification is one of the earliest signs of articular cartilage degeneration.[6]

◆ Morphologic Assessment of Cartilage

Cartilage-Sensitive Pulse Sequences

Traditional T1- and T2-weighted techniques are inadequate for the accurate assessment of articular cartilage. T1-weighted images result in poor delineation between the intermediate signal intensity cartilage and the low- to intermediate-signal intensity joint fluid. Conventional spin-echo or heavily T2-weighed fast spin echo (FSE) techniques result in poor delineation between the subchondral bone and the deep component of the articular cartilage, with consequent factitious thickening of the subchondral bone and thinning of the articular cartilage.[7] Several different cartilage-sensitive pulse sequences are available for MR imaging; of these, fat-suppressed three-dimensional (3D) gradient echo and FSE sequences are the most accurate and most widely used techniques.

Fat-suppressed 3D spoiled or T1-weighted gradient echo images obtained with isotropic voxels have the advantage of producing thin, contiguous slices that can be reformatted in any plane. The sharp contrast boundary between the low-signal intensity bone and the high-signal intensity articular cartilage makes these sequences amenable to semiautomated segmentation algorithms, allowing for accurate assessment of cartilage thickness and volume. This pulse sequence has been used for longitudinal assessment of cartilage volume in osteoarthritis trials and for quantitative assessment of focal cartilage defects and subsequent fill following repair.[8–10] This technique is, however, less sensitive to partial-thickness cartilage defects than FSE sequences and requires longer scan times. It is also limited by metal-induced susceptibility artifact, which may be a significant problem when imaging in the presence of metallic hardware or residual metallic debris following arthroscopy.[11]

An intermediate echo time (TE) two-dimensional FSE technique is one of the most popular pulse sequences for the assessment of articular cartilage. This technique provides good contrast between the intermediate signal intensity of articular cartilage, the low signal intensity of fibrocartilage and subchondral bone, and the high signal intensity of synovial fluid. On intermediate TE FSE images, articular cartilage demonstrates a normal gray-scale stratification, which corresponds to the cartilage zonal anatomy. Partial-thickness chondral lesions and chondral flaps are also well depicted with this technique (**Fig. 2.1**).[12] Use of an FSE technique with a wide receiver bandwidth minimizes susceptibility artifact, allowing accurate assessment of cartilage in the presence of metallic hardware or debris.[13]

Intermediate TE FSE sequences are subject to the "magic angle" effect, which must be

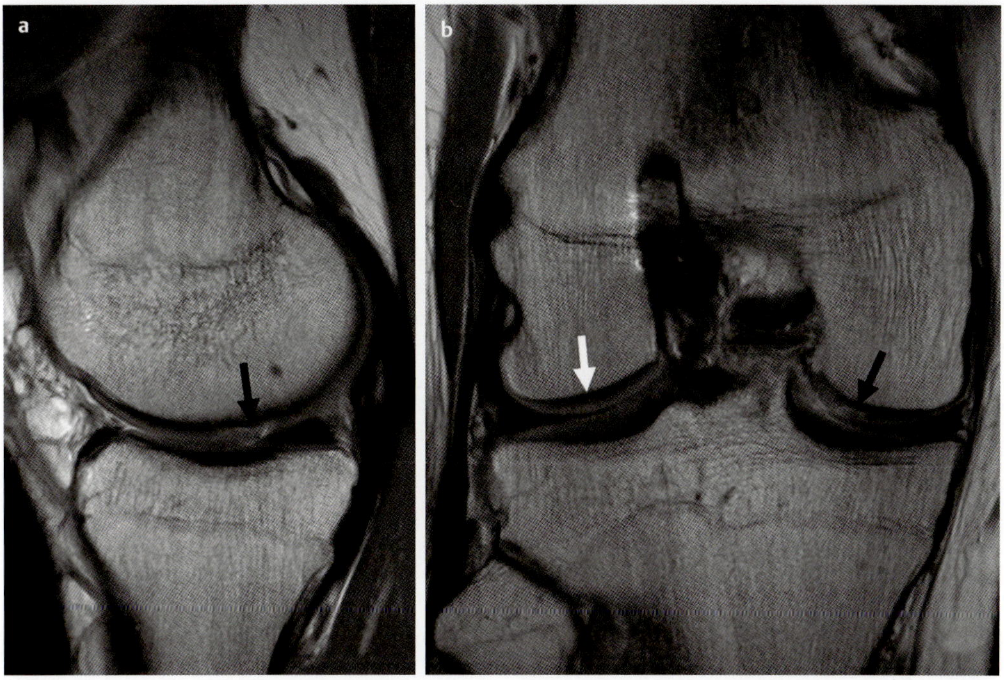

Fig. 2.1 Sagittal (**a**) and coronal (**b**) fast spin echo images in a 30-year-old man demonstrate delamination of cartilage over the medial femoral condyle with a flap extending to subchondral bone (*black arrow*). The cartilage over the lateral femoral condyle demonstrates normal gray-scale stratification (*white arrow*).

considered when interpreting these images. The signal intensity of ordered tissues, such as cartilage and tendons, depends on the orientation of the collagen fibers relative to the external magnetic field (B_0), which in a conventional MR machine runs parallel to the long axis of the patient's body. When highly structured tissues are imaged at 55 degrees to the external magnetic field using a short TE, there is a normal prolongation of T2 values, a phenomenon known as "magic angle" effect.[14,15] In the knee, magic angle effect commonly manifests as increased signal intensity in the cartilage over the anterior and posterior femoral condyles and in the submeniscal zone of the posterior tibial condyles.

The application of fat suppression to FSE images allows the detection of bone marrow edema and increases the contrast differences between cartilage, fluid, and synovium. However, fat suppression results in a lower signal-to-noise ratio; therefore, it is more difficult to achieve the high spatial resolution that is required to discern subtle fissures and surface fibrillation in the articular cartilage.[7,16] Thus, routine assessment of articular cartilage with FSE techniques should include high-resolution, non-fat-suppressed FSE images in at least two planes as well as a fat-suppressed FSE image in one plane.[12]

◆ Evaluation of Cartilage Biochemistry

Quantitative MR imaging techniques allow more sophisticated assessment of cartilage degeneration and cartilage repair. These techniques detect changes in the ultrastructure of cartilage and provide an assessment of cartilage biochemistry. They therefore have the potential to detect changes in cartilage biochemistry that may precede discernible cartilage thinning on traditional MR techniques. Quantitative MR imaging techniques are classified into those that detect alterations in collagen fiber orientation and those that detect alterations in the proteoglycan content.

Collagen

T2 mapping is the most commonly used quantitative MR technique for the assessment of collagen orientation. Additional techniques such as diffusion tensor imaging have been shown to be sensitive to collagen orientation[17–19] but are not currently in widespread clinical use.

T2 Mapping

The T2 (spin-spin) relaxation time reflects the loss of signal that occurs due to dephasing of the excited nuclei after the disturbing radio-frequency pulse is applied. T2 mapping is performed by acquiring several images at different TEs at the same slice location. The T2 calculation is performed on a pixel-by-pixel basis by fitting the signal intensity from each echo image and the corresponding TE to an exponential decay equation.

The T2 map of articular cartilage reflects the collagen fiber orientation and the mobile water content,[20] and is displayed using a color-coded map. In the radial zone, where the collagen is oriented perpendicular to the subchondral plate, short T2 values are obtained. In the transitional zone, where there is a more random orientation of collagen, longer T2 values are obtained. Prolongation of T2 relaxation times has been shown to be associated with osteoarthritis and breakdown in cartilage structure.[21–23] Quantitative T2 measurements demonstrate excellent inter- and intraobserver reliability,[24–26] thereby offering a tool for reproducible assessment of cartilage status over time.

Proteoglycan

Imaging strategies aimed at assessing the proteoglycan component of the extracellular matrix exploit the fixed-charge density property of the articular cartilage. The fixed-charge density in cartilage is largely due to the concentration and distribution of the negatively charged glycosaminoglycan chains within the proteoglycan macromolecules. Techniques used to assess proteoglycan content and distribution include sodium-23 imaging, delayed gadolinium-enhanced MR imaging of cartilage (dGEMRIC), and T1rho (ρ)-weighted MR imaging.

Sodium Imaging

Similar to hydrogen nuclei, sodium-23 is also a suitable nucleus for MR imaging. The presence of the negatively charged glycosaminoglycan molecules in articular cartilage generates an attraction toward the positively charged sodium-23, allowing for relative measurement of fixed-charge density.[27,28] Sodium imaging has been used as an imaging standard by which to assess proteoglycan distribution and content. However, this technique is limited in its clinical application by the lower concentration of sodium-23 in articular cartilage relative to hydrogen, resulting in low signal-to-noise ratio and requiring long scan times. Sodium imaging also requires the use of specialized coils and the ability to scan with multinuclear spectroscopy software, which is not widely available across clinical systems.[29,30]

dGEMRIC

dGEMRIC also exploits the fixed-charge property of articular cartilage through the use of an injection of negatively charged gadolinium contrast. Gadolinium is administered intravenously, the patient performs 10 minutes of exercise, and, following a 90-minute delay, T1-weighted maps are obtained, usually through the use of a specialized inversion recovery pulse sequence.[31] The gadolinium penetrates the articular cartilage, with the amount of penetration being inversely proportional to the glycosaminoglycan content. The gadolinium acts to shorten T1-relaxation times, allowing for the generation of T1 maps. In areas with depleted glycosaminoglycan content, there will be an increased distribution of gadolinium and therefore a higher T1 signal, which is reflected by a diminished "relative glycosaminoglycan index."

T1rho (ρ) Mapping

T1rho (ρ) is a technique used to assess the low-frequency interactions between hydrogen in macromolecules and free water.

Termed *spin-lattice relaxation in the rotating frame*, this technique uses clusters of radiofrequency pulses to "lock" magnetization in the transverse plane and limit dephasing of protons. After a specified period of time (the spin-lock time), the magnetization vector is realigned with B_0, and data are then acquired using an FSE or a spiral sequence.[32,33] Similar to T2 mapping, T1rho is calculated on a pixel-by-pixel basis by fitting the signal intensity from each spin-lock image and the corresponding spin-lock length to an exponential decay equation.

T1rho has been shown to reflect proteoglycan content in articular cartilage. The normalized T1rho rate is highly correlated to fixed-charge density in sodium-23 imaging, as well as to proteoglycan content and distribution on histology.[34–36] Subjects with osteoarthritis have longer T1rho values than asymptomatic controls, and T1rho may be even more sensitive to early cartilage degeneration than T2 mapping alone.[37] While clinically feasible at 1.5T and 3T,[37,38] T1rho is largely applied at 3T and is a promising technique to detect changes in proteoglycan content in early cartilage degeneration.

Due to the specialty coil requirements for sodium imaging as well as the logistical constraints of the contrast-enhanced dGEMRIC technique, the authors favor the use of noncontrast T1rho at 3T for assessment of proteoglycan content.

◆ Evaluation of Articular Cartilage Repair Techniques

In part due to its limited vascular supply, mature articular cartilage has little to no capacity for spontaneous repair.[39,40] A wide variety of cartilage repair techniques have been described; results following these procedures have varied widely. Most of the literature has relied on subjective clinical assessment or conventional radiographs to evaluate the success of repair techniques. The use of second-look arthroscopy with biopsy remains the gold standard but is limited by its invasive nature, potential for operative surgeon bias, and poor patient acceptance. With advancements in imaging techniques, MR imaging is increasingly recognized as an alternative method of noninvasive evaluation of the results of articular cartilage repair.[41]

The variables that should be assessed on MR imaging following cartilage repair differ according to the repair technique[2] and are summarized in **Table 2.1**. The expected MR

Table 2.1 Diagnostic checklist for MR imaging of cartilage repair techniques

Microfracture
• Signal intensity of the repair cartilage • Morphology of reparative tissue (flush, proud, depressed) • Volume or percent of fill by reparative tissue • Peripheral integration (fissures at repair–native cartilage interface) • Underlying subchondral bone (extent of bone marrow edema) • Overgrowth of subchondral bone • Assessment of host cartilage (adjacent/opposing surfaces) • Reactive synovitis
Autologous chondrocyte implantation (ACI)
• Signal intensity of the repair cartilage • Morphology of reparative tissue (flush, proud, depressed) • Volume or percent of fill by reparative tissue • Peripheral integration (fissures at repair–native cartilage interface) • Presence of delamination • Periosteum overlying the defect (periosteal hypertrophy) • Underlying subchondral bone (bone marrow edema) • Assessment of host cartilage (adjacent/opposing surfaces) • Reactive synovitis
Osteochondral transplantation
Osseous phase • Presence/absence of displacement of plugs • Restoration of radius of curvature of joint surface • Peripheral integration of osseous components • Presence of subchondral bone marrow edema
Articular phase • Signal intensity of the repair cartilage • Morphology of reparative tissue (flush, proud, depressed) • Volume or percent of fill by reparative tissue • Peripheral integration (fissures at repair–native cartilage interface)
Other features • Assessment of host cartilage (adjacent/opposing surfaces) • Reactive synovitis

imaging appearance following commonly used cartilage repair techniques is summarized in **Table 2.2**.

Table 2.2 Summary of imaging findings following cartilage repair

Microfracture
Early postoperative period • Reparative cartilage is hyperintense to native cartilage • Moderate subchondral bone marrow edema pattern
Late postoperative period • Signal intensity of reparative cartilage decreases as it matures • Subchondral bone marrow edema pattern dissipates
Specific complication • Overgrowth of subchondral bone
Autologous chondrocyte implantation (ACI)
Early postoperative period • Reparative cartilage is hyperintense to native cartilage and periosteal cover • Subchondral bone marrow edema pattern • ±Hypertrophy of periosteal cover
Late postoperative period • Signal intensity of reparative cartilage decreases as it matures • Subchondral bone marrow edema pattern dissipates
Specific complications • Graft delamination • Graft (periosteal) hypertrophy
Osteochondral autografts
Early postoperative period • Subchondral bone marrow edema pattern
Late postoperative period • Subchondral bone marrow edema pattern dissipates
Specific complications • Plug displacement/subsidence • Poor osseous integration
Osteochondral allograft (OCA)
Early postoperative period (0–3 mo) • Graft bone marrow edema pattern
Late postoperative period (3–6 mo) • Graft bone marrow edema pattern dissipates • Overgrowth of subchondral bone may occur
Specific complications • Graft rejection (persistent bone marrow edema, collapse, fluid undermining graft, global synovitis) • Graft collapse

Microfracture

Microfracture is a cartilage repair technique that is based on local bone marrow stimulation and the release of multipotential stem cells,[42] which, over time, differentiate into repair tissue, predominantly composed of fibrocartilage.[43,44]

MR imaging assessment following microfracture should include the assessment of (1) the signal intensity of the repair cartilage compared with native cartilage; (2) the morphology of the repair, being either flush, proud, or depressed with respect to the native cartilage; (3) the volume or percent fill of the defect in two imaging planes; (4) the degree of peripheral integration to the adjacent cartilage, including the absence/presence of fissures; (5) assessment of the underlying bone, including the presence of bone marrow edema; (6) the surface geometry of the bone, including the presence of proud subchondral bone; (7) the status of the articular cartilage in the adjacent and opposing surfaces; and (8) the presence of a reactive synovitis.[2]

The MR imaging appearance of a chondral defect treated by microfracture evolves over time. In the early postoperative period, the reparative tissue is less organized and has increased water content when compared with normal articular cartilage, and as such, it appears hyperintense compared with the native cartilage.[2,45] Over time, there is maturation of the signal characteristics of the repair tissue, which may appear hypointense to the native cartilage,[2,45] suggesting the presence of reparative fibrocartilage.[46–48]

The percent fill of the defect increases with time; however, the overall percent fill following microfracture may be less than that seen following autologous chondrocyte implantation.[2] Fissures between the area of microfracture and native cartilage appear commonly[45] and do not necessarily correlate to symptoms.[2] By 1 to 2 years after surgery, the treated defect should be filled with tissue that has a smooth, well-defined surface. Adverse functional scores at 2 years have been shown to correlate with poor percentage fill, indicating a correlation between objective MR imaging findings and subjective clinical outcome.[45]

Bone marrow edema may be seen in the subchondral bone at short-term follow-up but usually appears mild[2] and gradually diminishes over time.[44] In patients with failure of the microfracture procedure, the subchondral bone marrow edema does not diminish and may become more conspicuous over time.

Because microfracture involves the release of pluripotential stem cells, these cells may also differentiate into bone. Overgrowth of the subchondral bone has been noted in 25 to 49% of patients following microfracture[2,45] (**Fig. 2.2**). Osseous overgrowth may not be a negative prognostic factor and does not correlate with adverse clinical scores at short-term follow-up.[45] Osseous overgrowth may, however, result in a thinner layer of reparative tissue and inferior filling of the cartilage defect, which has been shown to correlate to inferior functional outcomes.[49]

Studies using quantitative MR imaging of microfracture have shown that at short-term follow-up (< 6 months) T2 values over the repair site are prolonged, mimicking the signal characteristics of the repair tissue observed on conventional MR imaging. Over time (> 12 months), there is progressive shortening and maturation of T2 values,[50] although T2 values of the repair tissue may never return to normal. At a follow-up time of 24 months after microfracture, T2 values have been shown to remain globally reduced compared with native cartilage.[51,52] The T2 index, calculated by the T2 value of the repair tissue compared with native cartilage, correlates with subjective functional scores, such as the Lysholm score and International Knee Documentation Committee (IKDC) form of subjective function,[51] further supporting the use of quantitative MR imaging as an objective tool to evaluate cartilage repair procedures.

Autologous Chondrocyte Implantation

Autologous chondrocyte implantation (ACI) is a two-stage surgical technique for the repair of symptomatic deep chondral defects[49] that entails growth of the patient's native chondrocytes in tissue culture followed by reimplantation and coverage by a periosteal flap.[53,54] Matrix-assisted chondrocyte transplantation (MACT) is a modification of this procedure, which uses biomaterials seeded with chondrocytes as carriers and scaffolds for cell growth, without the use of a periosteal cover.[55]

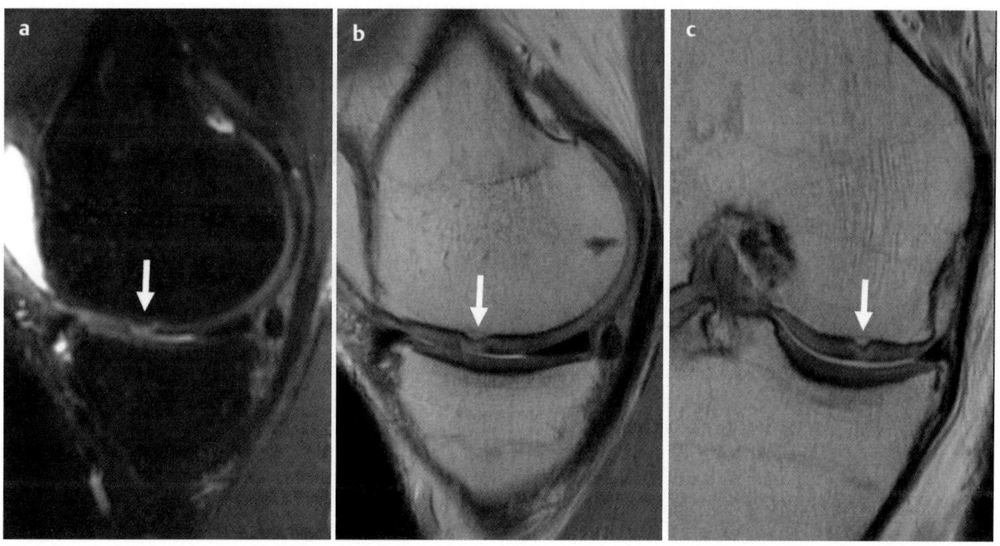

Fig. 2.2 Sagittal inversion recovery (**a**) and sagittal (**b**) and coronal (**c**) fast spin echo images in a 42-year-old man 5 months following microfracture of the medial femoral condyle. Proud bone formation is seen at the repair site (*arrow*), but there is good fill by hyperintense reparative tissue.

ACI is frequently used for larger cartilage defects than microfracture; areas up to 12 cm^2 have previously been transplanted.[42,54] ACI has been shown to provide better defect fill at all time periods when compared with microfracture[2]; however, the technique is associated with some specific potential complications, including delamination of the graft and periosteal hypertrophy,[49] which can be well demonstrated on MR imaging.

MR imaging assessment following ACI should include the assessment of (1) the signal intensity of the repair cartilage compared with native cartilage; (2) the morphology of the repair, being either flush, proud, or depressed with respect to the native cartilage; (3) the volume or percent fill of the defect in two imaging planes; (4) the degree of peripheral integration to the adjacent cartilage, including the absence/presence of fissures and the presence of delamination; (5) the periosteum overlying the defect, including the presence of periosteal hypertrophy; (6) the subchondral bone, including the presence of a bone marrow edema pattern; (7) the status of the articular cartilage in the adjacent and opposing surfaces; and (8) the presence of a reactive synovitis.[2]

Previous studies have investigated the correlation between MR imaging findings and clinical outcome measures following MACT. A study correlating MR imaging findings to the Knee Injury and Osteoarthritis Outcome Score (KOOS) and visual analog scale (VAS) score for pain and function demonstrated a significant correlation for the volume of repair fill, signal intensity of the repair tissue, structure of the repair tissue (presence or absence of fissures), and changes in the subchondral bone. Interclass coefficients demonstrated a strong agreement between observers, confirming the reproducibility of MR imaging in the longitudinal assessment of cartilage repair.[56]

A three-phase time course has been proposed in the healing of ACI. In the first 6 weeks, the "proliferative phase," the graft site fills with a soft, primitive repair tissue. From 7 weeks to 6 months, the "transition phase," there is expansion of the extracellular matrix, and the graft takes on the consistency of gelatin. The "remodeling phase" from 6 months to 3 years involves remodeling of the extracellular matrix to produce hyaline-like repair tissue.[57,58] The signal characteristics of the repair tissue on MR imaging reflect its histology. In the immediate postoperative period the reparative tissue is disorganized with increased water content and therefore appears hyperintense on MR images.[44,59,60] This allows the reparative tissue to be readily differentiated from the overlying periosteum, which appears hypointense. At 3 to 6 months following repair, there is a decline in the signal intensity of the reparative tissue as it becomes increasingly organized and integrated with the adjacent cartilage (**Fig. 2.3**). Complete integration can take up to 2 years.

Following ACI there should ideally be complete fill of the defect with repair tissue that restores the contour of the articular surface. Complete fill of the defect has been observed as early as 3 weeks postoperatively.[61] However, filling defects at the repair site in the early postoperative period (3 to 6 months) are common and often demonstrate progressive fill on follow-up examinations (6 to 12 months).[59,62] The interface with the adjacent cartilage is rarely smooth; hyperintense fissures less than 2 mm wide are common.[2] The long-term significance of small fissures is unknown, but larger fissures may represent failure of the repair cartilage to integrate with the native articular cartilage.

Delamination of the ACI graft is uncommon, occurring in fewer than 5% of patients,[62] and mostly occurs within the first 6 to 9 months following surgery.[44,49,58] A displaced delaminated graft appears as a defect within the repair site, and the displaced tissue may be seen as a loose body within the joint. In the case of in situ delamination, the appearance is similar to a cartilage flap, with a thin rim of fluid intensity between the base of the repair tissue and the underlying subchondral bone.[44] It must, however, be remembered that a fluid-like appearance to the graft in the first 4 weeks is normal due to immaturity of the repair tissue. This should not be misinterpreted as graft delamination; closer inspection will reveal the low-signal periosteum overlying the implant.[44]

Overgrowth of reparative tissue has been reported following ACI, which may be due to periosteal hypertrophy and/or thickening

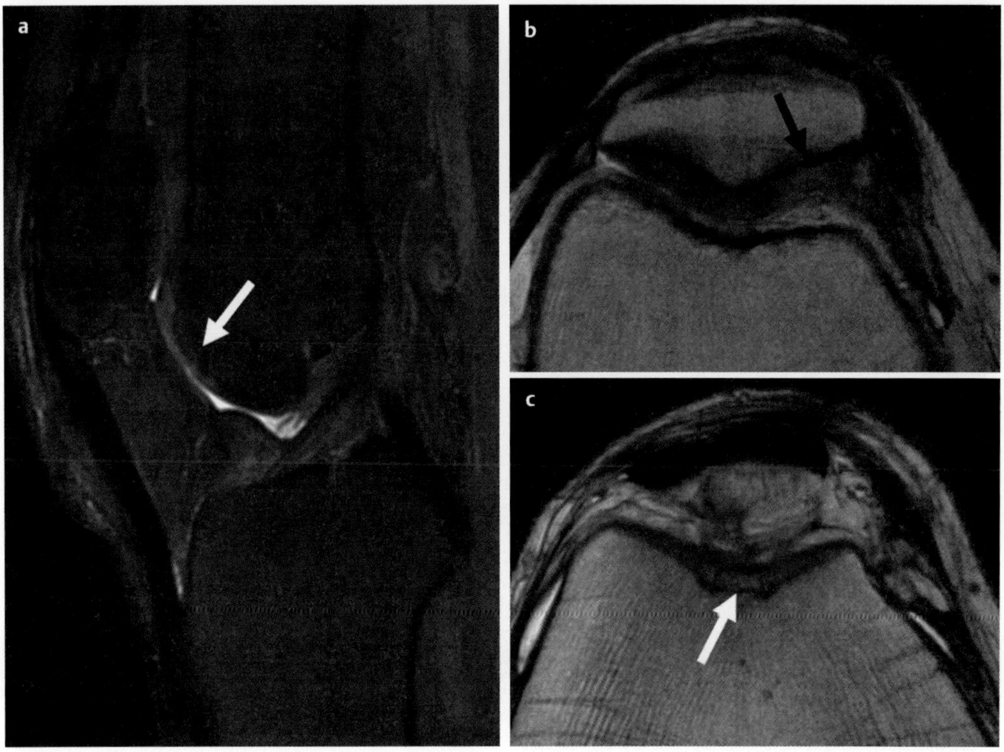

Fig. 2.3 Sagittal inversion recovery (**a**) and axial (**b, c**) fast spin echo images in a 43-year-old man 6 months following autologous chondrocyte implantation. There is good fill over the trochlea (*white arrow*) and moderate fill over the patella (*black arrow*) with hyperintense reparative tissue. The periosteal cover is well depicted over the trochlea. There is no bone marrow edema or reactive synovitis.

of the matrix.[2,63] Although the majority of patients remain asymptomatic, some patients report a sensation of catching, which can be painful.[49,58] This complication most commonly occurs at 3 to 9 months following surgery and may require arthroscopic resection of the hypertrophic periosteal cover.[54] In a study of 35 ACI procedures, graft hypertrophy occurred in 63% of lesions, accounting for moderate postoperative morbidity.[2] On MR imaging, periosteal hypertrophy appears as focal areas of repair tissue that extend above the expected contour of the articular surface. The hypertrophied tissue may grow beyond the margins of the graft, appearing as a flap of tissue extending over the adjacent native articular cartilage.[61]

The subchondral bone plate may appear irregular following ACI and often demonstrates a mild to moderate bone marrow edema pattern in the early postoperative period. The intensity and volume of the edema tend to diminish over time to regress to normal or a thin line of increased signal intensity deep to the subchondral bone. Persistent or increased edema in the marrow beneath an ACI suggests poor integration of the graft to the subchondral bone and warrants close clinical follow-up.[61]

Quantitative MR techniques have been used to provide insight into the biochemistry of repair tissue following ACI. In a cohort study of 15 patients treated with MACT, mean T2 values of the repair tissue were found to be significantly higher than control sites at short-term follow-up, but showed no significant difference at long-term follow-up (19 to 42 months).[64] In a study of 11 patients evaluated with dGEMRIC, investigators noted that at less than 12 months follow-up the relative glycosaminoglycan index in the ACI repair tissue was lower than that of native cartilage.

At follow-up of greater than 12 months, the glycosaminoglycan index was similar to that of native cartilage,[65] suggesting progressive maturation of the repair tissue. The studies suggest that quantitative MR imaging can provide information about the structure of the cartilage repair tissue.

In a study of 20 patients comparing microfracture and MACT, investigators noted that at 2 years following microfracture there was a global reduction in T2 values compared with control tissues; the repair tissue following microfracture also showed lack of the normal T2 stratification. In contrast, the MACT repair tissue demonstrated T2 values and T2 stratification similar to that of native cartilage,[52] suggesting that ACI results in more hyaline-like and mature repair tissue at the site of repair when compared with the microfracture technique.

Osteochondral Transplantation

Osteochondral transplantation may be performed using autogenous tissue, biphasic synthetic copolymer plugs, or fresh cryopreserved allograft tissue. Although some surgeons reserve these techniques for the repair of defects associated with bone and cartilage defects, such as osteonecrosis or osteochondritis dissecans, others may use osteochondral transplantation for isolated cartilage defects with intact subchondral bone.

MR imaging assessment following osteochondral repair includes the assessment of the osseous and articular phases.[2,50,66] Assessment of the osseous phase includes (1) the presence or absence of displacement of the plug or the presence of subchondral collapse in the setting of allograft transplantation; (2) the restoration of the radius of curvature of the subchondral bone; (3) the degree of osseous integration of the plug/allograft into the recipient site; (4) the presence of a subchondral bone marrow edema pattern. Assessment of the articular phase includes (5) the signal intensity of the repair cartilage compared with native cartilage; (6) the morphology of the repair, being either flush, proud, or depressed with respect to the native cartilage; (7) the volume or percent fill of the defect in two imaging planes: (8) the degree of peripheral integration to the adjacent cartilage, including the absence/presence of fissures and the presence of delamination. Additional features to assess include (1) the status of the articular cartilage in the adjacent and opposing surfaces and (2) the presence of a reactive synovitis.

Autologous Osteochondral Plugs

Autologous osteochondral transfer involves the harvesting of osteochondral plugs from a non-weight-bearing portion of the knee and transfer into cored holes created within the articular defect to be treated, using "press fit" fixation.[67,68]

Assessment of the osseous phase of incorporation involves assessment of plug position and restoration of the radius of curvature of the joint surface. Incongruity of the articular surface may be due to technical problems during graft placement or may occur later as a result of subsidence or degradation of the osteochondral plugs.[69] On MR imaging, an incongruent graft repair site can be seen as a step-off at the subchondral bone or articular surface (**Fig. 2.4**). Evaluation of restoration of the radius of curvature is clinically important because surface incongruities of as little as 0.5 mm have been shown to be associated with significant increases ($> 40\%$) in surface contact pressures.[70] This may explain why, in animal studies, elevated plugs tend to demonstrate poor integration with the surrounding articular cartilage when compared with plugs that are placed flush.[71,72]

Osseous integration of the plugs manifests as trabecular incorporation with the native bone and should be complete or partial in the majority of patients. When solid bony incorporation occurs, normal fatty marrow signal is seen within and around the plugs (**Fig. 2.5**). In animal studies, autologous grafts demonstrate solid osseous incorporation between 6 and 14 weeks.[73,74] Failure of osseous integration manifests as hyperintense signal at the native bone–graft interface and may be associated with a bone marrow edema pattern (**Fig. 2.6**).[66]

There are some potential pitfalls in the MR imaging assessment of the osseous phase, which should not be misinterpreted as a failure of osseous integration. The appearance of

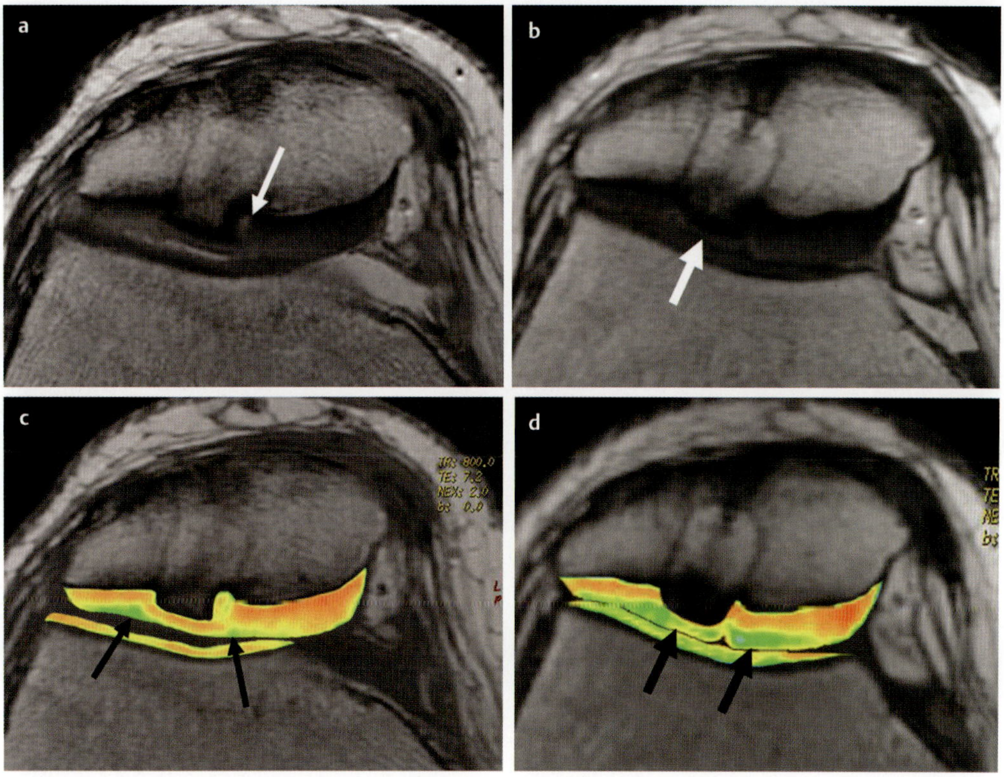

Fig. 2.4 Axial fast spin echo images in an 18-year-old woman 4 months (**a**) and 16 months (**b**) following autologous osteochondral repair over the lateral patellar facet. The plug is proud to the subchondral bone. On the initial scan there is a hyperintense fissure at the medial interface (*thin white arrow*). On the follow-up scan, the fissure has remodeled but there is progressive increased signal intensity in the cartilage over the plug (*thick white arrow*). Corresponding T2 map at 4 months (**c**) demonstrates normal T2 stratification over the repair site but prolongation of T2 values at the repair–native tissue interface (*thin black arrows*). At 16 months (**d**) there is more diffuse prolongation of T2 values over the repair site and adjacent cartilage (*thick black arrows*). (Adapted with permission from Nho SJ, Foo LF, Green DM, et al. Magnetic resonance imaging and clinical evaluation of patellar resurfacing with press-fit osteochondral autograft plugs. Am J Sports Med 2008;36(6):1101–1109.)

hypointense signal at the osseous interface is indicative of sclerosis due to the tight fit created with the "press-fit" technique used for plug placement. Over time, this tends to fill in with normal fatty marrow signal. Even in well-incorporated repairs, a small area of fat devoid of trabeculae may be identified deep to the base of the osteochondral plug, as the recipient tunnel is often made deeper than the length of the plug.[44] This should not be misinterpreted as failure of the repair.

Subchondral bone marrow edema is often present initially but is expected to resolve as the graft incorporates with the subchondral bone.[69] Detection of cystic cavities, fluid-like signal intensity, or the presence of a persistent bone marrow edema pattern suggests poor osseous integration of the graft and warrants close follow-up (**Fig. 2.7**).[44]

The degree of cartilage integration involves assessment of the signal intensity of the repair cartilage and percent fill. As the cartilage over the site of repair consists of normal articular cartilage, the signal in the cartilage over the plugs should ideally mimic that of native articular cartilage. The T2 values over autologous plugs also tend to show the expected normal T2 stratification, with shorter values in the radial zone and prolongation of values in the more superficial

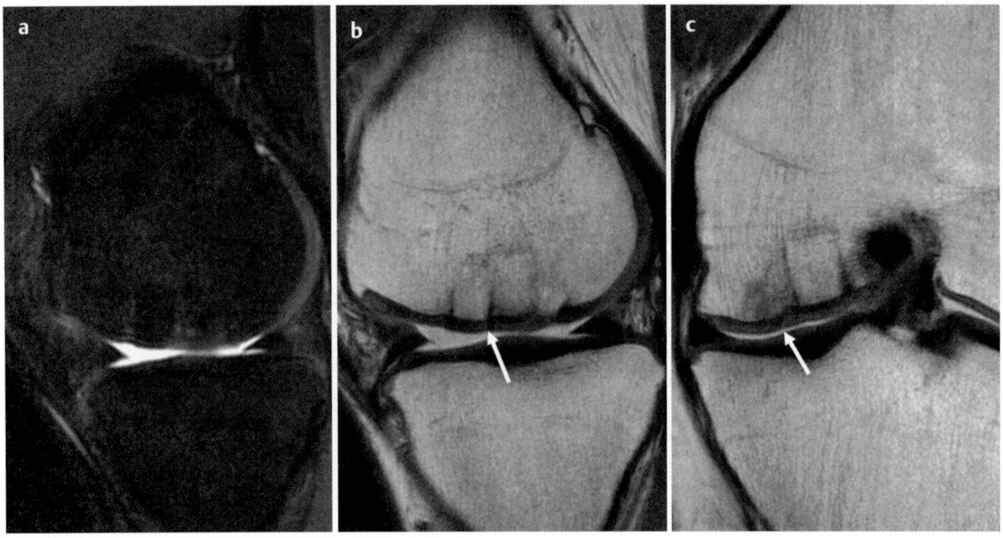

Fig. 2.5 Sagittal inversion recovery (**a**) and sagittal (**b**) and coronal (**c**) fast spin echo images in a 19-year-old man 6 months following autologous osteochondral repair over the medial femoral condyle. There is good osseous incorporation of the plugs and good restoration of the radius of curvature of the joint surface. Thin (< 2 mm) fissures are present at the area of peripheral integration with the native articular cartilage (*arrows*).

zones.[50] It is, however, common to observe hyperintense signal in the cartilage overlying the plugs and prolongation of T2 values when compared with native articular cartilage.[69]

While the osseous portion of the plug typically demonstrates excellent incorporation, persistent fissures at the cartilaginous level between the graft and native articular cartilage are common.[2,69,75] Second-look arthroscopies and biopsy specimens have shown that the cartilage transplanted into the recipient site remains hyaline-like and that a fibrocartilaginous bond, consisting of organized scar tissue, forms between the cartilage plugs and the native articular cartilage.[71]

Synthetic Biphasic Copolymer Plugs

Without the use of backfill, the defect created by harvest of an autologous osteochondral plug will typically fill in with reparative fibrocartilage and fibrous tissue. Because of concern about donor-site morbidity, there has been increased interest in the use of synthetic biphasic copolymer plugs, which can be used for "backfill" of the donor site and may also be used for primary cartilage repair.[76] These biphasic plugs have a scaffold impregnated with growth factors to promote bone growth in the deeper components, as well as the growth of repair cartilage over the superficial surface.

The signal characteristic of synthetic plugs on MR imaging are distinctly different from those of autologous tissue. In the early postoperative period (6 to 12 months) it is common to see unfavorable MR imaging findings, including depressed morphology of the plugs, resorption at the interface with the native bone, incomplete fill by reparative tissue, and prolongation of T2 values over the repair site. These findings do not necessarily reflect a failure of the repair technique but rather reflect the natural history of the synthetic plugs. With longer follow-up (> 16 months), there is usually a marked improvement in the MR imaging appearance with a high percentage of lesion fill, good osseous incorporation, and restoration of the radius of curvature of the joint surface. In addition, progressive shortening of T2 values is observed over the plugs, which come to approach relaxation times observed in native articular cartilage.[77]

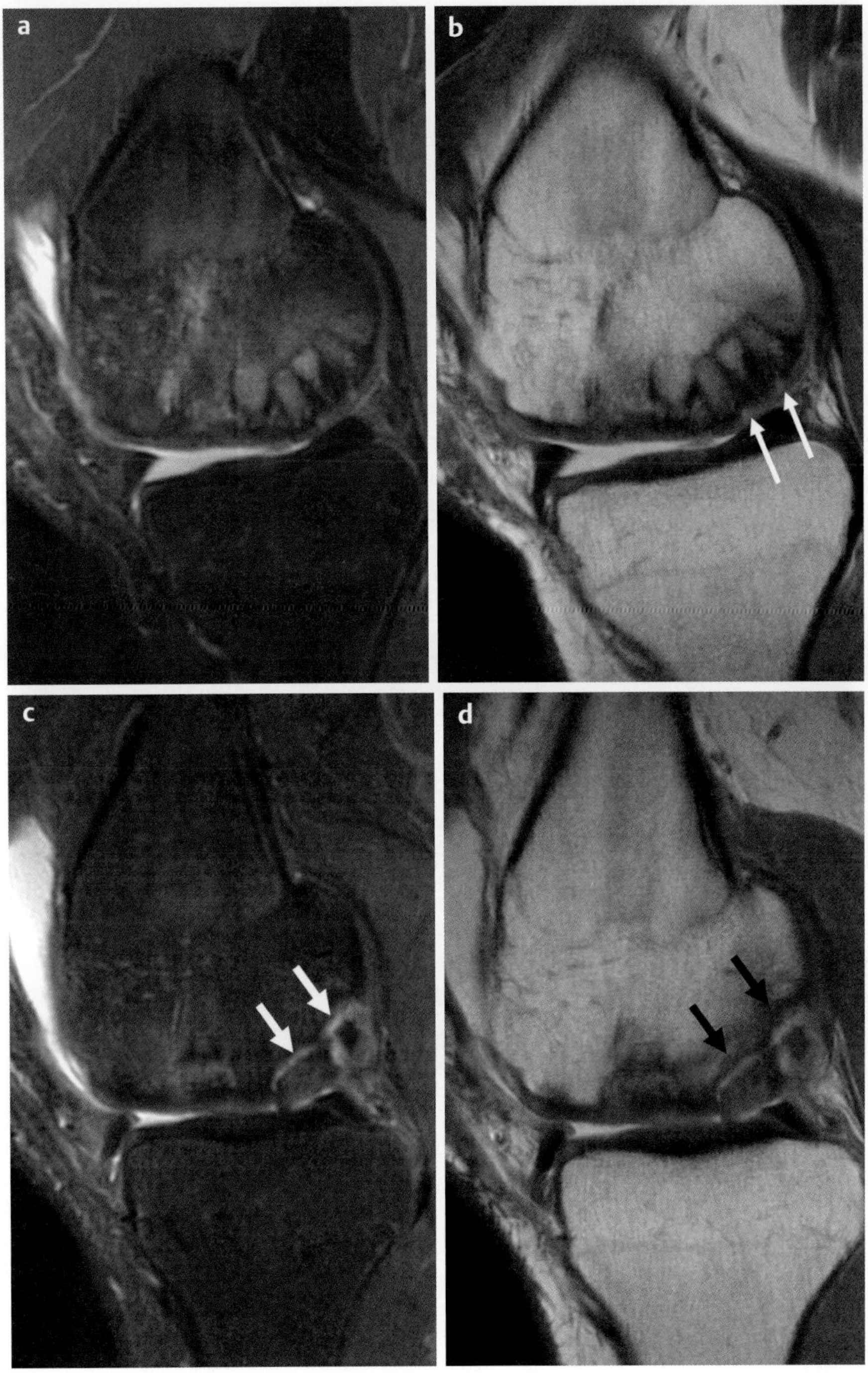

Fig. 2.6 Sagittal inversion recovery (left) and fast spin echo (right) images in a 16-year-old man following autologous osteochondral transplantation over the medial femoral condyle. At 2 months' follow-up (**a, b**) there is incomplete osseous incorporation of the plugs with a mild bone marrow edema pattern. Hyperintense fissures are seen at the area of peripheral integration with the native cartilage (*thin white arrows*). Images at 14 months postsurgery (**c, d**) demonstrate fluid signal undermining the posterior two plugs (*thick white arrows*) with sclerosis in the subchondral bone (*black arrows*), indicative of failure of osseous incorporation.

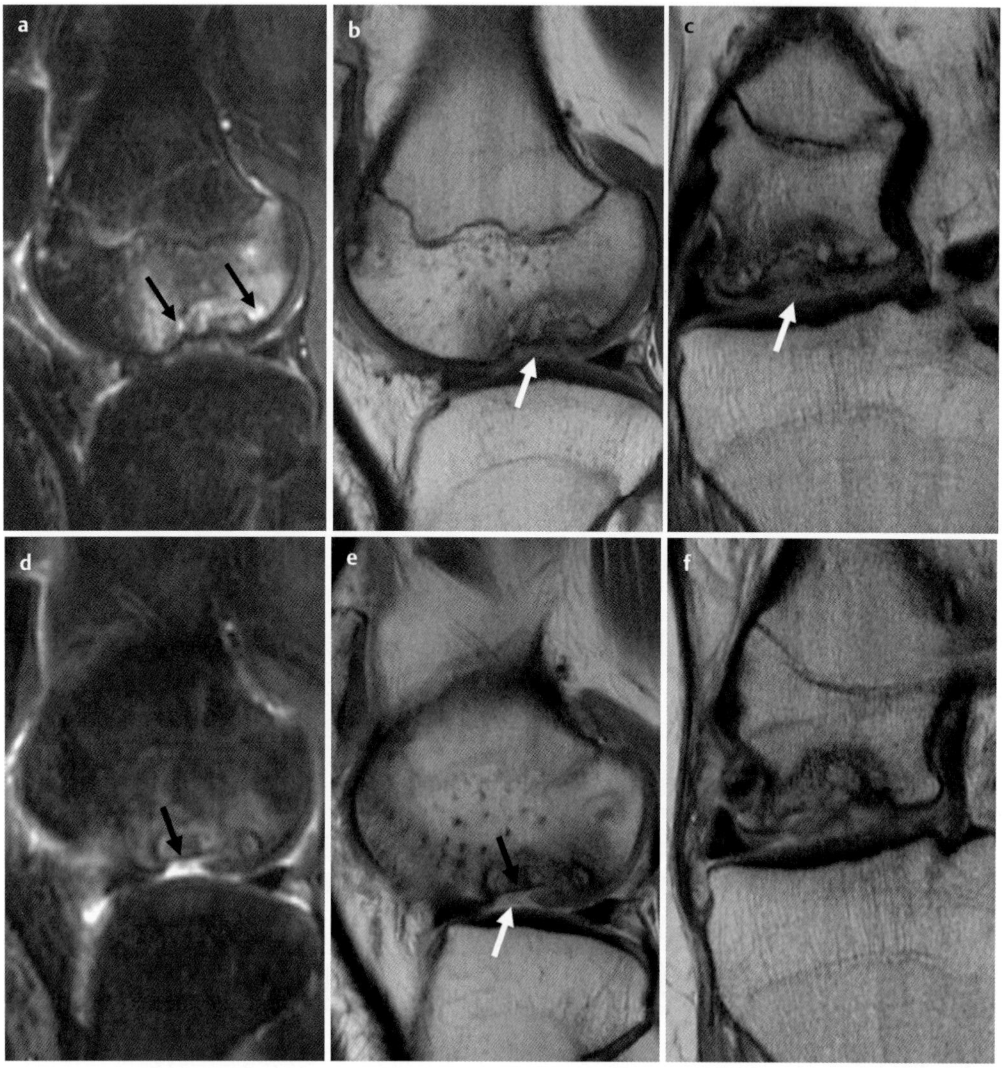

Fig. 2.7 Sagittal inversion recovery (left) and sagittal (middle) and coronal (right) fast spin echo images in a 12-year-old girl following autologous osteochondral transplantation over the lateral femoral condyle. At 29 months postsurgery (**a, b, c**) there is persistent bone marrow edema, cyst formation at the interface with the native bone (*black arrows*), and flattening of the subchondral surface. The articular cartilage is hyperintense but is intact (*white arrows*). At 36 months postsurgery (**d, e, f**) there is progressive subchondral collapse (*black arrows*) with delamination of the overlying cartilage and a flap extending down to bone (*white arrow*).

The delayed biologic incorporation of synthetic plugs when compared with autologous plugs must be recognized when interpreting MR images. Complete incorporation of synthetic plugs may take more than 2 years.[78] Increased signal in the plugs should therefore not be mistaken for delayed biologic incorporation (**Fig. 2.8**) and unfavorable MR imaging findings at early follow-up should be interpreted with caution.

Osteochondral Allograft Plugs

Osteochondral allograft (OCA) involves the harvesting of an osteochondral plug from a cadaver and transplantation into a donor site using press-fit technique, with or without pin fixation.[44] As with autologous osteochondral plugs, the MR imaging assessment of OCA plugs includes assessment of osseous integration, articular integration, and the

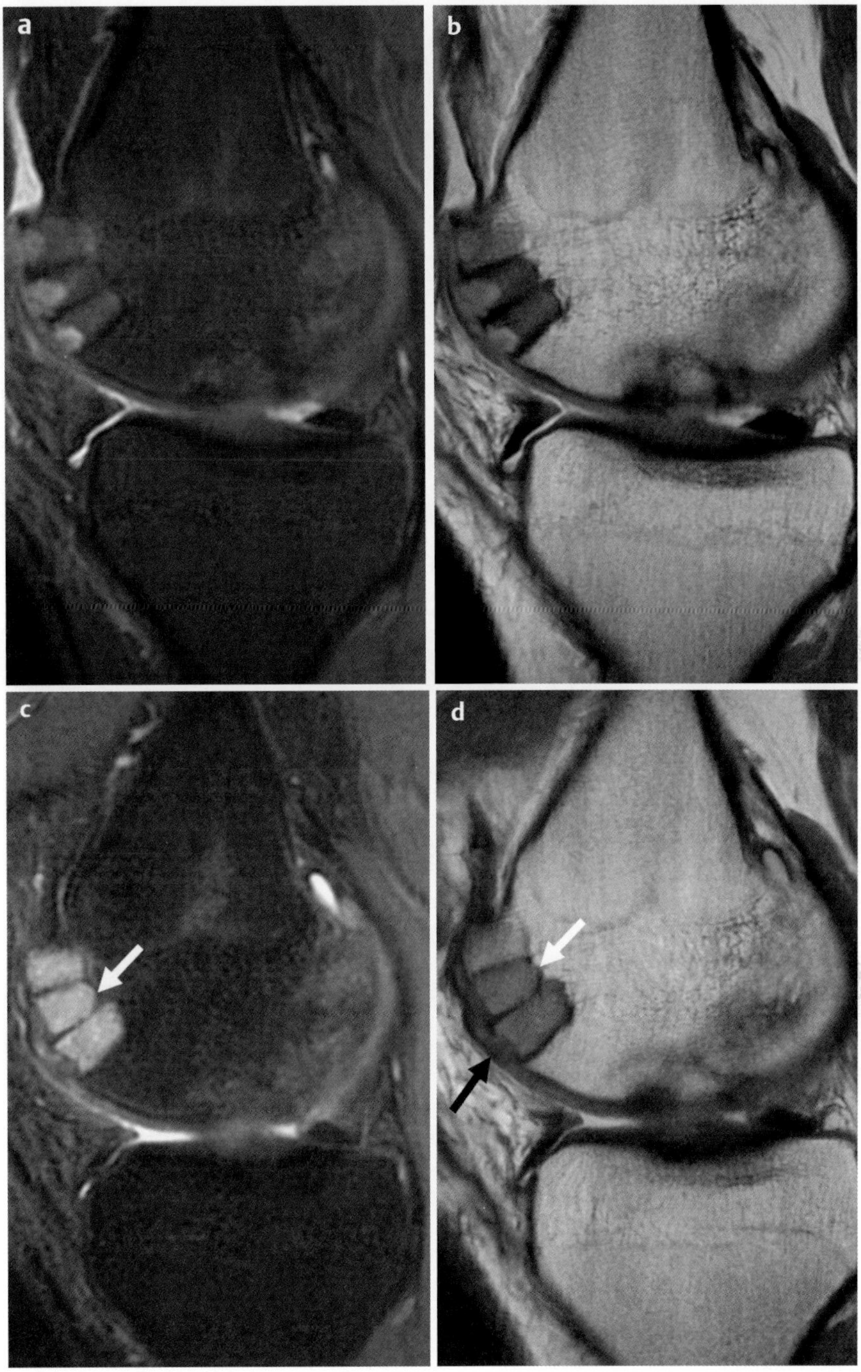

Fig. 2.8 Sagittal inversion recovery (left) and sagittal (right) fast spin echo images in a 17-year-old man following autologous osteochondral transfer into the medial femoral condyle with backfill over the donor site using synthetic biphasic copolymer plugs. At 6 months postsurgery (**a, b**) there is incomplete osseous and articular incorporation of the plugs. Incomplete osseous incorporation (*white arrows*) persists at 12 months postsurgery (**c, d**), but there is progressive low signal intensity of the overlying repair cartilage with good fill of the cartilage defect (*black arrow*). Delayed biologic incorporation of synthetic plugs is common when compared with autologous plugs, and may take up to 2 years.

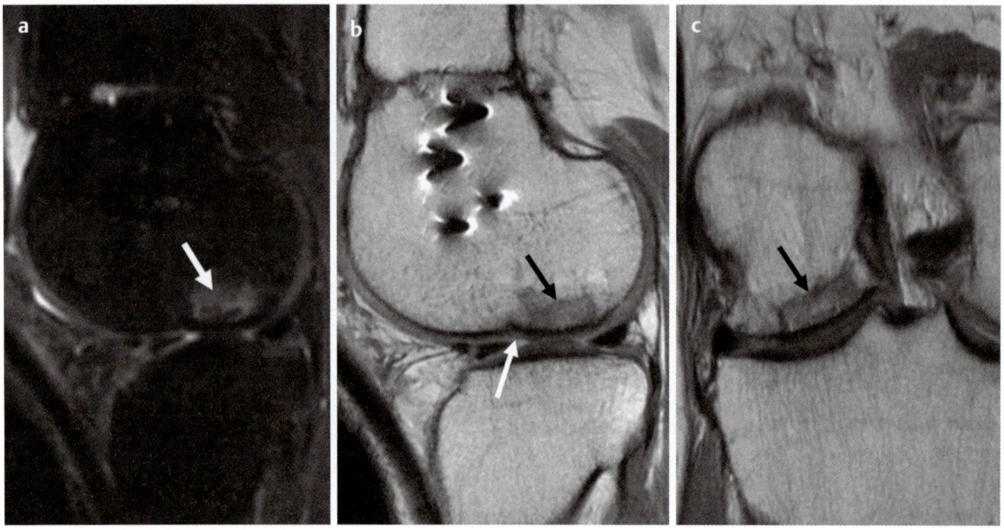

Fig. 2.9 Sagittal inversion recovery (**a**) and sagittal (**b**) and coronal (**c**) fast spin echo images in a 22-year-old woman 11 months following osteochondral allograft transplantation into the lateral femoral condyle. There is a mild bone marrow edema pattern (*thick white arrow*) with incomplete osseous incorporation (*black arrows*). There is overall good restoration of the radius of curvature of the joint. Small fissures are seen at the anterior margin of peripheral integration with the native articular cartilage (*thin white arrow*).

restoration of the radius of curvature of the joint surface.

Osseous incorporation is assessed by evaluating the graft marrow signal, graft–host interface, and graft congruity (**Fig. 2.9**). In the early postoperative period, mild or moderate bone marrow edema in the graft marrow and at the graft–host interface is common. Bone marrow edema decreases over time as graft incorporation occurs. Osseous trabecular incorporation manifests as normal fatty marrow signal extending within and around the plugs, and it has been shown to be complete or partial in the majority of patients and to correlate to clinical outcome measures, such as the Short Form-36 outcome score.[79] A graft–host interface that displays high signal on fluid-sensitive images and becomes thickened over time suggests the presence of granulation tissue, edema, or fluid due to incomplete incorporation and possible graft instability.[44] The presence of low signal intensity on all pulse sequences in the allograft bone strongly suggests loss of bone viability, which may lead to eventual implant failure.[79]

As the donor tissue is obtained from a foreign host, there is the potential for an adverse immunologic reaction to the OCA. Patients who express positive humoral immune responses demonstrate a decreased rate of bony incorporation and an increased rate of bone marrow edema pattern and surface collapse of the graft.[75] When the patient rejects an OCA, MR imaging usually demonstrates signal abnormalities in the graft marrow or at the graft–host interface before changes in the cartilage become evident. A reactive synovitis also appears to correlate to an adverse humoral response.[75]

While bony incorporation occurs in the majority of OCA transplants, fissures at the site of peripheral integration are common and histologically have been shown to represent fibrous tissue or fibrocartilage.[66] The cartilage over the allograft should display the normal signal characteristics and grayscale stratification of articular cartilage. Progressive cartilage degeneration over the allograft and cartilage delamination are potential complications that are well demonstrated on MR imaging.

◆ Conclusion

MR imaging is an increasingly used, noninvasive method for the assessment of articular cartilage defects and cartilage repair.

MR imaging may define the natural history of cartilage repair procedures and detect complications, potentially obviating the need for second-look arthroscopy. The continued application of quantitative MR imaging allows for assessment of tissue biochemistry. The future study of these pulse sequences may provide indirect insight into repair tissue mechanical properties, thus providing noninvasive insight into functional tissue capacity.

References

1. Mainil-Varlet P, Aigner T, Brittberg M, et al; International Cartilage Repair Society. Histological assessment of cartilage repair: a report by the Histology Endpoint Committee of the International Cartilage Repair Society (ICRS). J Bone Joint Surg Am 2003;85-A(Suppl 2):45–57
2. Brown WE, Potter HG, Marx RG, Wickiewicz TL, Warren RF. Magnetic resonance imaging appearance of cartilage repair in the knee. Clin Orthop Relat Res 2004;(422):214–223
3. Buckwalter JA, Hunziker E, Rosenberg J. Articular cartilage and knee joint function. In: Whit EJ ed. Articular Cartilage: Composition, Structure, Response to Injury and Methods of Facilitation of Repair. New York: Raven Press; 1990:19–56
4. Rosenberg L, Hellmann W, Kleinschmidt AK. Electron microscopic studies of proteoglycan aggregates from bovine articular cartilage. J Biol Chem 1975;250(5):1877–1883
5. Buckwalter JA, Mankin HJ. Articular cartilage: tissue design and chondrocyte-matrix interactions. Instr Course Lect 1998;47:477–486
6. Black BR, Chong R, Potter HG. Cartilage imaging in sports medicine. Sports Med Arthrosc 2009;17(1):68–80
7. Shindle MK, Foo LF, Kelly BT, et al. Magnetic resonance imaging of cartilage in the athlete: current techniques and spectrum of disease. J Bone Joint Surg Am 2006;88(Suppl 4):27–46
8. Eckstein F, Schnier M, Haubner M, et al. Accuracy of cartilage volume and thickness measurements with magnetic resonance imaging. Clin Orthop Relat Res 1998; (352):137–148
9. Eckstein F, Hudelmaier M, Wirth W, et al. Double echo steady state magnetic resonance imaging of knee articular cartilage at 3 Tesla: a pilot study for the Osteoarthritis Initiative. Ann Rheum Dis 2006;65(4):433–441
10. Glaser C, Tins BJ, Trumm CG, Richardson JB, Reiser MF, McCall IW. Quantitative 3D MR evaluation of autologous chondrocyte implantation in the knee: feasibility and initial results. Osteoarthritis Cartilage 2007;15(7):798–807
11. Potter HG, Foo LF. Magnetic resonance imaging of articular cartilage: trauma, degeneration, and repair. Am J Sports Med 2006;34(4):661–677
12. Potter HG, Linklater JM, Allen AA, Hannafin JA, Haas SB. Magnetic resonance imaging of articular cartilage in the knee. An evaluation with use of fast-spin-echo imaging. J Bone Joint Surg Am 1998;80(9):1276–1284
13. Potter HG, Foo LF. Magnetic resonance imaging of joint arthroplasty. Orthop Clin North Am 2006;37(3):361–373; vi–vii
14. Erickson SJ, Prost RW, Timins ME. The "magic angle" effect: background physics and clinical relevance. Radiology 1993;188(1):23–25
15. Fullerton GD, Rahal A. Collagen structure: the molecular source of the tendon magic angle effect. J Magn Reson Imaging 2007;25(2):345–361
16. Potter HG, Chong R. Magnetic resonance imaging assessment of chondral lesions and repair. J Bone Joint Surg Am 2009;91(Suppl 1):126–131
17. Glaser C. New techniques for cartilage imaging: T2 relaxation time and diffusion-weighted MR imaging. Radiol Clin North Am 2005;43(4):641–653, vii
18. de Visser SK, Bowden JC, Wentrup-Byrne E, et al. Anisotropy of collagen fibre alignment in bovine cartilage: comparison of polarised light microscopy and spatially resolved diffusion-tensor measurements. Osteoarthritis Cartilage 2008;16(6):689–697
19. Meder R, de Visser SK, Bowden JC, Bostrom T, Pope JM. Diffusion tensor imaging of articular cartilage as a measure of tissue microstructure. Osteoarthritis Cartilage 2006;14(9):875–881
20. Mosher TJ, Dardzinski BJ. Cartilage MRI T2 relaxation time mapping: overview and applications. Semin Musculoskelet Radiol 2004;8(4):355–368
21. Mosher TJ, Dardzinski BJ, Smith MB. Human articular cartilage: influence of aging and early symptomatic degeneration on the spatial variation of T2—preliminary findings at 3T. Radiology 2000;214(1):259–266
22. David-Vaudey E, Ghosh S, Ries M, Majumdar S. T2 relaxation time measurements in osteoarthritis. Magn Reson Imaging 2004;22(5):673–682
23. Alhadlaq HA, Xia Y, Moody JB, Matyas JR. Detecting structural changes in early experimental osteoarthritis of tibial cartilage by microscopic magnetic resonance imaging and polarised light microscopy. Ann Rheum Dis 2004;63(6):709–717
24. Glaser C, Mendlik T, Dinges J, et al. Global and regional reproducibility of T2 relaxation time measurements in human patellar cartilage. Magn Reson Med 2006;56(3):527–534
25. Raya JG, Dietrich O, Horng A, Weber J, Reiser MF, Glaser C. T2 measurement in articular cartilage: impact of the fitting method on accuracy and precision at low SNR. Magn Reson Med 2010;63(1):181–193
26. Mosher TJ, Zhang Z, Reddy R, et al. Knee articular cartilage damage in osteoarthritis: analysis of MR image biomarker reproducibility in ACRIN-PA 4001 multicenter trial. Radiology 2011;258(3):832–842
27. Lesperance LM, Gray ML, Burstein D. Determination of fixed charge density in cartilage using nuclear magnetic resonance. J Orthop Res 1992;10(1):1–13
28. Shapiro EM, Borthakur A, Gougoutas A, Reddy R. 23Na MRI accurately measures fixed charge density in articular cartilage. Magn Reson Med 2002;47(2):284–291
29. Wheaton AJ, Borthakur A, Shapiro EM, et al. Proteoglycan loss in human knee cartilage: quantitation with sodium MR imaging—feasibility study. Radiology 2004;231(3):900–905
30. Borthakur A, Shapiro EM, Akella SV, Gougoutas A, Kneeland JB, Reddy R. Quantifying sodium in the

human wrist in vivo by using MR imaging. Radiology 2002;224(2):598–602
31. Bashir A, Gray ML, Burstein D. Gd-DTPA2- as a measure of cartilage degradation. Magn Reson Med 1996;36(5):665–673
32. Wheaton AJ, Dodge GR, Borthakur A, Kneeland JB, Schumacher HR, Reddy R. Detection of changes in articular cartilage proteoglycan by T(1rho) magnetic resonance imaging. J Orthop Res 2005;23(1):102–108
33. Li X, Han ET, Ma CB, Link TM, Newitt DC, Majumdar S. In vivo 3T spiral imaging based multi-slice T(1rho) mapping of knee cartilage in osteoarthritis. Magn Reson Med 2005;54(4):929–936
34. Duvvuri U, Reddy R, Patel SD, Kaufman JH, Kneeland JB, Leigh JS. T1rho-relaxation in articular cartilage: effects of enzymatic degradation. Magn Reson Med 1997;38(6):863–867
35. Wheaton AJ, Casey FL, Gougoutas AJ, et al. Correlation of T1rho with fixed charge density in cartilage. J Magn Reson Imaging 2004;20(3):519–525
36. Regatte RR, Akella SV, Borthakur A, Kneeland JB, Reddy R. Proteoglycan depletion-induced changes in transverse relaxation maps of cartilage: comparison of T2 and T1rho. Acad Radiol 2002;9(12):1388–1394
37. Li X, Benjamin Ma C, Link TM, et al. In vivo T(1rho) and T(2) mapping of articular cartilage in osteoarthritis of the knee using 3T MRI. Osteoarthritis Cartilage 2007;15(7):789–797
38. Regatte RR, Akella SV, Wheaton AJ, et al. 3D-T1rho-relaxation mapping of articular cartilage: in vivo assessment of early degenerative changes in symptomatic osteoarthritic subjects. Acad Radiol 2004;11(7):741–749
39. Buckwalter JA. Articular cartilage: injuries and potential for healing. J Orthop Sports Phys Ther 1998;28(4):192–202
40. Curl WW, Krome J, Gordon ES, Rushing J, Smith BP, Poehling GG. Cartilage injuries: a review of 31,516 knee arthroscopies. Arthroscopy 1997;13(4): 456–460
41. Trattnig S, Winalski CS, Marlovits S, Jurvelin JS, Welsch GH, Potter HG. Magnetic resonance imaging of cartilage repair: a review. Cartilage 2011;2(1):5–26
42. Steadman JR, Rodkey WG, Rodrigo JJ. Microfracture: surgical technique and rehabilitation to treat chondral defects. Clin Orthop Relat Res 2001;(391, Suppl):S362–S369
43. Minas T, Nehrer S. Current concepts in the treatment of articular cartilage defects. Orthopedics 1997;20(6):525–538
44. Alparslan L, Winalski CS, Boutin RD, Minas T. Postoperative magnetic resonance imaging of articular cartilage repair. Semin Musculoskelet Radiol 2001;5(4):345–363
45. Mithoefer K, Williams RJ III, Warren RF, et al. The microfracture technique for the treatment of articular cartilage lesions in the knee. A prospective cohort study. J Bone Joint Surg Am 2005;87(9):1911–1920
46. Frisbie DD, Morisset S, Ho CP, Rodkey WG, Steadman JR, McIlwraith CW. Effects of calcified cartilage on healing of chondral defects treated with microfracture in horses. Am J Sports Med 2006;34(11):1824–1831
47. Frisbie DD, Oxford JT, Southwood L, et al. Early events in cartilage repair after subchondral bone microfracture. Clin Orthop Relat Res 2003;(407): 215–227
48. Frisbie DD, Trotter GW, Powers BE, et al. Arthroscopic subchondral bone plate microfracture technique augments healing of large chondral defects in the radial carpal bone and medial femoral condyle of horses. Vet Surg 1999;28(4):242–255
49. Peterson L, Minas T, Brittberg M, Nilsson A, Sjögren-Jansson E, Lindahl A. Two- to 9-year outcome after autologous chondrocyte transplantation of the knee. Clin Orthop Relat Res 2000;(374):212–234
50. Potter HG, Chong R, Sneag DB. Magnetic resonance imaging of cartilage repair. Sports Med Arthrosc 2008;16(4):236–245
51. Domayer SE, Kutscha-Lissberg F, Welsch G, et al. T2 mapping in the knee after microfracture at 3.0 T: correlation of global T2 values and clinical outcome—preliminary results. Osteoarthritis Cartilage 2008;16(8):903–908
52. Welsch GH, Mamisch TC, Domayer SE, et al. Cartilage T2 assessment at 3-T MR imaging: in vivo differentiation of normal hyaline cartilage from reparative tissue after two cartilage repair procedures—initial experience. Radiology 2008;247(1):154–161
53. Brittberg M, Lindahl A, Nilsson A, Ohlsson C, Isaksson O, Peterson L. Treatment of deep cartilage defects in the knee with autologous chondrocyte transplantation. N Engl J Med 1994;331(14):889–895
54. Minas T. Autologous chondrocyte implantation for focal chondral defects of the knee. Clin Orthop Relat Res 2001; (391, Suppl):S349–S361
55. Marlovits S, Zeller P, Singer P, Resinger C, Vécsei V. Cartilage repair: generations of autologous chondrocyte transplantation. Eur J Radiol 2006;57(1):24–31
56. Marlovits S, Singer P, Zeller P, Mandl I, Haller J, Trattnig S. Magnetic resonance observation of cartilage repair tissue (MOCART) for the evaluation of autologous chondrocyte transplantation: determination of interobserver variability and correlation to clinical outcome after 2 years. Eur J Radiol 2006;57(1):16–23
57. Breinan HA, Minas T, Hsu HP, Nehrer S, Sledge CB, Spector M. Effect of cultured autologous chondrocytes on repair of chondral defects in a canine model. J Bone Joint Surg Am 1997;79(10):1439–1451
58. Minas T, Peterson L. Advanced techniques in autologous chondrocyte transplantation. Clin Sports Med 1999;18(1):13–44; v–vi
59. Trattnig S, Ba-Ssalamah A, Pinker K, Plank C, Vecsei V, Marlovits S. Matrix-based autologous chondrocyte implantation for cartilage repair: noninvasive monitoring by high-resolution magnetic resonance imaging. Magn Reson Imaging 2005;23(7):779–787
60. Moriya T, Wada Y, Watanabe A, et al. Evaluation of reparative cartilage after autologous chondrocyte implantation for osteochondritis dissecans: histology, biochemistry, and MR imaging. J Orthop Sci 2007;12(3):265–273
61. Alparslan L, Minas T, Winalski CS. Magnetic resonance imaging of autologous chondrocyte implantation. Semin Ultrasound CT MR 2001;22(4): 341–351
62. Henderson IJ, Tuy B, Connell D, Oakes B, Hettwer WH. Prospective clinical study of autologous

chondrocyte implantation and correlation with MRI at three and 12 months. J Bone Joint Surg Br 2003;85(7):1060–1066

63. Tins BJ, McCall IW, Takahashi T, et al. Autologous chondrocyte implantation in knee joint: MR imaging and histologic features at 1-year follow-up. Radiology 2005;234(2):501–508

64. Trattnig S, Mamisch TC, Welsch GH, et al. Quantitative T2 mapping of matrix-associated autologous chondrocyte transplantation at 3 Tesla: an in vivo cross-sectional study. Invest Radiol 2007;42(6):442–448

65. Gillis A, Bashir A, McKeon B, Scheller A, Gray ML, Burstein D. Magnetic resonance imaging of relative glycosaminoglycan distribution in patients with autologous chondrocyte transplants. Invest Radiol 2001;36(12):743–748

66. Glenn RE Jr, McCarty EC, Potter HG, Juliao SF, Gordon JD, Spindler KP. Comparison of fresh osteochondral autografts and allografts: a canine model. Am J Sports Med 2006;34(7):1084–1093

67. Matsusue Y, Yamamuro T, Hama H. Arthroscopic multiple osteochondral transplantation to the chondral defect in the knee associated with anterior cruciate ligament disruption. Arthroscopy 1993;9(3):318–321

68. Bobić V. Arthroscopic osteochondral autograft transplantation in anterior cruciate ligament reconstruction: a preliminary clinical study. Knee Surg Sports Traumatol Arthrosc 1996;3(4):262–264

69. Nho SJ, Foo LF, Green DM, et al. Magnetic resonance imaging and clinical evaluation of patellar resurfacing with press-fit osteochondral autograft plugs. Am J Sports Med 2008;36(6):1101–1109

70. Koh JL, Wirsing K, Lautenschlager E, Zhang LO. The effect of graft height mismatch on contact pressure following osteochondral grafting: a biomechanical study. Am J Sports Med 2004;32(2): 317–320

71. Hangody L, Kish G, Kárpáti Z, Udvarhelyi I, Szigeti I, Bély M. Mosaicplasty for the treatment of articular cartilage defects: application in clinical practice. Orthopedics 1998;21(7):751–756

72. Pearce SG, Hurtig MB, Clarnette R, Kalra M, Cowan B, Miniaci A. An investigation of 2 techniques for optimizing joint surface congruency using multiple cylindrical osteochondral autografts. Arthroscopy 2001;17(1):50–55

73. Lane JM, Brighton CT, Ottens HR, Lipton M. Joint resurfacing in the rabbit using an autologous osteochondral graft. J Bone Joint Surg Am 1977;59(2): 218–222

74. Dew TL, Martin RA. Functional, radiographic, and histologic assessment of healing of autogenous osteochondral grafts and full-thickness cartilage defects in the talus of dogs. Am J Vet Res 1992;53(11):2141–2152

75. Sirlin CB, Brossmann J, Boutin RD, et al. Shell osteochondral allografts of the knee: comparison of MR imaging findings and immunologic responses. Radiology 2001;219(1):35–43

76. Williams RJ, Gamradt SC. Articular cartilage repair using a resorbable matrix scaffold. Instr Course Lect 2008;57:563–571

77. Bedi A, Foo LF, Williams RJ III, Potter HG; The Cartilage Study Group. The maturation of synthetic scaffolds for osteochondral donor sites of the knee: an MRI and T2 mapping analysis. Cartilage 2010;1(1):20–28

78. Foo LF, Jawetz ST, Williams RJ. MRI and quantitative T2 mapping of cartilage repair using synthetic biphasic acellular scaffold. ISMRM annual meeting (paper presentation, abstract 1030). Berlin, Germany; 2007

79. Williams RJ III, Ranawat AS, Potter HG, Carter T, Warren RF. Fresh stored allografts for the treatment of osteochondral defects of the knee. J Bone Joint Surg Am 2007;89(4):718–726

3

Evolving Biomarkers in Osteoarthritis

Debabrata Patra and Linda J. Sandell

Accurate and efficient diagnosis and prognosis are central to effective patient care. Localized pain, discomfort, or an annual check-up is usually the starting point for patient examination by a physician. But the route of patient care taken by a physician depends on several factors derived from blood tests, radiographs, magnetic resonance imaging (MRI) scans, or a physical examination, all of which generate a set of biomarkers advising the physician on the next step in patient care. Thus, conceptually a biomarker is any set of parameters that when out of the normal range would inform the physician of abnormal body physiology or injury in the patient. In the field of osteoarthritis (OA), however, a physician has very little on which to plan an effective, in-depth patient care stratagem to prevent OA progression. Patients are usually seen when the disease has advanced to a stage at which further progression cannot be prevented. At this stage, a physician's recommendations are limited to palliative or salvage procedures such as activity modification as in rest and/or joint immobilization, anti-inflammatories, analgesics, or total joint arthroplasty. These limitations in therapeutic options reflect a lack of clinically relevant diagnostic or disease-staging methodologies and the unavailability of any disease-modifying treatments, compounded by an incomplete understanding of OA pathogenesis in conjunction with the heterogeneity of the human population.

Like several other chronic diseases, OA is characterized by a silent phase of the disease when the individual is not aware that the disease process has begun. **Fig. 3.1** is a representation of the phases of OA in humans (modified from Kraus 2010)[1] where biomarker validation is desirable for phases 0 and 1 but is currently hypothetical. This has led to a concentrated movement in search of biomarkers to detect early OA. During early OA, though an individual may appear normal on radiographs or MRI scans, abnormalities in joint physiology may indicate OA initiation by elevations in the levels of cartilage breakdown products or markers of inflammation. These biochemical markers are the next wave in OA diagnosis, monitoring OA progression, understanding changes in body physiology, and determining efficacy of pharmaceutical interventions. An OA biochemical biomarker could be any molecule that would act as a surrogate to inform the physician of either disease initiation or progression and allow for accurate disease staging and prognosis for determining optimal preventive and therapeutic strategies. Ideally biomarkers for OA will be joint-specific and detectable during the silent, nonsymptomatic phase of OA. If possible, biomarkers will also be able to

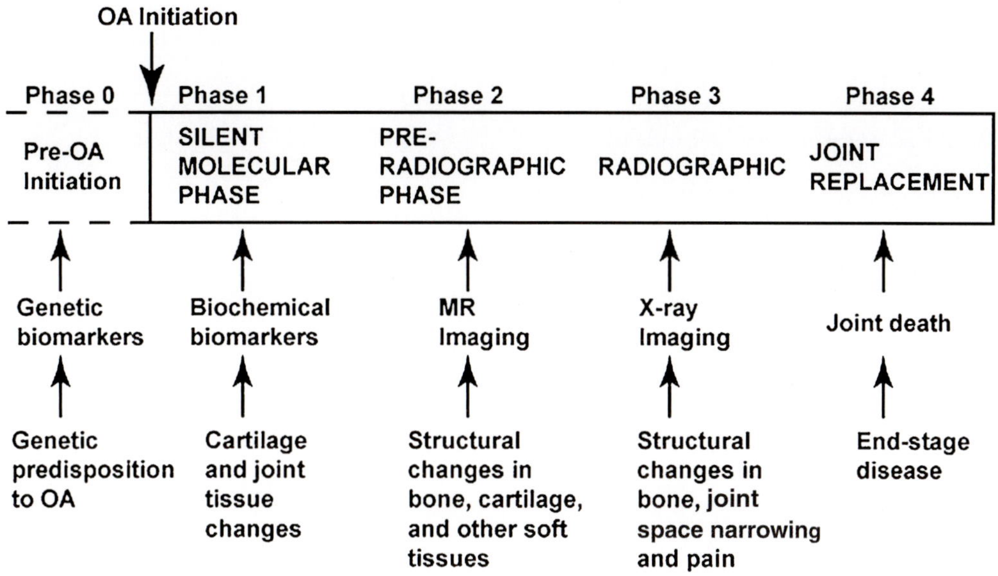

Fig. 3.1 A diagrammatic representation of the osteoarthritis (OA) disease process in humans showing the various stages of the disease and an accompanying diagnosing scheme. Patients usually enter this scheme at phase 3, but with advances in magnetic resonance (MR) imaging they are beginning to be detected in phase 2. Early OA detection in phases 0 and 1 is desirable but not in practice due to incomplete development of biomarkers for these stages.

distinguish age- and trauma-related changes to allow a physician to assess the degree of damage after injury.

To derive the maximum benefit from a biomarker it would be advisable to arrive at a consensus for a definition of the various stages of the disease process. In OA, often this is lacking. Many individuals are often reported to have radiographic OA as defined by joint space narrowing. However, most of these individuals often do not have symptomatic OA as defined by pain or discomfort. But joint space narrowing would be due to changes in articular cartilage structure that would reflect changes in articular cartilage composition. An individual at such a stage of OA, radiographic but nonsymptomatic, may not be denoted clinically as having OA and therefore would be bypassed for treatment. This puts the burden of proof on the biomarker as it should be validated as truly representative of the disease process, should be an irrefutable proof of disease initiation, and should be universally recognized as such. A late-stage biomarker more representative of disease progression is not likely to be of use in detecting the elusive, early OA phases. Thus, a primary goal of OA biomarker development is to catch the disease in the early stages to prevent the ravages of late-stage OA. Several biomarkers are currently in use but have varying degrees of acceptance for their utility in OA (**Table 3.1**). The OA Biomarkers Global Initiative has been organized by Osteoarthritis Research Society International (OARSI) in the recent past to accelerate OA biomarker development.[2] A documentation by the OARSI Food and Drug Administration (FDA) initiative provides a summary and guide for application of in vitro, soluble, biochemical biomarkers for monitoring OA and pharmacological trials and also provides a scheme for the classification of these biomarkers.[3]

◆ Biomarkers Derived from Type II Collagen

The fibrillar type II collagen (Col II) is the major collagen of articular cartilage. Col II is synthesized as a procollagen with the amino

Table 3.1 A list of biochemical biomarkers derived from cartilage, bone, and synovial tissue mentioned in this review that are under scrutiny for their usefulness in diagnosing, staging, and monitoring OA progression and in monitoring pharmacological interventions

Biochemical biomarker	Indicator of	Tissue origin	Potential for clinical application
PIINP, PIIANP, NPII	Type II collagen synthesis	Cartilage	Mild
PIICP (CPII)	Type II collagen synthesis	Cartilage	Mild
CTX-II	Type II collagen breakdown	Cartilage, bone	Exceptionally strong
TIINE, Coll 2–1, Coll 2–1NO_2	Type II collagen breakdown	Cartilage	Mild
ARGS	Aggrecan breakdown	Cartilage	Moderate
AGG1 (G1–1H11), AGG2 (6D6-G2), CS846	Aggrecan breakdown/ turnover	Cartilage	Mild
Keratan sulfate	Aggrecan breakdown	Cartilage	Moderate
COMP	COMP breakdown	Cartilage, meniscus, synovium	Mild
NTX-1, CTX-1, PINP	Bone resorption	Bone	Moderate
Osteocalcin	Bone synthesis	Bone	Moderate
HA, Glc-Gal-PYD	Synovial tissue breakdown	Synovium, though HA could also be from cartilage and meniscus	Moderate
MMP-3, MMP-13	Joint tissue breakdown	Cartilage, synovium	Moderately strong

Note: The potential for clinical application of these biochemical biomarkers is subjective and based on their popularity for use in following and monitoring OA progression in published studies, recent advances, and their utility and success in published pharmacological trials (see text for details).

(N)-terminal (PIINP and PIIANP) and carboxy (C)-terminal (PIICP; also referred to as CPII) domains removed during assembly of the molecule. Therefore, enzyme-linked immunosorbent assays (ELISA) to detect these in urine or serum using antibodies to PIINP/PIIANP/PIICP have been designed to indicate recent Col II synthetic activity or increased collagen synthesis as a sign of abnormal cartilage metabolism. In a 4-year study to investigate the prognostic value of PIICP, synovial fluid levels of PIICP were found to correlate well with radiographic progression of knee OA.[4] However, detection of PIINP (produced primarily by mature chondrocytes[5]) and PIIANP (identical to PIINP except that it has an additional 69 amino acid, cysteine-rich domain and is produced by chondroprogenitor cells and dedifferentiated, pathological chondrocytes) provides an additional dimension in that changes in PIIANP:PIINP ratios could be a reflection of the pathological state of the cartilage. The NPII assay (designed to measure peptides derived from the Col II N-propeptide with the antibody detecting different peptides than the PIIANP assay) demonstrated that N-terminal procollagen type II derived peptide levels are indeed higher in human plasma and urine from patients with radiographically confirmed OA,[6] suggesting that detecting Col II–derived peptides indicative of synthesis is a viable option for biomarkers. However, these potential biomarkers have not been validated for clinical use in OA diagnosis or monitoring treatment.

The biochemical biomarkers to date that have received a lot of attention and have shown promise in diagnosing OA effectively are primarily those that detect Col II breakdown. Of these, C-terminal cross-linked telopeptide of Col II (CTX-II), which can be detected in urine in humans by an ELISA, is the most characterized and appears to be the most promising as a noninvasive biomarker for monitoring OA.[7] It provides specific and sensitive data regarding Col II breakdown and has found acceptance in OA and rheumatoid arthritis (RA) as a biomarker for joint structure changes. It has been demonstrated to correlate well with power Doppler ultrasound synovitis and bone mineral density loss that are early markers of inflammatory arthritis.[8] However, as several OA and RA patients demonstrate normal levels of urinary (u) CTX-II, CTX-II alone may not be diagnostically useful. Furthermore, CTX-II was detected in calcified cartilage–bone interface besides the articular cartilage matrix,[9] suggesting that the tissue origins of CTX-II are not completely understood. However, levels of CTX-II have been demonstrated to correlate well with total body burden of osteophyte, a major pathological feature of OA, suggesting that CTX-II can be a component of a biomarker panel for clinical use in OA.[10] In practice, CTX-II changes are often monitored in combination with other biomarkers, and its use has found success in several pharmacological trials, though in monitoring primarily hip and knee OA. As such, by combining markers of both Col II synthesis (via use of PIIANP) and collagen breakdown (via use of CTX-II), it was shown that patients with knee OA who had the largest uncoupling between Col II synthesis (low levels of PIIANP) and Col II degradation (high levels for CTX-II) had an eightfold more rapid progression of joint damage than other patients,[11] suggesting that combining these two biomarkers would be effective in identifying subjects with high risk for progressive knee OA. CTX-II was also used to monitor the beneficial aspects of orally treating knee OA patients with salmon calcitonin (sCT).[12] In this study it was monitored in combination with other noteworthy biomarker candidates for OA such as matrix metalloproteinase (MMP)-3 (considered a significant predictor of joint space narrowing), along with the collagenase MMP-13 (enzyme that degrades Col II), all of which demonstrated significant decreases on sCT intake. CTX-II alone has also found success in well-controlled studies to identify OA patients with high cartilage turnover and also in monitoring pharmacological intervention of OA with glucosamine sulfate.[13]

Besides CTX-II, other Col II breakdown products have also been used as biomarkers in monitoring OA, such as TIINE, Coll 2–1, and Coll 2–1NO_2 in both serum and urine in immunoassays. Urinary TIINE is a Col II neoepitope and unlike CTX-II has the advantage that the exact nature of the immunoreactive epitope is known and has been characterized by mass spectrometry.[14] It is produced by the action of several MMPs, such as MMP-13, and this study demonstrated that an MMP-13 inhibitor can reduce the levels of the TIINE neoepitope, suggesting its utility as a biomarker in monitoring drug efficacy. Assays that detect Coll 2–1 and its nitrated form, Coll 2–1NO_2, provide information on oxidative related helical unwinding or further breakdown of the triple helical region of Col II and have been found to be increased in patients with primary knee OA.[15,16] These relatively new biomarkers have, however, found only limited success in monitoring pharmacological trials.

◆ Biomarkers Derived from Aggrecan

Although monitoring Col II–derived degradation products has been a primary focus of investigators, several aggrecan breakdown products informative of OA pathological conditions may have potential for application as biomarkers. Given the fact that loss of proteoglycan staining in the articular cartilage in models of OA is almost immediate, aggrecan-derived biomarkers could potentially be effective for early detection of OA. In fact, aggrecan breakdown resulting in production of the ARGS neoepitope sequence in aggrecanase-cleaved aggrecan[17] has recently been demonstrated to be detectable in human synovial fluid, serum, and urine, with the second-generation BC3-C2 antibody in a sensitive immunoassay[18] opening up avenues for aggrecan-derived neoepitopes being considered as worthy biomarkers for OA diagnosis and as clinical end points for disease-modifying OA drugs. The degenerative aggrecan breakdown products AGG1 (G1–1H11) and AGG2 (6D6-G2) have been used in combination with CTX-II to validate elevated levels of synovial fluid–derived visfatin as indicative of degenerative cartilage changes during knee OA.[19] Likewise, in female patients with knee OA, AGG1 and AGG2 correlated well with synovial fluid adiponectin, though no correlation was observed with CTX-II.[20] Immunoassays designed to detect a complex of fibronectin (another cartilage component) and aggrecan in synovial fluids have found applications in monitoring knee pain due to meniscal injury.[21] Fluctuations in the serum levels of CS846, a derivative of the chondroitin sulfate side chains of aggrecan, indicative of both aggrecan synthesis and turnover, have been shown to be associated with joint space narrowing, though it did not show promise in a pharmacological trial.[22] But significant decreases in serum levels of keratan sulfate (from the keratan sulfate side chains of aggrecan) in patients with knee OA treated with chondroitin sulfate have been demonstrated, suggesting the usefulness of keratan sulfate in monitoring OA pharmacological trials and in the utility of chondroitin sulfate in treating knee OA symptoms.[23]

◆ Bone, Synovium, and Other Sources of Biochemical Biomarkers

As OA is a disease of the joint as opposed to only the articular cartilage, bone and synovial tissue degenerative changes also contribute to the arsenal of biochemical biomarkers for OA. But bone markers have not gained the level of confidence that cartilage-derived biomarkers have achieved. This is due to the fact that bone markers are not likely to be exclusive and selective indicators of OA and may not be able to distinguish OA-related pathophysiology from other bone activities, such as those resulting from postmenopausal osteoporosis, age-related skeletal turnover, or other bone diseases, such as osteonecrosis.[24,25] However, bone markers have been used effectively in combination with cartilage turnover markers to assess OA and effectiveness of pharmacological interventions in OA. The urinary bone resorption marker N-telopeptide of type I collagen (NTX-1) in combination with the uCTX-II was used to determine the efficacy of risedronate in improving joint structure and reducing pain, stiffness, and joint function in patients with primary knee OA.[26,27] A significant drop in both uCTX-II and uNTX-I was observed in these patients with risedronate treatment (though found ineffective at reducing OA progression).

Very few study designs have based their conclusions on data from a single biomarker. In the vast majority of studies, combinatorial biomarker tracking has allowed investigators to distinguish between different stages of OA and definitions of OA. For example, in a recent study to test the relationship between biomarkers and early radiographically defined knee OA (based on the Kellgren-Lawrence [K/L] grading system), knee pain, and joint inflammation, uCTX-II levels alone did not correlate well with pain, but the ratio of uCTX-II/serum CPII and levels of the synovial marker hyaluronic acid (HA) increased with onset of OA irrespective of joint pain in patients with K/L grade 2 of OA.[28] On the other hand, in the same study, levels of uCTX-II alone as well as that of uNTX-I along with CPII and HA all increased significantly in patients if they had knee pain irrespective of the K/L grade. A combination of bone formation (PINP and osteocalcin) and CTX-I and NTX-I (both indicators of bone resorption activity) was used to classify patients with knee OA into subgroups who lose cartilage at different rates over 2 years.[29] This study showed that higher bone remodeling is associated with reduced cartilage loss. Thus, within the OA population it will be possible to subgroup individuals based on rates of cartilage loss when combining both bone resorption and formation markers.

Detection of the glycosylated analogue of pyridinoline, glucosyl galactosyl pyridinoline (Glc-Gal-PYD) in urine by high performance liquid chromatography reflects specifically synovial tissue degeneration.[30] Glc-Gal-PYD in combination with uCTX-II was found to have a strong association with disease severity and the presence of OA at the tibiofemoral and patellofemoral joints in men.[31] In the same study, changes in serum levels of CTX-I and osteocalcin, however, did not correlate well with the disease. Glc-Gal-PYD and uCTX-II was also used to study the efficacy of ibuprofen in reducing the symptoms of knee OA.[32] Several synovial fluid–derived biomarkers, such as interleukin (IL)-1, IL-6, IL-8, IL-11, leukemia inhibitory factor (LIF), cartilage oligomeric matrix protein (COMP), osteocalcin, and osteogenic protein-1 (OP-1), have been analyzed in the synovial fluid from OA and RA patients.[33] This study suggested that elevated levels of IL-11, LIF, and OP-1 would be more appropriate as biomarkers indicating OA, while elevated levels of IL-1, IL-6, IL-8, LIF, and OP-1 would be more likely to be indicative of RA.

The enzymes that break down Col II and aggrecan are themselves excellent indicators of disease states. Native fibrillar Col II is broken down first by the combined action of several MMPs such as MMP-1, -8, -13, and -14 and then broken down further by MMP-2, -3, and -9. Thus, the detection of these Col II modifying enzymes by immunoassays can also serve as biomarkers. As mentioned above, in studies to test the effectiveness of sCT in alleviating symptoms of OA, MMP-3 and MMP-13 along with CTX-II also served as biomarkers.[12] Significant reduction in levels of both MMP-3 and MMP-13 was observed, attesting to the utility of these enzymes as useful biomarkers for future drug efficacy studies. Thus, a panel of biomarkers will allow for rapid diagnosis and improved determination of OA. It should be noted, however, that a vast majority of these studies were conducted with patients suffering from knee OA. Therefore biomarkers that have found acceptance as indicative of OA are directly informative of knee OA as per these studies. In some cases, they are also indicative of hip OA. However, much work still needs to be done to validate these biomarkers as indicative of generalized OA or their relationship to other types of OA such as hand OA.

◆ Diurnal Variations in Biomarker Detection

Most studies using biomarkers are conducted within the framework of several variables that are often not identical across studies. Therefore unifying the diagnostic potential of a particular biomarker among different studies will need a standardization of the methodology to conduct these studies that does not exist currently. A critical limiting factor that prevents this unification is the diurnal variation among the biological samples that are collected. Serum and/or urine collected at different times of the day or night, before or after fasting, with or without physical activity, show significant differences in levels of some biomarkers. Levels of serum COMP are found to be constant during normal daytime activities but decrease during the night.[34] Thus, as long as clinical sampling is limited to daytime, no further standardization is necessary for analyzing serum COMP levels. However, in patients with radiographic knee OA, uCTX-II concentrations were found to vary with morning activity, decreasing significantly when sampled ~4 hours after arising from bed in the morning as compared with sampling before arising from bed.[35] On the other hand serum levels of HA increased significantly when analyzed 1 hour after arising from bed as compared with sampling before arising from bed.[36] Diurnal variation in CTX-I levels is also well established, decreasing significantly after food intake when compared with fasting levels.[37] These studies suggest that serum or urine sampling for any biomarker needs to be standardized and these standardization protocols should be applied both for diagnosis and for determining the efficacy of drugs in clinical trials. It would also be important to correlate the levels of biomarkers with different stages of the disease to make biological sense. For example, low levels of cartilage turnover markers in advanced stages of OA when there is little or no cartilage left need to be interpreted in combination with other diagnostic parameters before coming to a conclusion about the stage of the disease.

◆ Genotypes as Potential Biomarkers for OA

In addition to biochemical biomarkers, genetic biomarkers are also being developed to identify people with a genetic disposition to develop OA, significantly ahead of any radiographic changes. OA is a complex disease and its etiology is poorly understood. Besides the age-related degenerative cartilage changes, myriad factors come into play whose contribution to OA are difficult to quantify and enumerate. Genetic biomarkers are designed primarily to understand the etiology of OA with the perk being that they may also shed light on the pathophysiology of OA. The premise for this undertaking is the simple observation that OA tends to run in families. Several approaches have been taken to understand the heritability of OA; chief among them are candidate gene studies or the genome-wide linkage scans (GWAS). Candidate gene studies differ from GWAS studies in that one has advance knowledge of the association of the gene with cartilage and/or skeletal development; GWAS studies on the other hand were implemented to identify hitherto nonrelated OA susceptibility genes. Both these approaches have identified mutations and polymorphisms in several ECM genes and signaling molecules that show strong association with OA. In the candidate gene studies, several mutations in ECM genes resulting in various forms of skeletal dysplasias have been studied that demonstrate that these diseases are severe, early onset, hereditary forms of OA.[38] Identification of the disease genes by linkage analysis has uncovered mutations in the *COMP* gene to be responsible for pseudoachondroplasia (PSACH);[39] mutation in the type IX collagen (*COL9A1*)[40] or the matrilin-3 (*MATN3*) gene results in a dominant form of multiple epiphyseal dysplasia (MED);[41] spondyloepiphyseal dysplasia (SED) congenita results from a mutation in the type II collagen gene (*COL2A1*).[42] Patients suffering from these diseases have a short stature and invariably suffer from a highly specific form of OA as they get older. For example, patients with SED congenita exhibit severe hip OA, with moderately severe spine and knee OA but never hand OA. But patients with *COMP*-related MED demonstrate severe hip and hand OA but minimal knee OA. These differences occur despite the Col II and COMP proteins being widespread in cartilage. Thus these are examples of monogenic diseases showing familial OA that are essentially biomarkers for OA.

While candidate gene association studies target only a small portion of the human genome, GWAS studies on the other hand perform a genome-wide screen and have used primarily single nucleotide polymorphism (SNP) in genes to study their association with OA. These studies have identified several genes, but chief among them is the SNP at the + 104 position in the 5′ UTR of the growth and differentiation factor 5 (*GDF5*) gene that has shown a strong association to OA.[43] In humans, the nucleotide T at + 104, as opposed to C, results in higher incidence of OA and was first discovered in the Japanese and Chinese population with knee and hip OA. This SNP exerts allelic differences in transcriptional activity in chondrogenic cells with the OA-associated + 104T allele, resulting in reduced *GDF5* transcriptional activity. In a mouse model of dominant-negative *Gdf5* mutation, joint formation is impaired and mice develop severe OA endorsing the association between *GDF5* and OA in humans.[44] Currently, the *GDF5* gene has acquired a reputation that is close to that of a global OA gene as later studies have replicated this association in European populations as well.[45] Besides *GDF5*, an aspartic acid repeat polymorphism (the D14 allele) in the asporin (*ASPN*) gene has also shown strong association with hip and knee OA in different ethnic populations, and like the *GDF5* gene, it negatively affects chondrogenesis, the D14 allele resulting in a stronger inhibition of TGF-β induced expression of cartilage ECM genes such as aggrecan and type II collagen.[46] Interestingly, an SNP in the IL-1β gene, a chemokine known to drive the pathobiology of OA, is associated with severe erosive hand OA.[47] Thus, these studies hint at the possibility to screen individuals genetically and identify them to be at risk for OA even before incidents of radiographic OA or pain, effectively behaving as genetic biomarkers for OA.

It is important to realize, however, that ethnic differences can still come into play, and different genetic biomarkers may need to be validated for different ethnic populations. Even though GWAS studies are often replicated using different ethnic populations to increase the reliability of a disease-associated SNP, there is still considerable diversity among different ethnic populations and even among individuals of the same nationality. In some replication studies, differences were also noted between the sexes.

In the Rotterdam study, SNPs in *GDF5* gene correlated with knee OA, but not with hip OA, primarily in women; no associations were found in men.[48] These differences may be explained by the inclusion criteria of the population used in the study (patients had less severe radiographic OA as compared with previous studies); the differences, however, could also be explained by hormonal differences between men and women. Thus, a common genetic biomarker for both men and women may not be appropriate to diagnose OA. Furthermore, the arcOGEN consortium stage 1 study suggested that OA is a highly polygenic disease with multiple genes making contributions to the OA disease phenotype.[49] Thus, a genetic biomarker based on a single gene may not have the diagnostic power to identify individuals of different ethnic populations at risk for OA. The standardization of inclusion criteria of the population cohorts for GWAS studies, as well as a standardization of the definitions of different stages of OA, would help in unifying data from different laboratories. Furthermore, like biochemical biomarkers, combinatorial genetic biomarker tracking might need to be explored to increase the level of confidence in their usefulness for OA.

◆ Future in OA Biomarker Development

Though a significant body of work exists on biomarker development in OA, none of these biomarkers can yet be considered to be the gold standard for diagnosing, staging, monitoring treatment, or prognosticating for OA. Part of this problem is due to the complex nature of OA, the lack of complete understanding of its pathophysiology, and the heterogeneity of the human population. With the current crop of biomarkers, the pitfalls in diagnosis may be avoided by tracking a panel of biomarkers and/or by expressing them in ratios. In the recent past, several worthy efforts have been undertaken to apply the latest state-of-the-art proteomic technologies to improve understanding of disease mechanisms of OA while also generating new biomarkers for potential clinical application. The use of liquid chromatography coupled with mass spectrometry has allowed for the detection of altered lipid metabolism in plasma in humans with OA.[50] Thus, abnormal lipid profiling, through lipidomics, is a worthy addition as a biomarker for a more accurate OA diagnosis. Likewise, the use of metabolomics has identified abnormal metabolite ratios of valine to histidine and of leucine (or isoleucine) to histidine in human sera as significantly associated with knee OA.[51] It seems likely that these "thinking out of the box" approaches to OA will allow for a better understanding of the disease process with consequent identification and validation of a specific set of biomarkers to diagnose OA earlier, and to allow for better treatment options in OA.

Acknowledgments

We thank Virginia Kraus and Bentham Science (publishers of Current Drug Targets) for permission to modify their figure on which our Fig. 3.1 is based. This work was supported by National Institutes of Health grants RO1 AR50847, RO1 AR45550, and RO1 AR36994 to L. J. Sandell.

References

1. Kraus VB. Waiting for action on the osteoarthritis front. Curr Drug Targets 2010;11(5):518–520
2. Kraus VB, Nevitt M, Sandell LJ. Summary of the OA biomarkers workshop 2009—biochemical biomarkers: biology, validation, and clinical studies. Osteoarthritis Cartilage 2010;18(6):742–745
3. Kraus VB, Burnett B, Coindreau J, et al; OARSI FDA Osteoarthritis Biomarkers Working Group. Application of biomarkers in the development of drugs intended for the treatment of osteoarthritis. Osteoarthritis Cartilage 2011;19(5):515–542
4. Sugiyama S, Itokazu M, Suzuki Y, Shimizu K. Procollagen II C propeptide level in the synovial fluid as a predictor of radiographic progression in early knee osteoarthritis. Ann Rheum Dis 2003;62(1):27–32
5. Sandell LJ, Morris N, Robbins JR, Goldring MB. Alternatively spliced type II procollagen mRNAs define distinct populations of cells during vertebral development: differential expression of the amino-propeptide. J Cell Biol 1991;114(6):1307–1319
6. Nemirovskiy OV, Sunyer T, Aggarwal P, Abrams M, Hellio Le Graverand MP, Mathews WR. Discovery and development of the N-terminal procollagen type II (NPII) biomarker: a tool for measuring collagen type II synthesis. Osteoarthritis Cartilage 2008;16(12):1494–1500

7. Christgau S, Garnero P, Fledelius C, et al. Collagen type II C-telopeptide fragments as an index of cartilage degradation. Bone 2001;29(3):209–215
8. Freeston JE, Garnero P, Wakefield RJ, Hensor EM, Conaghan PG, Emery P. Urinary type II collagen C-terminal peptide is associated with synovitis and predicts structural bone loss in very early inflammatory arthritis. Ann Rheum Dis 2011; 70(2):331–333
9. Bay-Jensen AC, Andersen TL, Charni-Ben Tabassi N, et al. Biochemical markers of type II collagen breakdown and synthesis are positioned at specific sites in human osteoarthritic knee cartilage. Osteoarthritis Cartilage 2008;16(5):615–623
10. Kraus VB, Kepler TB, Stabler T, Renner J, Jordan J. First qualification study of serum biomarkers as indicators of total body burden of osteoarthritis. PLoS ONE 2010;5(3):e9739
11. Garnero P, Ayral X, Rousseau JC, et al. Uncoupling of type II collagen synthesis and degradation predicts progression of joint damage in patients with knee osteoarthritis. Arthritis Rheum 2002;46(10):2613–2624
12. Manicourt DH, Azria M, Mindeholm L, Thonar EJ, Devogelaer JP. Oral salmon calcitonin reduces Lequesne's algofunctional index scores and decreases urinary and serum levels of biomarkers of joint metabolism in knee osteoarthritis. Arthritis Rheum 2006;54(10):3205–3211
13. Christgau S, Henrotin Y, Tankó LB, et al. Osteoarthritic patients with high cartilage turnover show increased responsiveness to the cartilage protecting effects of glucosamine sulphate. Clin Exp Rheumatol 2004;22(1):36–42
14. Nemirovskiy OV, Dufield DR, Sunyer T, Aggarwal P, Welsch DJ, Mathews WR. Discovery and development of a type II collagen neoepitope (TIINE) biomarker for matrix metalloproteinase activity: from in vitro to in vivo. Anal Biochem 2007;361(1):93–101
15. Deberg MA, Labasse AH, Collette J, Seidel L, Reginster JY, Henrotin YE. One-year increase of Coll 2-1, a new marker of type II collagen degradation, in urine is highly predictive of radiological OA progression. Osteoarthritis Cartilage 2005;13(12):1059–1065
16. Deberg M, Labasse A, Christgau S, et al. New serum biochemical markers (Coll 2-1 and Coll 2-1 NO2) for studying oxidative-related type II collagen network degradation in patients with osteoarthritis and rheumatoid arthritis. Osteoarthritis Cartilage 2005;13(3):258–265
17. Sandy JD, Flannery CR, Neame PJ, Lohmander LS. The structure of aggrecan fragments in human synovial fluid. Evidence for the involvement in osteoarthritis of a novel proteinase which cleaves the Glu 373-Ala 374 bond of the interglobular domain. J Clin Invest 1992;89(5):1512–1516
18. Swearingen CA, Carpenter JW, Siegel R, et al. Development of a novel clinical biomarker assay to detect and quantify aggrecanase-generated aggrecan fragments in human synovial fluid, serum and urine. Osteoarthritis Cartilage 2010;18(9): 1150–1158
19. Duan Y, Hao D, Li M, et al. Increased synovial fluid visfatin is positively linked to cartilage degradation biomarkers in osteoarthritis. Rheumatol Int 2011; January 19 (Epub ahead of print)
20. Hao D, Li M, Wu Z, Duan Y, Li D, Qiu G. Synovial fluid level of adiponectin correlated with levels of aggrecan degradation markers in osteoarthritis. Rheumatol Int 2010; May 13 (Epub ahead of print)
21. Scuderi GJ, Golish SR, Cook FF, Cuellar JM, Bowser RP, Hanna LS. Identification of a novel fibronectin-aggrecan complex in the synovial fluid of knees with painful meniscal injury. J Bone Joint Surg Am 2011;93(4):336–340
22. Mazzuca SA, Poole AR, Brandt KD, Katz BP, Lane KA, Lobanok T. Associations between joint space narrowing and molecular markers of collagen and proteoglycan turnover in patients with knee osteoarthritis. J Rheumatol 2006;33(6):1147–1151
23. Uebelhart D, Thonar EJ, Delmas PD, Chantraine A, Vignon E. Effects of oral chondroitin sulfate on the progression of knee osteoarthritis: a pilot study. Osteoarthritis Cartilage 1998;6(Suppl A):39–46
24. Garnero P, Delmas PD. Noninvasive techniques for assessing skeletal changes in inflammatory arthritis: bone biomarkers. Curr Opin Rheumatol 2004;16(4):428–434
25. Marx RE, Cillo JE Jr, Ulloa JJ. Oral bisphosphonate-induced osteonecrosis: risk factors, prediction of risk using serum CTX testing, prevention, and treatment. J Oral Maxillofac Surg 2007;65(12):2397–2410
26. Bingham CO III, Buckland-Wright JC, Garnero P, et al. Risedronate decreases biochemical markers of cartilage degradation but does not decrease symptoms or slow radiographic progression in patients with medial compartment osteoarthritis of the knee: results of the two-year multinational knee osteoarthritis structural arthritis study. Arthritis Rheum 2006;54(11):3494–3507
27. Spector TD, Conaghan PG, Buckland-Wright JC, et al. Effect of risedronate on joint structure and symptoms of knee osteoarthritis: results of the BRISK randomized, controlled trial [ISRCTN01928173]. Arthritis Res Ther 2005;7(3): R625–R633
28. Ishijima M, Watari T, Naito K, et al. Relationships between biomarkers of cartilage, bone, synovial metabolism and knee pain provide insights into the origins of pain in early knee osteoarthritis. Arthritis Res Ther 2011;13(1):R22
29. Berry PA, Maciewicz RA, Cicuttini FM, Jones MD, Hellawell CJ, Wluka AE. Markers of bone formation and resorption identify subgroups of patients with clinical knee osteoarthritis who have reduced rates of cartilage loss. J Rheumatol 2010;37(6):1252–1259
30. Gineyts E, Garnero P, Delmas PD. Urinary excretion of glucosyl-galactosyl pyridinoline: a specific biochemical marker of synovium degradation. Rheumatology (Oxford) 2001;40(3):315–323
31. Jordan KM, Syddall HE, Garnero P, et al. Urinary CTX-II and glucosyl-galactosyl-pyridinoline are associated with the presence and severity of radiographic knee osteoarthritis in men. Ann Rheum Dis 2006;65(7):871–877
32. Gineyts E, Mo JA, Ko A, et al. Effects of ibuprofen on molecular markers of cartilage and synovium turnover in patients with knee osteoarthritis. Ann Rheum Dis 2004;63(7):857–861
33. Kokebie R, Aggarwal R, Lidder S, et al. The role of synovial fluid markers of catabolism and anabolism in osteoarthritis, rheumatoid arthritis and

asymptomatic organ donors. Arthritis Res Ther 2011;13(2):R50

34. Andersson ML, Petersson IF, Karlsson KE, et al. Diurnal variation in serum levels of cartilage oligomeric matrix protein in patients with knee osteoarthritis or rheumatoid arthritis. Ann Rheum Dis 2006;65(11):1490–1494

35. Kong SY, Stabler TV, Criscione LG, Elliott AL, Jordan JM, Kraus VB. Diurnal variation of serum and urine biomarkers in patients with radiographic knee osteoarthritis. Arthritis Rheum 2006;54(8):2496–2504

36. Criscione LG, Elliott AL, Stabler T, Jordan JM, Pieper CF, Kraus VB. Variation of serum hyaluronan with activity in individuals with knee osteoarthritis. Osteoarthritis Cartilage 2005;13(9):837–840

37. Qvist P, Christgau S, Pedersen BJ, Schlemmer A, Christiansen C. Circadian variation in the serum concentration of C-terminal telopeptide of type I collagen (serum CTx): effects of gender, age, menopausal status, posture, daylight, serum cortisol, and fasting. Bone 2002;31(1):57–61

38. Ikegawa S. Genetic analysis of skeletal dysplasia: recent advances and perspectives in the post-genome-sequence era. J Hum Genet 2006;51(7): 581–586

39. Briggs MD, Hoffman SM, King LM, et al. Pseudoachondroplasia and multiple epiphyseal dysplasia due to mutations in the cartilage oligomeric matrix protein gene. Nat Genet 1995;10(3):330–336

40. Czarny-Ratajczak M, Lohiniva J, Rogala P, et al. A mutation in COL9A1 causes multiple epiphyseal dysplasia: further evidence for locus heterogeneity. Am J Hum Genet 2001;69(5):969–980

41. Chapman KL, Mortier GR, Chapman K, Loughlin J, Grant ME, Briggs MD. Mutations in the region encoding the von Willebrand factor A domain of matrilin-3 are associated with multiple epiphyseal dysplasia. Nat Genet 2001;28(4):393–396

42. Tiller GE, Rimoin DL, Murray LW, Cohn DH. Tandem duplication within a type II collagen gene (COL2A1) exon in an individual with spondyloepiphyseal dysplasia. Proc Natl Acad Sci U S A 1990;87(10):3889–3893

43. Miyamoto Y, Mabuchi A, Shi D, et al. A functional polymorphism in the 50 UTR of GDF5 is associated with susceptibility to osteoarthritis. Nat Genet 2007;39(4):529–533

44. Daans M, Luyten FP, Lories RJ. GDF5 deficiency in mice is associated with instability-driven joint damage, gait and subchondral bone changes. Ann Rheum Dis 2011;70(1):208–213

45. Chapman K, Takahashi A, Meulenbelt I, et al. A meta-analysis of European and Asian cohorts reveals a global role of a functional SNP in the 5′ UTR of GDF5 with osteoarthritis susceptibility. Hum Mol Genet 2008;17(10):1497–1504

46. Kizawa H, Kou I, Iida A, et al. An aspartic acid repeat polymorphism in asporin inhibits chondrogenesis and increases susceptibility to osteoarthritis. Nat Genet 2005;37(2):138–144

47. Stern AG, de Carvalho MR, Buck GA, et al; I-NODAL Network. Association of erosive hand osteoarthritis with a single nucleotide polymorphism on the gene encoding interleukin-1 beta. Osteoarthritis Cartilage 2003;11(6):394–402

48. Vaes RB, Rivadeneira F, Kerkhof JM, et al. Genetic variation in the GDF5 region is associated with osteoarthritis, height, hip axis length and fracture risk: the Rotterdam study. Ann Rheum Dis 2009;68(11):1754–1760

49. Panoutsopoulou K, Southam L, Elliott KS, et al; arcOGEN Consortium. Insights into the genetic architecture of osteoarthritis from stage 1 of the arcOGEN study. Ann Rheum Dis 2011;70(5):864–867

50. Castro-Perez JM, Kamphorst J, DeGroot J, et al. Comprehensive LC-MS E lipidomic analysis using a shotgun approach and its application to biomarker detection and identification in osteoarthritis patients. J Proteome Res 2010;9(5):2377–2389

51. Zhai G, Wang-Sattler R, Hart DJ, et al. Serum branched-chain amino acid to histidine ratio: a novel metabolomic biomarker of knee osteoarthritis. Ann Rheum Dis 2010;69(6):1227–1231

4

Using Animal Models in Osteoarthritis Biomarker Research

Bridget C. Garner, Aaron M. Stoker, Keiichi Kuroki, Richard Evans, Cristi Reeves Cook, and James L. Cook

Osteoarthritis (OA) is characterized by morphologic, molecular, biochemical, and biomechanical abnormalities in the articular cartilage and other articular tissues. It has been estimated that over 27 million adults have a clinical diagnosis of OA; therefore, it is the most common form of arthritis.[1] OA was historically regarded as a noninflammatory arthritis, but there is now cumulative evidence that inflammation has a prominent role in OA.[2–5] Signals, including those associated with proinflammatory cytokines, can be sent and received among articular tissues, thus creating a "cross-talk" environment. A complex interaction involving cytokines,[3,4,6–13] proteolytic enzymes,[7,10,11,13,14] leukocytes,[15] synoviocytes,[16,17] and chondrocytes[3,11–13,18] likely initiates and exacerbates pathologic changes in osteoarthritic joints. Even with this knowledge much of the pathogenesis of OA remains unknown, and diagnosis early in the disease course is rare. Consequently the desire to develop an early OA diagnostic biomarker panel, a process that could have profound ramifications on treatment and prevention of this disease, has exploded.

Animal models are routinely used to study human OA as these models provide significant advantages over clinical human research. The most relevant models have been described in the dog, horse, sheep, goat, guinea pig, rat, rabbit, and mouse, and for more details on each of these models the reader is directed to a series of initiatives.[19–27] Highlights of each animal model, as well as a concise list of their advantages and disadvantages (**Table 4.1**), are included here.

Given that the dog is considered a nearly ideal species for translational investigation of human OA and at least 20% of dogs over 1 year of age in the United States are affected by naturally occurring clinical OA,[28] the dog is the most used animal model for investigation of OA.[20] Surgical transection of the cranial (anterior) cruciate ligament[29] and meniscal transection[30] are useful methods of surgical induction, but spontaneous or chemical induction models and additional surgical methods are also described.[20] Surgical and chemical induction methods are employed in rabbits and guinea pigs, whereas only surgical induction methods have been reported in sheep and goats.[22,23,27] The spontaneous OA-prone Dunkin-Hartley guinea pig[31] also offers a unique opportunity to study disease mechanisms. A commonly used surgical model of OA induction in the horse is the osteochondral fragment-exercise model,[32–34] which, unlike most other surgical induction models, does not induce joint instability. The medial meniscal tear model with or without concurrent exercise is most often used to induce OA in the rat.[21,35,36] While mice can

Table 4.1 Advantages and disadvantages of each of the most relevant animal models for translational investigation of human OA

Model(ref)	Advantages	Disadvantages
Dog[20]	Anatomically similar to humans	Can be expensive to acquire if research bred
	Arthroscopy possible	Larger facilities needed compared with rodents
	Clinical progression and treatment similarities	Variability between breeds
	Degenerative, trauma, and overuse etiologies occur	
	Genome sequence available	
G. pig[22]	Histologically similar to human OA	Not great for exercise model since sedentary
	Husbandry practices well developed	Not large enough for arthroscopy
	Imaging such as CT and MRI can be used	
	Sedentary lifestyle ideal for some models	
	Similar risk factors for development of OA	
	Sufficient tissue to harvest	
Horse[24]	Abundant tissue to collect	Need special facilities
	Arthroscopy possible	Trained personnel needed for safety
	Can control activity level	
	Progressive OA occurs	
Mouse[26]	Easy to handle	Difficult tissue and fluid collection due to size
	Genome sequence available	Tissue sectioning requires trained personnel
	Histologically similar to human OA	
	Transgenic and knockout models available	
Rabbit[23]	Animals readily available for low cost	Gait impedes functional studies
	Ease of handling	Not large enough for arthroscopy
	Lesions develop rapidly	
	More tissue than mice, rats, guinea pigs	
Rat[21]	Animals readily available for low cost	Iatrogenic injuries during surgery common
	Ease of handling	Minimal tissue/body fluids for collection
	Genome sequence available	Not large enough for arthroscopy
	Lesions develop rapidly	Transgenic models not available
Sheep/goat[27]	Biomechanically similar to humans	Genome not fully sequenced
	Can house on pasture in some areas	Need to validate oral treatments in ruminants
	Easily managed and handled	
	Large enough for arthroscopy	
	Sufficient tissue for collection	

Abbreviations: OA, osteoarthritis; G. pig, guinea pig; CT, computed tomography; MRI, magnetic resonance imaging.

be challenging to study for reasons related to their small size, the availability of genetically modified or knockout mice is invaluable to learning more about OA.[26] As an example of animal modeling, the authors will focus on the dog to illustrate the level and limitations of modeling within a specific species.

Biomarkers are "objective indicators of normal biologic processes, pathogenic processes, or pharmacologic responses to therapeutic interventions,"[37] and they have been classified under the BIPED scheme: Burden of disease, Investigative, Prognostic, Efficacy of intervention, and Diagnostic.[38] Expansion of this classification system is already under way as the Osteoarthritic Research Society International Food and Drug Administration (OARSI FDA) initiative Biomarkers Working Group recently elected to modify this acronym to BIPEDS to include a safety category.[39] Another method of biomarker classification involves whether each is a direct or an indirect indicator of OA.[40] Direct biomarkers originate from the articular tissues, whereas indirect biomarkers do not arise from the joint tissues but can ultimately affect the joint's microenvironment.[41] Both types provide useful information; therefore, the study of both direct and indirect biomarkers of OA is warranted.

Changes in direct markers reflect extracellular matrix (ECM) catabolism or anabolism and may reveal the status of joint degeneration. Several biomarker candidates causing or resulting from chondrocyte,[13,42–46] collagen,[11,47–51] or aggrecan[52–55] degradation have been assessed, and these include matrix metalloproteases (MMPs), keratan sulfate, cartilage oligomeric matrix protein (COMP), and the collagenase-generated cleavage epitope of type II collagen (C2C or Col2–3/4C). Studies have shown strong correlations between these degradative markers and magnetic resonance imaging (MRI) findings.[56–60] Markers associated with bone turnover have also been identified, but they are less commonly considered to be diagnostic markers as these are often altered only during advanced stages of disease.[61–64]

MMPs are zinc- and calcium-dependent enzymes that degrade specific substrates of the ECM and include stromelysins, collagenases, gelatinases, and membrane metalloproteases. MMPs have a role in the normal turnover of the connective tissue matrix that occurs during growth and development, but the unchecked production of MMPs is a commonly documented cause of cartilage matrix degradation in OA in animals.[48,49,54,65–69] Increased collagen 2 denaturation and fragmentation are often present in OA articular cartilage,[48] and MMP13, which is increased in OA chondrocytes, appears to be directly involved.[50] Stimulation of canine chondrocytes with interleukin 1β (IL-1β) to represent an in vitro OA environment led to an increase in MMP13 versus controls.[3] MMP3 has also been shown to increase in cartilage and synovial fluid from canines with experimentally induced OA,[70] and cultured canine chondrocytes stimulated with IL-1β demonstrated increased matrix degradation and MMP3 expression.[71]

Keratan sulfate, a glycosaminoglycan constituent of the cartilage ECM, has been primarily investigated in animal models using the dog. Synovial fluid keratan sulfate concentrations have been shown to increase[72,73] or decrease[74,75] in dogs with OA with no clear correlation to factors such as manner of disease initiation (induced vs. spontaneous) or chronicity of disease. Given the discordant results, keratan sulfate may not be ideal for detecting or prognosticating OA, but it may prove useful for serial monitoring of known OA individuals.[75]

COMP is one of the most comprehensively studied diagnostic biomarker candidates in the synovial fluid, serum, and urine of veterinary patients. Elevations in COMP have been reported in the serum and synovial fluid of naturally occurring OA dogs compared with controls, but elevations due to induced synovitis were also reported, suggesting a lack of specificity (SP) for cartilage damage.[46] Another canine study supported these conclusions by showing that a variety of articular tissues, including ligament, tendon, meniscus, and articular cartilage, produced COMP.[76] Regardless, COMP may be diagnostically useful as serum and synovial fluid levels have been shown to correlate with canine knee MRI grades of cartilage lesions.[60] Urinary COMP was also predictive of the presence of osteophytes in racehorses with carpal fractures.[77] Serum COMP concentration was lower in osteoarthritic horses versus normal.[78]

The C2C or Col2–3/4C has also been investigated in several animal studies with some discordant results. Synovial fluid Col2 levels increased in dogs following induced anterior cruciate rupture.[42,79] Serum increases were also detected in the Pond-Nuki model, and these increases correlated with a marker of lipid peroxidation, suggesting a link between oxidative stress and cartilage degeneration.[80] In contrast, differences were not detected in synovial fluid, serum, or urine concentrations in dogs with naturally occurring anterior cruciate rupture compared with controls even though lameness, joint effusion, and osteophytes were present.[81]

The pathophysiology of OA centers on an imbalance of cartilage degradation and synthesis; therefore, several anabolic markers are also under investigation. Anabolic markers reflect molecules that are present in small quantities or entirely absent in health, or they can exhibit an altered structure in which epitopes usually hidden are revealed during OA.[51,55,82–86] Chondroitin sulfate (epitopes 3-B-3[−] and 846) is one example of an anabolic marker.

Like keratan sulfate, chondroitin sulfate comprises a large portion of the proteoglycan aggrecan. A monoclonal antibody (3-B-3) recognizes chondroitin sulfate epitopes with (3-B-3[+]) or without (3-B-3[−]) chondroitinase pretreatment. The 3-B-3(−) epitope is identified in the growth plate cartilage during normal growth and development and in the early stages of OA, but it is absent in healthy adult canine cartilage, suggesting anabolic production of an altered form of chondroitin sulfate as an attempted repair response.[85,86] Furthermore, it is expressed in superficial zone articular cartilage from destabilized canine knee joints in early stages of disease, and it is detectable before a loss of matrix and proteoglycans is identifiable by toluidine blue staining.[86] Synovial fluid values are increased in dogs with naturally occurring OA and experimental OA as early as 4 weeks after surgical destabilization.[73,79,85,87] In contrast to synovial fluid data, serum concentrations in dogs with hip dysplasia were lower than in those without joint disease.[88]

While direct markers may more closely represent the status of joint degeneration, indirect markers, including but not limited to cytokines, can be used to learn more about the processes preceding or leading to the development of OA. Cytokine and chemokine fluctuations within the synovial fluid of osteoarthritic patients have been documented, but comprehensive assessment of the potential clinical significance of those alterations is largely lacking in the literature.[9,16,89,90] Cytokines and chemokines have also demonstrated roles in the pathogenesis of OA, including induction of proteinase expression and inhibition of proteoglycan synthesis.[91,92] A more thorough exploration of these roles may provide significant information about the pathogenesis of OA and lead to the identification of early OA diagnostic biomarkers. Following stimulation with IL-1β, the up-regulation of cytokines, chemokines, and MMPs was more rapid than the down-regulation of matrix gene expression (*COL2A1* and aggrecan), suggesting that these types of molecules may be the first changes identifiable in early OA.[12] As a result, the authors were interested in learning more about the cytokine and chemokine profiles of dogs with and without OA and the relationships of these profiles to MMP concentrations.

Our objectives for this study were (1) to delineate the temporal alterations of cytokine, chemokine, and MMP concentrations in synovial fluid, serum, and urine in induced and naturally occurring osteoarthritic dogs in comparison to healthy dogs and (2) to assess the diagnostic value of particular markers through evaluation of sensitivity (SN), SP, and receiver operating characteristic (ROC) curve analysis. We hypothesized there would be strong correlations between cytokine and chemokine fluctuations and the status of the joint with respect to OA.

◆ Materials and Methods

Sample Collection

Part 1: Model Dogs

All procedures were approved by the institution's animal care and use committee. In this study, 21 adult, intact female, purpose-bred hound dogs > 20 kg were included. No more than 24 hours before surgery, blood was drawn from the cephalic vein into serum separator tubes. The serum was harvested within 2 hours of collection, and the specimens were stored in individual, airtight containers at −80°C. Urine was obtained from each dog by aseptic cystocentesis or by manual bladder expression if cystocentesis was not successful. The urine specimens were kept on ice until processing for storage in individual, airtight containers at −20°C. Serum and urine were collected again at 4, 8, and 12 weeks after surgery using the same protocol.

On the day of surgery, each dog was premedicated, anesthetized, and aseptically prepared for arthroscopic surgery of the right stifle (knee). Synovial fluid was collected via aseptic arthrocentesis of the knee joint, and the samples were kept on ice until they were aliquoted and frozen at −80°C for subsequent analysis. Using standard arthroscopic technique and instrumentation for the canine knee,[93] one of four surgical procedures was performed on each dog: transection of the anterior cruciate ligament[29] (ACL-T; $n = 5$), complete radial transection of the meniscus (MR)[30] ($n = 5$), creation of two 6.0- to 8.0-mm-long full-thickness grooves (GRs) in the cartilage of the weight-bearing portion of the medial femoral condyle (GR; $n = 6$) using a 3 mm OD (outside diameter) arthroscopic curette, or manipulation of all the aforementioned intra-articular structures without insult using an arthroscopic probe (SHAM; $n = 5$). Transections were visually confirmed, and GRs were measured with a calibrated probe. The nonoperated, contralateral hind limb served as an internal control for each dog, although synovial fluid was not collected at the time of surgery (baseline). Twelve weeks later, synovial fluid was collected by arthrocentesis from both the operated and contralateral control knees, and aliquots were frozen at −80°C. A second arthroscopic evaluation of the operated joint was performed and articular tissues were collected for histology as part of a concurrent study.

Part 2a: Client-Owned OA Dogs

Informed client consent was obtained for each dog. Blood and synovial fluid were obtained from 10 adult medium and large breed dogs presenting for surgical intervention of unilateral stifle OA (Pre-sx OA; $n = 10$). These dogs ranged from 3 to 8 years old (median 4.5 years) and included five male castrated and five female spayed dogs. Synovial fluid was obtained from the affected stifle via routine aseptic arthrocentesis, and blood was collected via jugular venipuncture. Synovial fluid samples were kept on ice until they were aliquoted and frozen at −80°C for subsequent analysis. Clinical OA was confirmed in each dog by a board-certified veterinary orthopaedic surgeon as determined by knee physical examination based on the presence of effusion, periarticular fibrosis, pain upon flexion and extension, and a lameness evaluation based on the visual examination of gait at a walk and trot. Radiographic evidence of knee OA including signs of osteophytosis, effusion, and sclerosis was confirmed by a board-certified veterinary radiologist.

All dogs underwent surgery for assessment, lavage, and stabilization of cruciate ligament deficiency and recovered uneventfully. Eight to 12 weeks later, the dogs returned for a postoperative recheck, and blood and synovial fluid were collected again to assess changes in markers after surgical intervention. Two female dogs did not return for follow-up; therefore, their postoperative blood and synovial fluid samples could not be obtained (Post-sx OA; $n = 8$). These two dogs were subsequently excluded from paired statistical analysis.

Part 2b: Control Group

The normal control group comprised seven medium and large breed adult dogs ranging from 2 to 5 years old (median 2.75 years). There were three castrated males, two spayed females, and two intact males. These dogs had no clinical history of joint trauma, were not lame, and were deemed to be free of clinical OA as determined by a board-certified veterinary orthopaedic surgeon. Radiographic evaluation of the shoulders, knees, and hips verified the absence of OA. Blood and synovial fluid were collected in a similar manner to the OA dogs at a time convenient to the clients, and synovial fluid samples were kept on ice until they were aliquoted and frozen at −80°C for subsequent analysis.

Multiplex Analysis

An aliquot (25 μL) from each synovial fluid, serum, and urine sample was thawed. The urine and synovial fluid samples were centrifuged at 14,000 rpm for 10 minutes to pellet debris, and the supernatant was removed. The synovial fluid was incubated with hyaluronidase (MP Biomedicals, LLC, Solon, Ohio) at 37°C for 60 minutes to decrease viscosity. Each aliquot was subsequently analyzed in duplicate using a multiplex canine cytokine and chemokine immunoassay (Millipore Corp., St. Louis, MO) based on the xMAP platform (Qiagen Inc., Valencia, CA) for IL-2, IL-4, IL-7, IL-8, IL-15, IL-18, IP-10, interferon gamma (INF-γ), tumor necrosis factor-alpha (TNF-α), monocyte chemoattractant protein 1 (MCP1), keratinocyte-derived chemoattractant (KC), and granulocyte-macrophage colony-stimulating factor (GM-CSF) according to the manufacturers' directions. The synovial fluid and serum from the client-owned OA dogs and their control group were also analyzed for IL-6 and IL-10 using the same assay. A multiplex human MMP immunoassay (R&D Systems, Minneapolis, MN) based on the xMAP platform for four MMPs: MMP2, MMP3, MMP9, and MMP13 was also used to analyze the client-owned dogs and their control group in duplicate. This MMP assay had been previously shown within our laboratory to cross-react with samples of canine origin.[94] Briefly, for each of the xMAP assays the synovial fluid, serum, and urine samples were admixed with antichemokine, cytokine, or MMP monoclonal antibody–charged, small (5.6 micron), polystyrene microspheres in a 96-well plate. Following an overnight incubation at 4°C, a biotinylated polyclonal secondary antibody was added, as well as streptavidin-phycoerythrin. The median fluorescence intensity was determined for each sample. The urine creatinine concentration was measured with an in-house chemistry analyzer (AU400; Olympus America Inc., Irving, TX), and the urine cytokine and chemokine values were standardized to this concentration (pg/mg).

Statistics

Planned comparisons between preoperative and postoperative samples (synovial fluid, serum, and urine) and between operated and contralateral limbs (synovial fluid only) were performed with the paired *t*-test. Comparisons between surgery model groups or between OA and normal individuals were performed with the unpaired *t*-test or the Mann–Whitney rank sum test (SigmaStat 3.5; Systat Software, Inc., San Jose, CA). Significance was set at $p < 0.05$. SN and SP were calculated for select markers of interest using the histopathological data as the reference test for the model dogs and the clinical examination and radiographic data for the hospital patients. When possible, Youden's index was used to select optimal concentration cut-offs that maximized SN and SP for each marker. When this index did not provide a balance between SN and SP, concentration cut-offs that selectively led to higher SN were chosen. The 95% confidence intervals (CIs) were calculated using the Clopper–Pearson method.

ROC Curves

ROC curves based on the logistic regression model were developed for markers of interest to the authors following initial statistical analysis. These curves were created and area under the curve (AUC) calculated for assessment of diagnostic value of certain parameters using JMP 7.0.2 software (SAS Institute, Cary, NC).

◆ Results

Part 1: Induction of OA

Arthroscopic findings (**Fig. 4.1a, b**), lameness scoring performed by a board-certified veterinary surgeon, and histologic scoring performed by a board-certified veterinary pathologist confirmed the induction of clinically significant OA in the ACL-T, GR, and MR groups, whereas evidence of OA was not identified in the SHAM dogs. The ACL-T and MR dogs exhibited the most severe joint pathologic changes (although characteristically different from each other), but GR dogs also had articular cartilage damage and synovitis. Both investigators were blinded with respect to the group during evaluation.

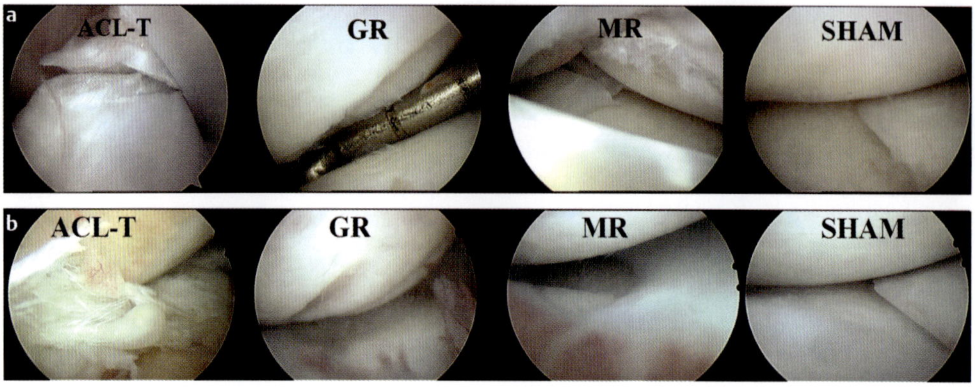

Fig. 4.1 (**a**, **b**) Initial and 12-week postoperative arthroscopic views of one dog from each surgical induction model. ACL-T, anterior cruciate ligament transection; GR, groove model; MR, meniscal release; SHAM, manipulation without insult.

Part 1: Synovial Fluid

Individuals without sufficient volume for analysis in operated (baseline and +12 weeks) and contralateral hind limbs were excluded from paired statistical analysis. There were no significant differences between the baseline and the +12-week SHAM samples or between operated and nonoperated hind limbs in the SHAM dogs for any analytes. MCP1 (**Fig. 4.2**) was significantly increased in ACL-T joints ($n = 4$) 12 weeks after surgery compared with baseline ($p = 0.036$) and nonoperated joints at 12 weeks ($p = 0.018$). MCP1 trended upward in the GR ($n = 5$) and MR groups ($n = 3$) at 12 weeks compared with baseline and nonoperated limbs, but statistical significance was not reached (GR: $p = 0.24$ and $p = 0.076$, MR: $p = 0.11$ and $p = 0.310$). MCP1 was not significantly higher in the ACL-T, GR, or MR joints compared with the SHAM joints 12 weeks after surgery.

IL-8 (**Fig. 4.3**) was significantly increased at 12 weeks in the GR dogs compared with baseline ($p = 0.024$), and in the +12-week MR group compared with the SHAM dogs ($p = 0.006$) at 12 weeks. KC (**Fig. 4.4**) was significantly decreased in the GR group operated and nonoperated limbs at 12 weeks ($n = 4$) compared with baseline ($p = 0.011$, $p = 0.017$). There were no other significant differences in cytokine or chemokine expression in ACL-T, GR, MR, and SHAM dogs between time zero samples and the +12- week nonoperated hindlimbs. Statistically significant differences were not detected for any of the remaining analytes.

Comparing the synovial fluid concentrations of SHAM joints with nonoperated hind

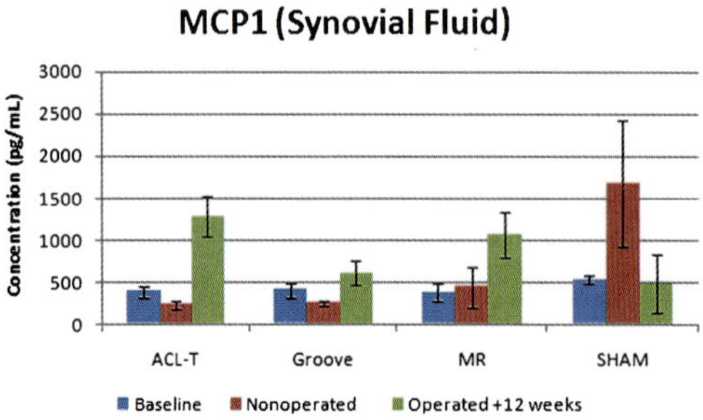

Fig. 4.2 Mean synovial fluid monocyte chemoattractant protein 1 concentrations (pg/mL) for surgery model dogs at initiation and conclusion of the study. Bars represent standard error of the mean.

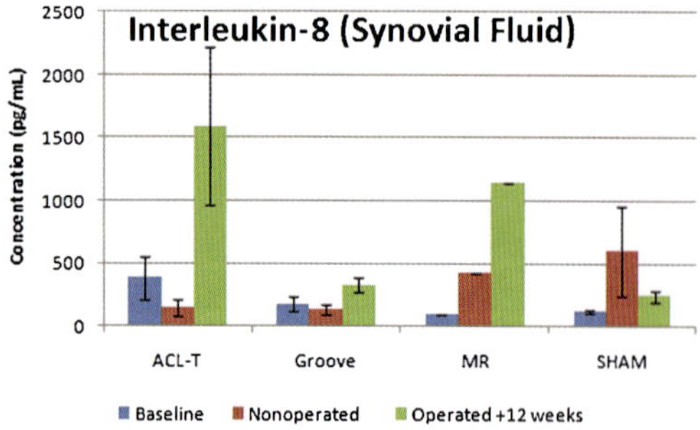

Fig. 4.3 Mean synovial fluid interleukin-8 concentrations (pg/mL) for surgery model dogs at initiation and conclusion of the study. Bars represent standard error of the mean.

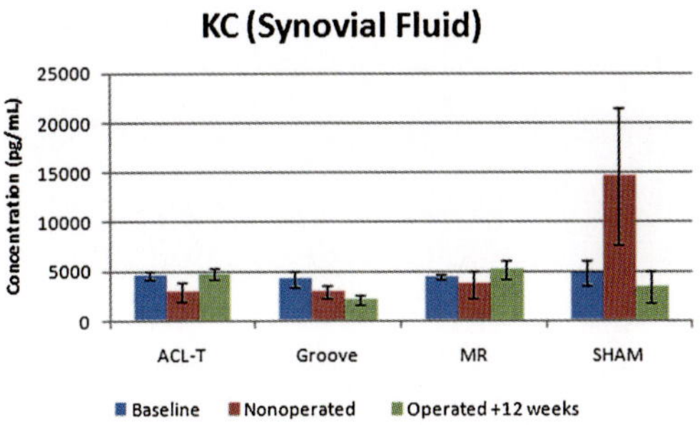

Fig. 4.4 Mean synovial fluid keratinocyte-derived chemoattractant (KC) concentrations (pg/mL) for surgery model dogs at initiation and conclusion of the study. Bars represent standard error of the mean.

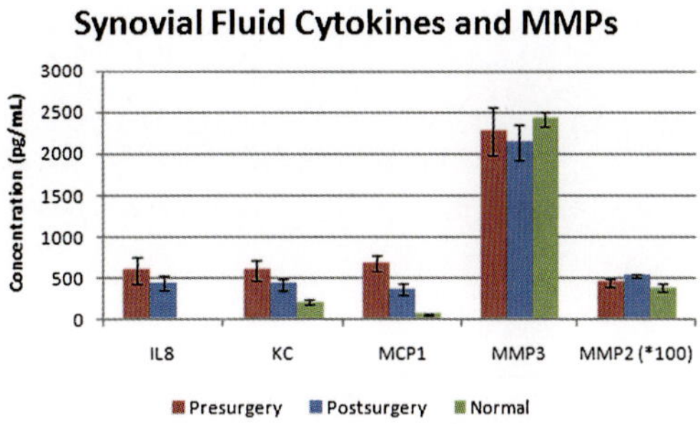

Fig. 4.5 Mean synovial fluid cytokine, chemokine, and MMP concentrations (pg/mL) for spontaneous OA dogs at surgery and 8 to 12 weeks after surgical stabilization compared with normal dogs. Bars represent standard error of the mean. MMP, matrix metalloprotease; OA, osteoarthritis.

limbs and operated hind limbs revealed significant concentration overlap (**Table 4.2**). However, IL-8 had the highest SN and SP with an SN of 100% (95% CI, 76.8 to 100) and an SP of 80% (95% CI, 28.4 to 99.5) at a concentration cut-off between 144 and 189 pg/mL. MCP1 and KC were less sensitive at 78% (95% CI, 49.2 to 95.3), but MCP1 was the most specific at 100% (95% CI, 47.8 to 100).

Part 1: Serum and Urine

IL-8 (data not shown) was significantly lower in the serum of ACL-T dogs 12 weeks after surgery compared with those at baseline ($p = 0.049$). IL-8 was significantly increased in the 8- and 12-week postoperative urine samples from each of the four treatment groups compared with day 0 and the 4-week samples ($p < 0.001$). There was not a statistically significant difference in IL-8 between the baseline and 4-week samples, nor was there a difference between the 8- and 12-week samples. Urine MCP1 was increased in the 8-week postoperative SHAM group compared with baseline ($p = 0.009$). No other statistically significant changes were detected in the serum or urine samples.

Part 2: Synovial Fluid

MCP1 was significantly higher in the Pre-sx OA dogs and Post-sx OA dogs compared with normal dogs ($p < 0.001$, $p = 0.009$), and there was a significant decrease in MCP1 following surgery compared with presurgery values ($p = 0.016$) (**Fig. 4.5**). IL-8 and KC were

Table 4.2 Median (range) synovial fluid concentrations from model dogs including the optimal concentration cut-offs and associated SN and SP percentages (with 95% CI)

Marker	SHAM ($n = 5$)	Non-op ($n = 15^a$)	Op ($n = 15^a$)	Cut-off[b] (pg/mL)	SN % (95% CI)	SP % (95% CI)
IL-8	119 (73–1,197)	202 (49–753)	781 (189–2,825)	144–189	100 (76.8–100.0)	80 (28.4–99.5)
MCP1	506 (345–619)	278 (55–933)	879 (168–1,981)	620	78 (49.2–95.3)	100 (47.8–100)
KC	4209 (2,995–5,396)	3137 (273–6,361)	4041 (1,697–6,624)	3000	78 (49.2–95.3)	20 (0.5–71.6)

[a]One individual did not have sufficient postoperative synovial fluid volume and is therefore not included in the non-op or op analyses.
[b]Values below cut-off concentration denote SHAM dogs.
Abbreviations: SN, sensitivity; SP, specificity; CI, confidence interval; SHAM, values for time zero; non-op, nonoperated hind limbs at 12 weeks after surgery; Op, operated hind limbs at 12 weeks after surgery; IL-8, interleukin-8; MCP1, monocyte chemoattractant protein 1; KC, keratinocyte-derived chemoattractant.

Table 4.3 Median (range) synovial fluid concentrations including the optimal concentration cut-offs and associated SN and SP percentages (with 95% CI) from clinical OA patients and normal dogs

Biomarker	NL	Pre-sx OA	Cut-off[a] (pg/mL)	SN % (95% CI)	SP % (95% CI)
IL-8	0 (0)	438 (112–4,004)	0–112	100 (69–100)	100 (59–100)
MCP1	73 (0–91)	734 (263–2,278)	91.5–263.3	100 (69–100)	100 (59–100)
KC	201 (141–377)	584 (181–8,603)	275	90 (56–100)	86 (42–100)
MMP2	3.95 (2.24–5.67)[b]	4.34 (2.76–6.66)[b]	3.72[b]	80 (44.4–97.4)	57 (18.4–90.1)
MMP3	2569 (1,950–2,700)	2150 (1,097–5,081)	2,451	40 (12.1–73.7)	42.8 (9.9–81.6)

Abbreviations: SN, sensitivity; SP, specificity; CI, confidence interval; OA, osteoarthritis; NL, normal dogs; Pre-sx OA, dogs presenting for surgical stabilization of knee osteoarthritis before surgery; IL-8, interleukin-8; MCP1, monocyte chemoattractant protein 1; KC, keratinocyte-derived chemoattractant; MMP2, matrix metalloprotease 2; MMP3, matrix metalloprotease 3.
[a]Values below cut-off concentration denote normal dogs.
[b]($\times 10^4$).

significantly higher in the Pre-sx OA dogs compared with normal dogs ($p < 0.001$, $p = 0.014$) and in the Post-sx OA dogs compared with normal dogs ($p = 0.002$, $p = 0.029$). Both analytes were numerically different (lower) in Post-sx OA dogs compared with Pre-sx OA dogs ($p = 0.340, 0.115$). IL-18 (INF-γ inducing factor) was significantly higher in the Pre-sx OA dogs compared with Post-sx OA dogs ($p = 0.016$) and with normal dogs ($p = 0.002$), but the remaining cytokines and chemokines were below the limit of detection. MMP2 was highest in the Post-sx OA dogs, and this was significantly higher than in the normal dogs ($p = 0.010$), whereas no significant difference was found between the Pre-sx OA dogs and the normal dogs. MMP3 was highest in the normal dogs and declined in the OA dogs after surgery, but these changes did not reach statistical significance. MMP9 and MMP13 were below the limit of detection of the assay.

Synovial fluid concentration data were compared between normal and Pre-sx OA dogs (**Table 4.3**). Concentration ranges did not overlap for IL-8 and MCP1, but they did for KC, MMP2, and MMP3. Optimal concentration cut-offs were selected as previously described, and at these particular cut-offs IL-8 and MCP1 performed well individually (SN and SP of 100% [SN 95% CI, 69 to 100; SP 95% CI, 59 to 100]). KC also performed reasonably well with an SN of 90% (95% CI, 56 to 100) and an SP of 86% (95% CI, 42 to 100). The MMPs were less strong differentiators.

Part 2: Serum

MMP2 and MMP3 were highest in the normal dogs and lowest in the Pre-sx OA dogs (**Fig. 4.6**). MMP2 was significantly higher in normal dogs and Post-sx OA dogs compared with Pre-sx OA dogs ($p = 0.005$, $p = 0.027$), but there was not a significant difference between normal and Post-sx OA dogs ($p = 0.270$). MMP3 was significantly higher in normal dogs and Post-sx OA dogs compared with Pre-sx OA dogs ($p = 0.002$, $p = 0.044$), and MMP3 was significantly higher in normal dogs compared with Post-sx OA dogs ($p = 0.025$). SN and SP with associated cut-off values for MMP2 and MMP3 are shown

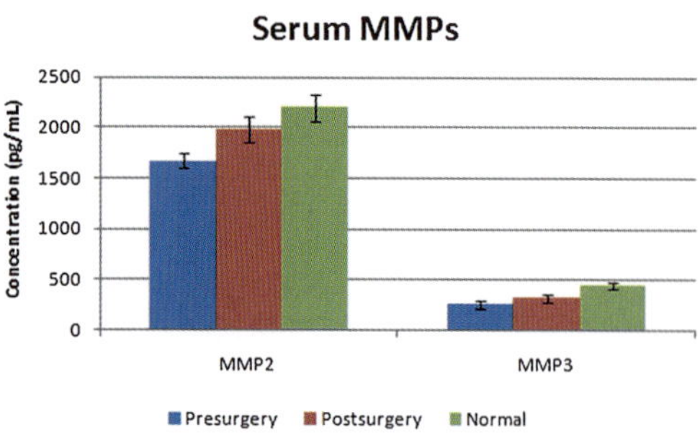

Fig. 4.6 Mean serum MMP2 and MMP3 concentrations (pg/mL) for spontaneous OA dogs at surgery and 8 to 12 weeks after surgical stabilization compared with normal dogs. Bars represent standard error of the mean. MMP2, matrix metalloprotease 2; MMP3, matrix metalloprotease 3, OA, osteoarthritis.

Table 4.4 Median (range) serum concentrations including the optimal concentration cut-offs and associated SN and SP percentages (with 95% CI) from clinical OA patients and normal dogs

Biomarker	NL	Pre-sx OA	Cut-off[a] (pg/mL)	SN % (95% CI)	SP % (95% CI)
MMP2	2,174.4 (1,771–3,002)	1,684.9 (1,281–1,945)	2,000	100 (69.2–100)	56 (21.2–86.3)
MMP3	413.8 (2,93.5–676)	256.7 (93.5–421.3)	400	80 (44.4–97.5)	67 (29.9–92.5)

Abbreviations: SN, sensitivity; SP, specificity; CI, confidence interval; OA, osteoarthritis; NL, normal dogs; Pre-sx OA, dogs presenting for surgical stabilization of knee osteoarthritis before surgery; MMP2, matrix metalloprotease 2; MMP3, matrix metalloprotease 3.
[a]Values above cut-off concentration denote normal dogs.

in **Table 4.4**. Significant differences were not detected between groups for serum IL-8, KC, MCP1, IL-18, IL-2, IL-7, or GM-CSF, and the remaining analytes were below the limit of detection for the assay.

ROC Analysis

For the surgery model dogs (data not shown), calculation of the AUC confirmed IL-8 was the single marker with the strongest ability to differentiate between SHAM and all OA dogs. The combination of MCP1, IL-8, and KC demonstrated similarly strong discriminatory ability, but the addition of MCP1 and KC did not greatly improve the performance of IL-8 alone. ROC curve analysis was also performed on the hospital patient data (**Fig. 4.7**).

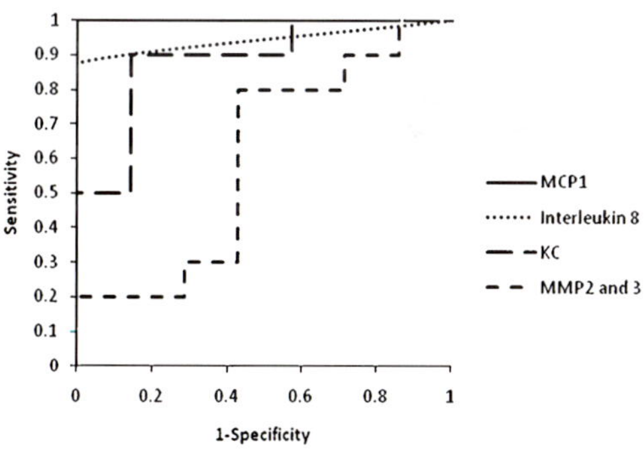

Fig. 4.7 ROC curve of MCP1, IL-8, KC, and MMP2 and MMP3 from dogs with (Pre-sx OA) and without (normal dog) spontaneously occurring OA. IL-8, interleukin 8, KC, keratinocyte-derived chemoattractant; MCP1, monocyte chemoattractant protein 1; MMP2, matrix metalloprotease 2; MMP3, matrix metalloprotease 3; OA, osteoarthritis; ROC, receiver operating characteristic.

IL-8 was a strong discriminator between normal and Pre-sx OA dogs, but the AUC for MCP1 was slightly higher. Combining MMP2 and MMP3 slightly improved their individual performance, but their AUC remained lower than MCP1, IL-8, and KC.

◆ Discussion

We have shown that changes in cytokine and chemokine concentrations occur within canine synovial fluid following surgical destabilization of the knee joint and may be useful for differentiating osteoarthritic versus normal knees. Furthermore, we have confirmed changes in the same three cytokines and chemokines occur in synovial fluid from dogs with naturally occurring OA, and the elevations in these markers decline once surgical stabilization has occurred. These findings suggest the incorporation of chemokines into a diagnostic or even treatment efficacy biomarker panel may prove useful, especially given their apparent early involvement in clinical OA.[91,92]

One of the significantly altered chemokines, MCP1; also known as chemokine (C-C motif) ligand 2, is responsible for selectively attracting mononuclear cells such as monocytes and memory T cells, but not neutrophils.[95–99] Arthropathies associated with monocytic infiltrates such as rheumatoid arthritis have been historically linked to alterations in MCP1.[17,100,101] MCP1 is elevated in synovial fluid and serum from RA sufferers, and the synovial tissue macrophages are the dominant source of this cytokine.[100] Evidence suggests MCP1 expression may be altered in OA as well, although reports in the veterinary literature are far less numerous than those in the human literature. Human OA articular chondrocytes express higher levels of MCP1 mRNA than normal chondrocytes, and OA synoviocytes also produce MCP1.[91,102] Stimulation with IL-1β, a cytokine used in in vitro models of OA, greatly increases the expression of MCP1.[12,91] MCP1 can also subsequently augment MMP3 expression, inhibit proteoglycan synthesis, and enhance proteoglycan release from chondrocytes in vitro.[91] A canine study confirms that MCP1 serum levels are significantly higher in critically ill dogs compared with healthy dogs and dogs recently undergoing surgery, including orthopedic procedures.[103] As a difference was not noted between healthy dogs and postoperative patients, it was concluded that surgery alone would not significantly alter serum levels of MCP1 or interfere with analysis in postoperative patients.[103] Synovial fluid data were not available from those patients, but the lack of significant increase in our SHAM dogs from baseline to +12 weeks suggests that there is either a brief increase or no significant change in synovial fluid MCP1 following surgical manipulation of the joint.

MCP1 exhibits high SN and SP for the client-owned dogs compared with normal, and the concentration ranges did not overlap between these two groups. As a result, synovial fluid MCP1 was considered one of the best markers to distinguish between non-OA and OA dogs. When absolute synovial fluid concentrations were analyzed, the client-owned OA dogs had very similar synovial fluid MCP1 concentrations to the +12-week postoperative model dogs. The most striking difference in MCP1 concentrations between the two segments of this study was the discrepancy between the SHAM dogs at time zero and the control individuals. At baseline the SHAM dogs had much higher synovial fluid concentrations of MCP1 compared with both the nonoperated hind limbs 12 weeks later and the normal dog group. At this time, there is no clear explanation for this apparent elevation at the initiation of the surgery model, but this may have contributed to the lack of significant differences between the SHAM and other surgery model groups 12 weeks after surgery.

While MCP1 may be useful as a diagnostic biomarker, it may also be helpful in evaluating treatment efficacy. For example, synovial fluid MCP1 was significantly lower in the Post-sx OA dogs compared with their Pre-sx OA values, whereas IL-8 and KC did not significantly decline after treatment. Furthermore, two client-owned dogs that developed minor postoperative complications had the highest MCP1 values of all the postoperative dogs. These findings suggest MCP1 has potential for clinical use in both diagnostic and treatment monitoring applications.

Neutrophil chemoattractants such as IL-8, also known as chemokine (C-X-C motif) ligand 8 (CXCL8), and KC or CXCL1[15,104–106] are both members of the CXC cytokine family, and phylogenetic analysis has shown KC may be similar to growth regulated oncogene-alpha (GROα) found in humans.[107] IL-8 and KC have been evaluated in joint diseases, with the majority of literature focusing on humans. Specifically, IL-8 is increased in humans with rheumatoid arthritis, OA, and other arthritides,[108,109] but in dogs it has been linked with Lyme disease.[110] The synovium appears to play a significant role as synovial expression of IL-8 is high in humans with Lyme arthritis, and RA and OA synoviocytes constitutively express IL-8.[102,111,112] This localization is helpful for the diagnostician as the degree of synovial expression appears to be reflected in the synovial fluid. For example, IL-8 mRNA expression is higher in canine OA knee synovial fluid than normal synovial fluid.[90] Additionally, increased numbers of functional receptors have been identified in OA chondrocytes, and the ligand—receptor interactions in these tissues have been shown to induce matrix-degrading enzymes such as MMP3.[7] IL-8 and KC promote cartilage hypertrophy, which can ultimately lead to dysregulated matrix repair and pathologic calcification in OA.[113]

Synovial fluid IL-8 performed well as an individual marker of OA with high SN and SP in both portions of the study, and the IL-8 concentration was increased in a majority of our surgery model dogs and spontaneous OA dogs. Furthermore, the concentration ranges did not overlap between normal and Pre-sx OA dogs, and the optimal cut-off concentrations were very similar between the model dogs and the spontaneous OA dogs. This suggests the synovial fluid IL-8 cut-off concentration identified here may be successfully extrapolated to other populations to rule in or out OA.

Although synovial fluid KC was not as sensitive or specific as IL-8 and MCP1 when used individually, it did promote an interesting question. A significant decrease was noted in the GR model dogs at the time of sacrifice compared with the initiation of the study, but there was no significant change between baseline and the end point for ACL-T dogs. In contrast, the Pre-sx OA dogs had higher KC synovial fluid concentrations than the client-owned dogs deemed free of disease. This disagreement may suggest that the alteration in KC is dependent on the type and/or chronicity of pathologic changes. All of the client-owned dogs had cruciate disease with most similarities to those included in the ACL-T model group, but one limitation of the second half of the study was that the chronicity of OA was variable or even unknown in the client-owned dogs. The inciting cause of OA was not specifically selected for or against in the spontaneous OA dogs, but the resulting cruciate-disease-heavy population was determined to be representative of spontaneous OA patients presenting to the hospital. Additional studies investigating dogs with noncruciate disease knee instability may provide insight as to whether or not KC may be helpful in determining the tissue(s) affected. Until that time, KC can be retained as a potential diagnostic biomarker of interest.

Unfortunately, the evaluation of cytokines and chemokines in the serum and urine was less rewarding than in the synovial fluid. This is not entirely surprising since other systemic conditions not related to OA can dramatically affect cytokine and chemokine concentrations. Urinary IL-8 can increase in the presence of a urinary tract infection, likely associated with its involvement in the recruitment of neutrophils.[114,115] However, elevations in urinary IL-8 have been detected in individuals with inflammation of nonurinary tract origin as well.[116] Urinalyses were not performed on the model dogs in this study to assess for urinary tract infection, but such a condition must be considered as a potential cause for the elevations seen. IL-8 was the only cytokine to show significant differences within the urine of OA dogs, but since similar elevations were also found in the SHAM dogs the increase was not likely associated with OA.

In contrast to the cytokines and chemokines, MMPs exhibited changes in the serum and the synovial fluid, although the serum concentrations showed a clearer trend. Perhaps the lack of a clear trend in synovial fluid MMP2 and MMP3 was related to the variability in the chronicity of disease of the hospital patients. The concentrations of serum MMP2 and MMP3 overlapped between the normal and Pre-sx OA individuals, and the selected cut-off concentrations had to sacrifice high SP to get a higher SN. However, serum MMP2 and MMP3 were highest in the normal patients. Ling et al also showed serum MMP2 was lower in individuals who subsequently developed OA and suggested altered ECM metabolism, including a decline in more constitutively expressed MMPs such as MMP2, played a role in the initiation and progression of OA.[117] Unfortunately, it cannot be ruled out that serum MMP2 and MMP3 values in the control individuals in this study may have been elevated due to unknown concurrent processes not directly related to OA.

To the authors' knowledge, the combination of synovial fluid MCP1, IL8, and KC for use in an OA diagnostic biomarker panel is unreported in dogs or humans. It has been reported that human knee articular cartilage produces MCP1, IL-8, and GROα (the human counterpart to KC), but this production did not differ significantly between OA and other joint diseases.[118] Further investigation is necessary to evaluate how reliably this particular combination of markers differentiates between types of arthritides in human and veterinary patients. Prospective studies to more closely follow the trends in these cytokine, chemokine, and MMP fluctuations in earlier stages of OA are also warranted.

This study has characterized the cytokine, chemokine, and MMP changes that occur in dog models of OA and in dogs with spontaneously occurring OA. As many of the biomarker studies performed in dogs to date have focused solely on direct indicators of OA, the indirect cytokine markers will be useful in further understanding the mechanisms leading up to the development of OA. More specifically, the evaluation of synovial fluid, serum, and urine-derived chemokine, cytokine, and MMP concentrations revealed potential additional biomarker candidates for the early diagnosis of OA in dogs. While synovial fluid IL-8 was the most sensitive marker for both portions of this study, the authors feel MCP1 and KC contribute additional information and should be retained within the tentative biomarker panel until further investigations can be performed. Although these markers were not as useful in the serum, the authors suggest that use of synovial fluid biomarkers has important clinical application based on the relative ease in obtaining samples, the associated costs, and the joint specific nature of these evaluations. MMP2 and MMP3 were not as sensitive or specific in the synovial fluid, but they may prove useful in serum evaluations and should be studied further. Studies are currently under way in our laboratory to determine the specific tissue source(s) of these cytokines and to see if synovial fluid from other joints possesses a similar profile.

In conclusion, animal models are routinely used for the translational investigation of human OA biomarkers. Each of the most relevant models was mentioned here, and a brief list of their advantages and disadvantages was compiled. Dogs are often used in studies such as the one described here, but species such as the guinea pig, the horse, and others have significantly contributed to the level of OA research currently available. Given the various etiologies, risk factors, and presentations of OA, it is unlikely a single animal model will ever be sufficient to study all aspects of this disease.

References

1. Helmick CG, Felson DT, Lawrence RC, et al; National Arthritis Data Workgroup. Estimates of the prevalence of arthritis and other rheumatic conditions in the United States. Part I. Arthritis Rheum 2008;58(1):15–25
2. Ray A, Kuroki K, Cook JL, et al. Induction of matrix metalloproteinase 1 gene expression is regulated by inflammation-responsive transcription factor SAF-1 in osteoarthritis. Arthritis Rheum 2003;48(1):134–145
3. Kuroki K, Stoker AM, Cook JL. Effects of proinflammatory cytokines on canine articular chondrocytes in a three-dimensional culture. Am J Vet Res 2005;66(7):1187–1196
4. Goldring MB. The role of cytokines as inflammatory mediators in osteoarthritis: lessons from animal models. Connect Tissue Res 1999;40(1):1–11
5. Scanzello CR, Plaas A, Crow MK. Innate immune system activation in osteoarthritis: is osteoarthritis a chronic wound? Curr Opin Rheumatol 2008;20(5):565–572
6. Borzi RM, Mazzetti I, Macor S, et al. Flow cytometric analysis of intracellular chemokines in chondrocytes in vivo: constitutive expression and enhancement in osteoarthritis and rheumatoid arthritis. FEBS Lett 1999;455(3):238–242
7. Borzì RM, Mazzetti I, Cattini L, Uguccioni M, Baggiolini M, Facchini A. Human chondrocytes express functional chemokine receptors and release matrix-degrading enzymes in response to C-X-C and C-C chemokines. Arthritis Rheum 2000;43(8):1734–1741
8. Fernandes JC, Martel-Pelletier J, Pelletier JP. The role of cytokines in osteoarthritis pathophysiology. Biorheology 2002;39(1-2):237–246
9. Hegemann N, Wondimu A, Kohn B, Brunnberg L, Schmidt MF. Cytokine profile in canine immune-mediated polyarthritis and osteoarthritis. Vet Comp Orthop Traumatol 2005;18(2):67–72
10. Hulejová H, Baresová V, Klézl Z, Polanská M, Adam M, Senolt L. Increased level of cytokines and matrix metalloproteinases in osteoarthritic subchondral bone. Cytokine 2007;38(3):151–156
11. Klatt AR, Paul-Klausch B, Klinger G, et al. A critical role for collagen II in cartilage matrix degradation: collagen II induces pro-inflammatory cytokines and MMPs in primary human chondrocytes. J Orthop Res 2009;27(1):65–70
12. Sandell LJ, Xing X, Franz C, Davies S, Chang LW, Patra D. Exuberant expression of chemokine genes by adult human articular chondrocytes in response to IL-1beta. Osteoarthritis Cartilage 2008;16(12):1560–1571
13. Tetlow LC, Adlam DJ, Woolley DE. Matrix metalloproteinase and proinflammatory cytokine production by chondrocytes of human osteoarthritic cartilage: associations with degenerative changes. Arthritis Rheum 2001;44(3):585–594
14. Borden P, Solymar D, Sucharczuk A, Lindman B, Cannon P, Heller RA. Cytokine control of interstitial collagenase and collagenase-3 gene expression in human chondrocytes. J Biol Chem 1996; 271(38):23577–23581
15. Schröder JM, Persoon NL, Christophers E. Lipopolysac-charide-stimulated human monocytes secrete, apart from neutrophil-activating peptide 1/interleukin 8, a second neutrophil-activating protein. NH2-terminal amino acid sequence identity with melanoma growth stimulatory activity. J Exp Med 1990;171(4):1091–1100
16. Venn G, Nietfeld JJ, Duits AJ, et al. Elevated synovial fluid levels of interleukin-6 and tumor necrosis factor associated with early experimental canine osteoarthritis. Arthritis Rheum 1993;36(6):819–826
17. Villiger PM, Terkeltaub R, Lotz M. Production of monocyte chemoattractant protein-1 by inflamed synovial tissue and cultured synoviocytes. J Immunol 1992;149(2):722–727
18. Tiku K, Thakker-Varia S, Ramachandrula A, Tiku ML. Articular chondrocytes secrete IL-1, express membrane IL-1, and have IL-1 inhibitory activity. Cell Immunol 1992;140(1):1–20

19. Aigner T, Cook JL, Gerwin N, et al. Histopathology atlas of animal model systems—overview of guiding principles. Osteoarthritis Cartilage 2010;18(Suppl 3):S2–S6
20. Cook JL, Kuroki K, Visco D, Pelletier JP, Schulz L, Lafeber FP. The OARSI histopathology initiative—recommendations for histological assessments of osteoarthritis in the dog. Osteoarthritis Cartilage 2010;18(Suppl 3):S66–S79
21. Gerwin N, Bendele AM, Glasson S, Carlson CS. The OARSI histopathology initiative—recommendations for histological assessments of osteoarthritis in the rat. Osteoarthritis Cartilage 2010;18(Suppl 3):S24–S34
22. Kraus VB, Huebner JL, DeGroot J, Bendele A. The OARSI histopathology initiative—recommendations for histological assessments of osteoarthritis in the guinea pig. Osteoarthritis Cartilage 2010;18(Suppl 3):S35–S52
23. Laverty S, Girard CA, Williams JM, Hunziker EB, Pritzker KP. The OARSI histopathology initiative—recommendations for histological assessments of osteoarthritis in the rabbit. Osteoarthritis Cartilage 2010;18(Suppl 3):S53–S65
24. McIlwraith CW, Frisbie DD, Kawcak CE, Fuller CJ, Hurtig M, Cruz A. The OARSI histopathology initiative—recommendations for histological assessments of osteoarthritis in the horse. Osteoarthritis Cartilage 2010;18(Suppl 3):S93–S105
25. Poole R, Blake S, Buschmann M, et al. Recommendations for the use of preclinical models in the study and treatment of osteoarthritis. Osteoarthritis Cartilage 2010;18(Suppl 3):S10–S16
26. Glasson SS, Chambers MG, Van Den Berg WB, Little CB. The OARSI histopathology initiative—recommendations for histological assessments of osteoarthritis in the mouse. Osteoarthritis Cartilage 2010;18(Suppl 3):S17–S23
27. Little CB, Smith MM, Cake MA, Read RA, Murphy MJ, Barry FP. The OARSI histopathology initiative—recommendations for histological assessments of osteoarthritis in sheep and goats. Osteoarthritis Cartilage 2010;18(Suppl 3):S80–S92
28. Johnston SA. Osteoarthritis. Joint anatomy, physiology, and pathobiology. Vet Clin North Am Small Anim Pract 1997;27(4):699–723
29. Pond MJ, Nuki G. Experimentally-induced osteoarthritis in the dog. Ann Rheum Dis 1973;32(4): 387–388
30. Luther JK, Cook CR, Cook JL. Meniscal release in cruciate ligament intact stifles causes lameness and medial compartment cartilage pathology in dogs 12 weeks postoperatively. Vet Surg 2009;38(4):520–529
31. Bendele AM, Hulman JF. Spontaneous cartilage degeneration in guinea pigs. Arthritis Rheum 1988;31(4):561–565
32. Frisbie DD, Al-Sobayil F, Billinghurst RC, Kawcak CE, McIlwraith CW. Changes in synovial fluid and serum biomarkers with exercise and early osteoarthritis in horses. Osteoarthritis Cartilage 2008;16(10):1196–1204
33. Frisbie DD, Kawcak CE, McIlwraith CW. Evaluation of the effect of extracorporeal shock wave treatment on experimentally induced osteoarthritis in middle carpal joints of horses. Am J Vet Res 2009;70(4):449–454
34. Frisbie DD, Kawcak CE, Trotter GW, Powers BE, Walton RM, McIlwraith CW. Effects of triamcinolone acetonide on an in vivo equine osteochondral fragment exercise model. Equine Vet J 1997;29(5):349–359
35. Janusz MJ, Bendele AM, Brown KK, Taiwo YO, Hsieh L, Heitmeyer SA. Induction of osteoarthritis in the rat by surgical tear of the meniscus: Inhibition of joint damage by a matrix metalloproteinase inhibitor. Osteoarthritis Cartilage 2002;10(10):785–791
36. Moore EE, Bendele AM, Thompson DL, et al. Fibroblast growth factor-18 stimulates chondrogenesis and cartilage repair in a rat model of injury-induced osteoarthritis. Osteoarthritis Cartilage 2005;13(7):623–631
37. De Gruttola VG, Clax P, DeMets DL, et al. Considerations in the evaluation of surrogate endpoints in clinical trials: Summary of a National Institutes of Health workshop. Control Clin Trials 2001;22(5):485–502
38. Bauer DC, Hunter DJ, Abramson SB, et al; Osteoarthritis Biomarkers Network. Classification of osteoarthritis biomarkers: a proposed approach. Osteoarthritis Cartilage 2006;14(8):723–727
39. Kraus VB. Osteoarthritis year 2010 in review: biochemical markers. Osteoarthritis Cartilage 2011;19(4):346–353
40. McIlwraith CW. Use of synovial fluid and serum biomarkers in equine bone and joint disease: a review. Equine Vet J 2005;37(5):473–482
41. Thonar EJ. Molecular markers of metabolic changes in osteoarthritis. Osteoarthritis Cartilage 1999;7(3):338– 339
42. Matyas JR, Atley L, Ionescu M, Eyre DR, Poole AR. Analysis of cartilage biomarkers in the early phases of canine experimental osteoarthritis. Arthritis Rheum 2004;50(2):543–552
43. Cawston T, Billington C, Cleaver C, et al. The regulation of MMPs and TIMPs in cartilage turnover. Ann N Y Acad Sci 1999;878:120–129
44. Bonassar LJ, Frank EH, Murray JC, et al. Changes in cartilage composition and physical properties due to stromelysin degradation. Arthritis Rheum 1995;38(2):173–183
45. Lohmander LS, Saxne T, Heinegård DK. Release of cartilage oligomeric matrix protein (COMP) into joint fluid after knee injury and in osteoarthritis. Ann Rheum Dis 1994;53(1):8–13
46. Misumi K, Vilim V, Carter SD, Ichihashi K, Oka T, Sakamoto H. Concentrations of cartilage oligomeric matrix protein in dogs with naturally developing and experimentally induced arthropathy. Am J Vet Res 2002;63(4):598–603
47. Barksby HE, Milner JM, Patterson AM, et al. Matrix metalloproteinase 10 promotion of collagenolysis via procollagenase activation: implications for cartilage degradation in arthritis. Arthritis Rheum 2006;54(10):3244–3253
48. Billinghurst RC, Dahlberg L, Ionescu M, et al. Enhanced cleavage of type II collagen by collagenases in osteoarthritic articular cartilage. J Clin Invest 1997;99(7):1534–1545
49. Huebner JL, Otterness IG, Freund EM, Caterson B, Kraus VB. Collagenase 1 and collagenase 3 expression in a guinea pig model of osteoarthritis. Arthritis Rheum 1998;41(5):877–890
50. Shlopov BV, Lie WR, Mainardi CL, Cole AA, Chubinskaya S, Hasty KA. Osteoarthritic lesions: involvement of three different collagenases. Arthritis Rheum 1997;40(11):2065–2074

51. Conrozier T, Poole AR, Ferrand F, et al. Serum concentrations of type II collagen biomarkers (C2C, C1, 2C and CPII) suggest different pathophysiologies in patients with hip osteoarthritis. Clin Exp Rheumatol 2008;26(3):430–435
52. Arican M, Carter SD, Bennett D, May C. Measurement of glycosaminoglycans and keratan sulphate in canine arthropathies. Res Vet Sci 1994;56(3):290–297
53. Campion GV, McCrae F, Schnitzer TJ, Lenz ME, Dieppe PA, Thonar EJ. Levels of keratan sulfate in the serum and synovial fluid of patients with osteoarthritis of the knee. Arthritis Rheum 1991;34(10):1254–1259
54. Sondergaard BC, Schultz N, Madsen SH, Bay-Jensen AC, Kassem M, Karsdal MA. MAPKs are essential upstream signaling pathways in proteolytic cartilage degradation: divergence in pathways leading to aggrecanase and MMP-mediated articular cartilage degradation. Osteoarthritis Cartilage 2010;18(3):279–288
55. Poole AR, Ionescu M, Swan A, Dieppe PA. Changes in cartilage metabolism in arthritis are reflected by altered serum and synovial fluid levels of the cartilage proteoglycan aggrecan. Implications for pathogenesis. J Clin Invest 1994;94(1):25–33
56. Chaganti RK, Kelman A, Lui L, et al; Study Of Osteoporotic Fractures Research Group (SOF). Change in serum measurements of cartilage oligomeric matrix protein and association with the development and worsening of radiographic hip osteoarthritis. Osteoarthritis Cartilage 2008; 16(5):566–571
57. Hunter DJ, Li J, LaValley M, et al. Cartilage markers and their association with cartilage loss on magnetic resonance imaging in knee osteoarthritis: the Boston Osteoarthritis Knee Study. Arthritis Res Ther 2007;9(5):R108
58. Bruyere O, Collette J, Kothari M, et al. Osteoarthritis, magnetic resonance imaging, and biochemical markers: a one year prospective study. Ann Rheum Dis 2006;65(8):1050–1054
59. King KB, Lindsey CT, Dunn TC, Ries MD, Steinbach LS, Majumdar S. A study of the relationship between molecular biomarkers of joint degeneration and the magnetic resonance-measured characteristics of cartilage in 16 symptomatic knees. Magn Reson Imaging 2004;22(8):1117–1123
60. Qi C, Changlin H. Levels of biomarkers correlate with magnetic resonance imaging progression of knee cartilage degeneration: a study on canine. Knee Surg Sports Traumatol Arthrosc 2007;15(7):869–878
61. Schmidt-Rohlfing B, Thomsen M, Niedhart C, Wirtz DC, Schneider U. Correlation of bone and cartilage markers in the synovial fluid with the degree of osteoarthritis. Rheumatol Int 2002; 21(5):193–199
62. Wong PK, Young L, Vaile JH, et al. Telopeptides as markers of bone turnover in rheumatoid arthritis and osteoarthritis. Intern Med J 2004;34(9-10):539–544
63. Bruyere O, Collette JH, Ethgen O, et al. Biochemical markers of bone and cartilage remodeling in prediction of longterm progression of knee osteoarthritis. J Rheumatol 2003;30(5):1043–1050
64. Arican M, Carter SD, Bennett D. Osteocalcin in canine joint diseases. Br Vet J 1996;152(4): 411–423
65. Billinghurst RC, Buxton EM, Edwards MG, McGraw MS, McIlwraith CW. Use of an antineoepitope antibody for identification of type-II collagen degradation in equine articular cartilage. Am J Vet Res 2001;62(7):1031–1039
66. van Meurs J, van Lent P, Stoop R, et al. Cleavage of aggrecan at the Asn341-Phe342 site coincides with the initiation of collagen damage in murine antigen-induced arthritis: a pivotal role for stromelysin 1 in matrix metalloproteinase activity. Arthritis Rheum 1999;42(10):2074–2084
67. Freemont AJ, Hampson V, Tilman R, Goupille P, Taiwo Y, Hoyland JA. Gene expression of matrix metalloproteinases 1, 3, and 9 by chondrocytes in osteoarthritic human knee articular cartilage is zone and grade specific. Ann Rheum Dis 1997;56(9):542–549
68. Karran EH, Young TJ, Markwell RE, Harper GP. In vivo model of cartilage degradation—effects of a matrix metalloproteinase inhibitor. Ann Rheum Dis 1995;54(8):662–669
69. Mehraban F, Kuo SY, Riera H, Chang C, Moskowitz RW. Prostromelysin and procollagenase genes are differentially up-regulated in chondrocytes from the knees of rabbits with experimental osteoarthritis. Arthritis Rheum 1994;37(8):1189–1197
70. Panula HE, Lohmander LS, Rönkkö S, Agren U, Helminen HJ, Kiviranta I. Elevated levels of synovial fluid PLA2, stromelysin (MMP-3) and TIMP in early osteoarthrosis after tibial valgus osteotomy in young beagle dogs. Acta Orthop Scand 1998;69(2):152–158
71. Cook JL, Anderson CC, Kreeger JM, Tomlinson JL. Effects of human recombinant interleukin-1beta on canine articular chondrocytes in three-dimensional culture. Am J Vet Res 2000;61(7): 766–770
72. Ratcliffe A, Beauvais PJ, Saed-Nejad F, Shurety W, Caterson B. Synovial fluid analyses detect and differentiate proteoglycan metabolism in canine experimental models of osteoarthritis and disuse atrophy. Agents Actions Suppl 1993;39:63–67
73. Lindhorst E, Vail TP, Guilak F, et al. Longitudinal characterization of synovial fluid biomarkers in the canine meniscectomy model of osteoarthritis. J Orthop Res 2000;18(2):269–280
74. Budsberg SC, Lenz ME, Thonar EJ. Serum and synovial fluid concentrations of keratan sulfate and hyaluronan in dogs with induced stifle joint osteoarthritis following cranial cruciate ligament transection. Am J Vet Res 2006;67(3):429–432
75. Innes JF, Sharif M, Barr AR. Relations between biochemical markers of osteoarthritis and other disease parameters in a population of dogs with naturally acquired osteoarthritis of the genual joint. Am J Vet Res 1998;59(12):1530–1536
76. Müller G, Michel A, Altenburg E. COMP (cartilage oligomeric matrix protein) is synthesized in ligament, tendon, meniscus, and articular cartilage. Connect Tissue Res 1998;39(4):233–244
77. Arai K, Tagami M, Hatazoe T, et al. Analysis of cartilage oligomeric matrix protein (COMP) in synovial fluid, serum and urine from 51 racehorses with carpal bone fracture. J Vet Med Sci 2008;70(9):915–921
78. Misumi K, Vilim V, Hatazoe T, et al. Serum level of cartilage oligomeric matrix protein (COMP) in equine osteoarthritis. Equine Vet J 2002;34(6):602–608

79. Chu Q, Lopez M, Hayashi K, et al. Elevation of a collagenase generated type II collagen neoepitope and proteoglycan epitopes in synovial fluid following induction of joint instability in the dog. Osteoarthritis Cartilage 2002;10(8):662–669
80. Goranov NV. Serum markers of lipid peroxidation, antioxidant enzymatic defense, and collagen degradation in an experimental (Pond-Nuki) canine model of osteoarthritis. Vet Clin Pathol 2007;36(2):192–195
81. Hayashi K, Kim SY, Lansdowne JL, Kapatkin A, Déjardin LM. Evaluation of a collagenase generated osteoarthritis biomarker in naturally occurring canine cruciate disease. Vet Surg 2009;38(1):117–121
82. Carney SL, Billingham ME, Caterson B, et al. Changes in proteoglycan turnover in experimental canine osteoarthritic cartilage. Matrix 1992;12(2):137–147
83. Caterson B, Mahmoodian F, Sorrell JM, et al. Modulation of native chondroitin sulphate structure in tissue development and in disease. J Cell Sci 1990;97(Pt 3):411–417
84. Sorrell JM, Mahmoodian F, Schafer IA, Davis B, Caterson B. Identification of monoclonal antibodies that recognize novel epitopes in native chondroitin/dermatan sulfate glycosaminoglycan chains: their use in mapping functionally distinct domains of human skin. J Histochem Cytochem 1990;38(3):393–402
85. Ratcliffe A, Shurety W, Caterson B. The quantitation of a native chondroitin sulfate epitope in synovial fluid lavages and articular cartilage from canine experimental osteoarthritis and disuse atrophy. Arthritis Rheum 1993;36(4):543–551
86. Visco DM, Johnstone B, Hill MA, Jolly GA, Caterson B. Immunohistochemical analysis of 3-B-(-) and 7-D-4 epitope expression in canine osteoarthritis. Arthritis Rheum 1993;36(12):1718–1725
87. Johnson KA, Hay CW, Chu Q, Roe SC, Caterson B. Cartilage-derived biomarkers of osteoarthritis in synovial fluid of dogs with naturally acquired rupture of the cranial cruciate ligament. Am J Vet Res 2002;63(6):775–781
88. Nganvongpanit K, Itthiarbha A, Ong-Chai S, Kongtawelert P. Evaluation of serum chondroitin sulfate and hyaluronan: biomarkers for osteoarthritis in canine hip dysplasia. J Vet Sci 2008;9(3):317–325
89. Hay CW, Chu Q, Budsberg SC, Clayton MK, Johnson KA. Synovial fluid interleukin 6, tumor necrosis factor, and nitric oxide values in dogs with osteoarthritis secondary to cranial cruciate ligament rupture. Am J Vet Res 1997;58(9):1027–1032
90. de Bruin T, de Rooster H, van Bree H, Cox E. Interleukin-8 mRNA expression in synovial fluid of canine stifle joints with osteoarthritis. Vet Immunol Immunopathol 2005;108(3-4):387–397
91. Yuan GH, Masuko-Hongo K, Sakata M, et al. The role of C-C chemokines and their receptors in osteoarthritis. Arthritis Rheum 2001;44(5):1056–1070
92. Rollins BJ. Monocyte chemoattractant protein 1: a potential regulator of monocyte recruitment in inflammatory disease. Mol Med Today 1996;2(5): 198–204
93. Mahn MM, Cook JL, Cook CR, Balke MT. Arthroscopic verification of ultrasonographic diagnosis of meniscal pathology in dogs. Vet Surg 2005;34(4):318–323
94. Breshears LA, Cook JL, Stoker AM, Fox DB, Luther JK. The Effect of Uniaxial Cyclic Tensile Load on Gene Expression in Canine Cranial Cruciate Ligamentocytes. Vet Surg 2010;39(4):433–443
95. Yoshimura T, Robinson EA, Tanaka S, Appella E, Kuratsu J, Leonard EJ. Purification and amino acid analysis of two human glioma-derived monocyte chemoattractants. J Exp Med 1989;169(4):1449–1459
96. Matsushima K, Larsen CG, DuBois GC, Oppenheim JJ. Purification and characterization of a novel monocyte chemotactic and activating factor produced by a human myelomonocytic cell line. J Exp Med 1989;169(4):1485–1490
97. Yoshimura T, Robinson EA, Tanaka S, Appella E, Leonard EJ. Purification and amino acid analysis of two human monocyte chemoattractants produced by phytohemagglutinin-stimulated human blood mononuclear leukocytes. J Immunol 1989;142(6):1956–1962
98. Carr MW, Roth SJ, Luther E, Rose SS, Springer TA. Monocyte chemoattractant protein 1 acts as a T-lymphocyte chemoattractant. Proc Natl Acad Sci U S A 1994;91(9):3652–3656
99. Taub DD, Proost P, Murphy WJ, et al. Monocyte chemotactic protein-1 (MCP-1), -2, and -3 are chemotactic for human T lymphocytes. J Clin Invest 1995;95(3):1370–1376
100. Koch AE, Kunkel SL, Harlow LA, et al. Enhanced production of monocyte chemoattractant protein-1 in rheumatoid arthritis. J Clin Invest 1992;90(3):772–779
101. Stankovic A, Slavic V, Stamenkovic B, Kamenov B, Bojanovic M, Mitrovic DR. Serum and synovial fluid concentrations of CCL2 (MCP-1) chemokine in patients suffering rheumatoid arthritis and osteoarthritis reflect disease activity. Bratisl Lek Listy (Tlacene Vyd) 2009;110(10):641–646
102. Seitz M, Loetscher P, Dewald B, Towbin H, Ceska M, Baggiolini M. Production of interleukin-1 receptor antagonist, inflammatory chemotactic proteins, and prostaglandin E by rheumatoid and osteoarthritic synoviocytes—regulation by IFN-gamma and IL-4. J Immunol 1994;152(4): 2060–2065
103. Duffy AL, Olea-Popelka FJ, Eucher J, Rice DM, Dow SW. Serum concentrations of monocyte chemoattractant protein-1 in healthy and critically ill dogs. Vet Clin Pathol 2010;39(3):302–305
104. Moser B, Clark-Lewis I, Zwahlen R, Baggiolini M. Neutrophil-activating properties of the melanoma growth-stimulatory activity. J Exp Med 1990;171(5):1797–1802
105. Lin M, Carlson E, Diaconu E, Pearlman E. CXCL1/KC and CXCL5/LIX are selectively produced by corneal fibroblasts and mediate neutrophil infiltration to the corneal stroma in LPS keratitis. J Leukoc Biol 2007;81(3):786–792
106. Fujiwara K, Ohkawara S, Takagi K, Yoshinaga M, Matsukawa A. Involvement of CXC chemokine growth-related oncogene-alpha in monosodium urate crystal-induced arthritis in rabbits. Lab Invest 2002;82(10):1297–1304
107. Modi WS, Yoshimura T. Isolation of novel GRO genes and a phylogenetic analysis of the CXC chemokine subfamily in mammals. Mol Biol Evol 1999;16(2):180–193
108. Remick DG, DeForge LE, Sullivan JF, Showell HJ. Profile of cytokines in synovial fluid specimens

from patients with arthritis. Interleukin 8 (IL-8) and IL-6 correlate with inflammatory arthritides. Immunol Invest 1992;21(4):321–327

109. Brennan FM, Zachariae CO, Chantry D, et al. Detection of interleukin 8 biological activity in synovial fluids from patients with rheumatoid arthritis and production of interleukin 8 mRNA by isolated synovial cells. Eur J Immunol 1990;20(9):2141–2144

110. Straubinger RK, Straubinger AF, Härter L, et al. Borrelia burgdorferi migrates into joint capsules and causes an up-regulation of interleukin-8 in synovial membranes of dogs experimentally infected with ticks. Infect Immun 1997; 65(4):1273–1285

111. Straubinger RK, Straubinger AF, Summers BA, Erb HN, Härter L, Appel MJ. Borrelia burgdorferi induces the production and release of proinflammatory cytokines in canine synovial explant cultures. Infect Immun 1998;66(1):247–258

112. Straubinger RK, Straubinger AF, Summers BA, Jacobson RH, Erb HN. Clinical manifestations, pathogenesis, and effect of antibiotic treatment on Lyme borreliosis in dogs. Wien Klin Wochenschr 1998;110(24):874–881

113. Merz D, Liu R, Johnson K, Terkeltaub R. IL-8/CXCL8 and growth-related oncogene alpha/CXCL1 induce chondrocyte hypertrophic differentiation. J Immunol 2003;171(8):4406–4415

114. Ko YC, Mukaida N, Ishiyama S, et al. Elevated interleukin-8 levels in the urine of patients with urinary tract infections. Infect Immun 1993;61(4):1307–1314

115. Zaki MelS. Interleukin 8 is a surrogate marker for rapid diagnosis of bacteriuria. Immunol Invest 2008;37(7):694–703

116. Taha AS, Grant V, Kelly RW. Urinalysis for interleukin-8 in the non-invasive diagnosis of acute and chronic inflammatory diseases. Postgrad Med J 2003;79(929):159–163

117. Ling SM, Patel DD, Garnero P, et al. Serum protein signatures detect early radiographic osteoarthritis. Osteoarthritis Cartilage 2009;17(1):43–48

118. Pulsatelli L, Dolzani P, Piacentini A, et al. Chemokine production by human chondrocytes. J Rheumatol 1999;26(9):1992–2001

II

Science and Techniques for Cartilage Repair

5

The Cartilage–Bone Interface

Caroline D. Hoemann, Charles-Hubert Lafantaisie-Favreau, Viorica Lascau-Coman, Gaoping Chen, and Jessica Guzmán-Morales

All Cartilage–Bone Interfaces Are Derived from an Initially Pure Cartilage Structure

Long bones develop first from embryonic mesenchymal stem cells that coalesce to form a "blastema," with a scant but uniform type I collagen matrix.[1] The blastema transforms at early fetal stages into a cartilaginous structure, or cartilage "anlage," with collagen type II as the main extracellular component.[2-6] The cartilage anlage contains a mixture of fusiform and round chondroblast cells,[7] which upon terminal differentiation will hypertrophy (become unusually large) and begin to express proteins that attract blood vessels and facilitate biomineralization.[8]

Mineralized bone begins to form when the fetal cartilage undergoes focal hypertrophy, which launches a process of endochondral ossification (EO). The very first cartilage–bone interfaces to form in the body are at the primary ossification centers in the shaft of developing long bones. These areas are marked by remodeling and vascular invasion in parallel with the deposition of a mineralizing collagen type I matrix that ensheathes and mechanically protects the blood vessels.[4,9] After the secondary ossification centers appear in the distal tibia and femur, three types of dynamic cartilage–bone interface are established, as illustrated in the immature rabbit knee (**Fig. 5.1a–c**). Articular cartilage sits on top of the epiphyseal subchondral bone (**Fig. 5.1b, e, h**). The growth plate is a cartilage structure firmly sandwiched between two layers of bone: the epiphyseal and the metaphyseal bone (**Fig. 5.1c, f, i**).

Growth Plate Cartilage–Bone Interface during Postnatal Development

In the developing knee, epiphyseal bone will continue to expand into the cartilage anlage until the cartilage interface forms a thin calcified layer that arrests vascular invasion. Calcified cartilage forms at the base of the articular cartilage and in certain growth plate reserve zones (**Fig. 5.2**), through mechanisms that are still not fully understood. Haines[10] previously noticed that growth plate reserve zones fused to "permanent" epiphyseal lines develop a thin layer of calcified cartilage/tidemark (**Fig. 5.2a**, proximal trochlea) while other reserve zones do not form tidemarks (**Fig. 5.2b**, distal trochlea) and eventually close without leaving a scar.

Growth plate hypertrophic cartilage (HTC) does not form a tidemark. This interface is actually a mixture of cartilage and bone, by definition of the primary spongiosa, where

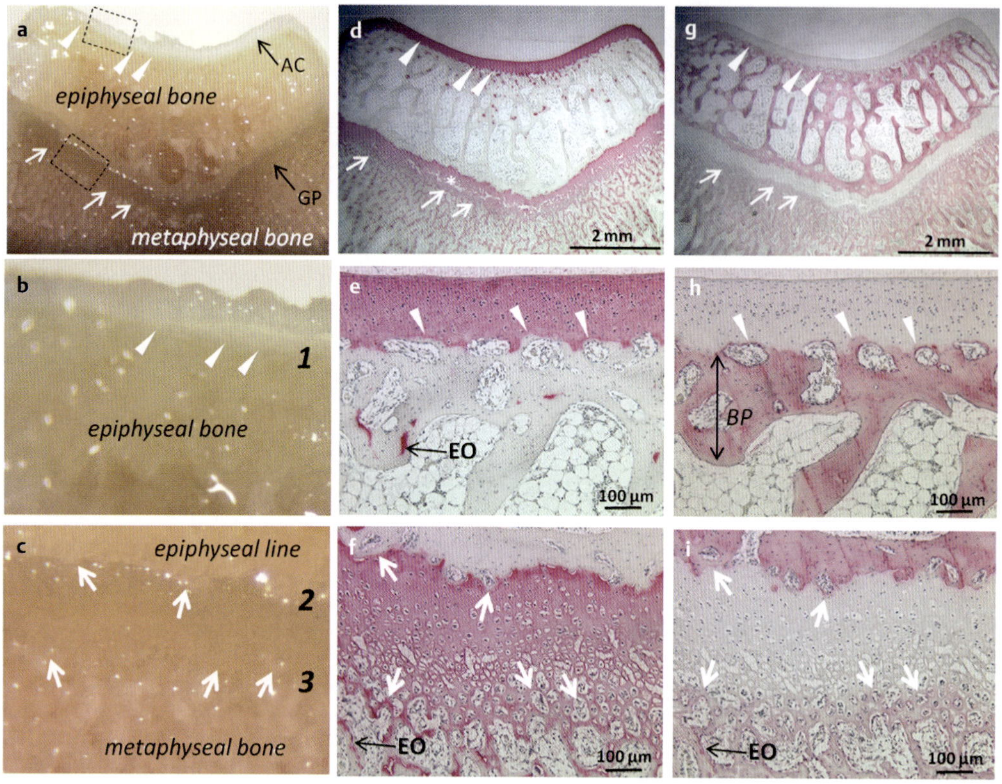

Fig. 5.1 Three dynamic cartilage–bone interfaces are established in the postnatal knee femoral end: (1) articular cartilage–epiphyseal bone, (2) growth plate–epiphyseal bone, and (3) growth plate–metaphyseal bone. A macroscopic transverse view of a skeletally immature ~ 4-month rabbit knee trochlea shows the articular cartilage (AC, *white arrowheads*, **a**, **b**) and growth plate (GP, *white arrows*, **a**, **c**), and distribution of collagen type II (**d**, **e**, **f**) and collagen type I (**g**, **h**, **i**) in serial sections from the area of the dashed squares in **a**. The distal femur was fixed in 4% paraformaldehyde/100 mM cacodylate, decalcified in EDTA with a Milestone microwave (Milestone, Shelton, CT), cryoembedded and cryosectioned using the CryoJane tape system (Leica Microsystems, Buffalo Grove, IL); the sections were predigested in hyaluronidase and protease to remove glycosaminoglycans and immunostained for collagen type II (monoclonal II6B3, DSHB, USA) or collagen type I (monoclonal antibody I-H85, VWR, Canada), using secondary biotinylated goat anti-mouse and avidin-alkaline phosphatase red substrate detection with iron hematoxylin counterstain.[91,92] In (**a**) and (**b**), the rough articular surface is a cutting artifact from the isomet diamond saw. The tear/crack in the growth plate indicated by a white asterisk in (**d**) is a cryosectioning artifact.[92] *Abbreviations:* EO, endochondral ossification; BP, subchondral bone plate; AC, articular cartilage; GP, growth plate.

new bone is deposited on the cartilage trabeculae carved out by invading blood vessels and marrow (**Figs. 5.1f, i,** and **5.3**). In trabecular bone maturing below the growth plate, an initially pure collagen type II-glycosaminoglycan (GAG) extracellular matrix is slowly incorporating collagen type I (EO, **Fig. 5.3**). Vascular invasion of the hypertrophic zone spurs a continual endochondral expansion of the distal femur (arrows, **Fig. 5.4a**), in tandem with appositional cartilage and bone growth (**Fig. 5.4b, c**).

The growth cartilage–metaphyseal bone interface is a dynamic and ever-expanding front of HTC undergoing vascular invasion and ossification. Interestingly, in newly formed endochondral bone, hypertrophic chondrocytes express zymogen forms of enzymes capable of remodeling collagen matrix, including matrix metalloproteinases (MMP-13, MMP-9) and complement C1s.[11,12] In knockout mice for MMP-13 or MMP-9, conversion of collagen type II HTC to collagen type I trabecular bone is inhibited.[13–15]

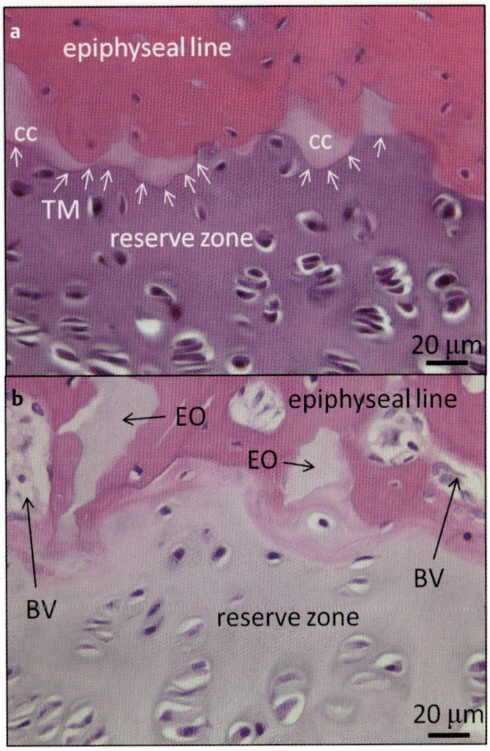

Fig. 5.2 The growth plate–epiphyseal bone interface sometimes includes a layer of calcified cartilage and a tidemark in the reserve zone (**a**, proximal trochlea), and in other areas is devoid of calcified cartilage or tidemark and fused to a more vascular bone (**b**, distal trochlea). Representative decalcified transverse sections from ~ 4-month-old rabbit trochlear growth plates stained with hematoxylin and eosin are shown, from $N = 7$ distinct New Zealand white rabbit femurs, ~ 4 months old. *Abbreviations:* TM, tidemark (*white arrows*); BV, blood vessels; CC, calcified cartilage; EO, endochondral ossification (cartilage remnant).

Remodeling of the HTC front by osteoclasts, chondroclasts, and bone-marrow-derived metalloproteinases drives the replacement of HTC with vascularized bone.[13–15] Growth plate cell proliferation and vascular invasion can be diminished by nutritional deprivation, ischemia, or supraphysiologic loading.[16–18] Blood vessel invasion of the HTC layer is believed to be naturally driven by hypertrophic chondrocyte secretion of angiogenic factors,[19] MMP-13,[11,13] and gelatinases capable of untethering matrix-bound vascular endothelial growth factor.[15] Osteoclasts that remodel the base of endochondral bone are also known to release angiogenic factors and can also promote vascular invasion and osteogenesis.[20,21] Apoptosis of hypertrophic chondrocytes is also implicated as an important driver of the endochondral growth process.[15,22]

Articular Cartilage–Bone Interface during Postnatal Development

A distinct and more advanced EO process is going on during postnatal articular cartilage growth (**Fig. 5.1e** and **h**). In 3- to 6-month-old rabbit articular cartilage, most chondrocytes are no longer proliferating, and a tidemark has formed at the base of the hypertrophic zone.[23] Bone is not being deposited along cartilage trabeculae; it has developed layer by layer to form a thick osteoid around blood vessels subjacent to the calcified cartilage layer. Only small patches of cartilage persist in the subchondral bone (EO, **Fig. 5.1e**). The remnants of GAG and collagen type II in trabecular bone are the hallmarks of EO.

Neonatal articular cartilage is relatively thick; it is filled with a system of endothelial-lined canals distinct from the normal vasculature.[7] Cartilage canals have been described in immature articular cartilage in a variety of large animals and in human (fetal ovine, 2-week-old calf, 2-year-old human).[7,24] Postnatal weight-bearing activity is associated with regression of the canals and a thinning and anisotropic organization of the articular cartilage layer. The articular layer continues to grow postnatally through an appositional or asymmetric layer-by-layer expansion, through cell division near the superficial zone[6,25,26] (**Fig. 5.4**). In the deep zone near the articular cartilage–bone interface, chondrocytes terminally differentiate into hypertrophic chondrocytes, cease to proliferate, and express collagen type II, collagen type X, alkaline phosphatase, and osteopontin, a highly phosphorylated hydroxylapatite-binding protein.[27–31] Like articular cartilage, the growth plate hypertrophic zone also contains collagen type X and alkaline phosphatase, but a tidemark is notably absent.[32,33] The tidemark that forms at the base of mature articular cartilage develops slightly below the region of chondrocytes expressing collagen

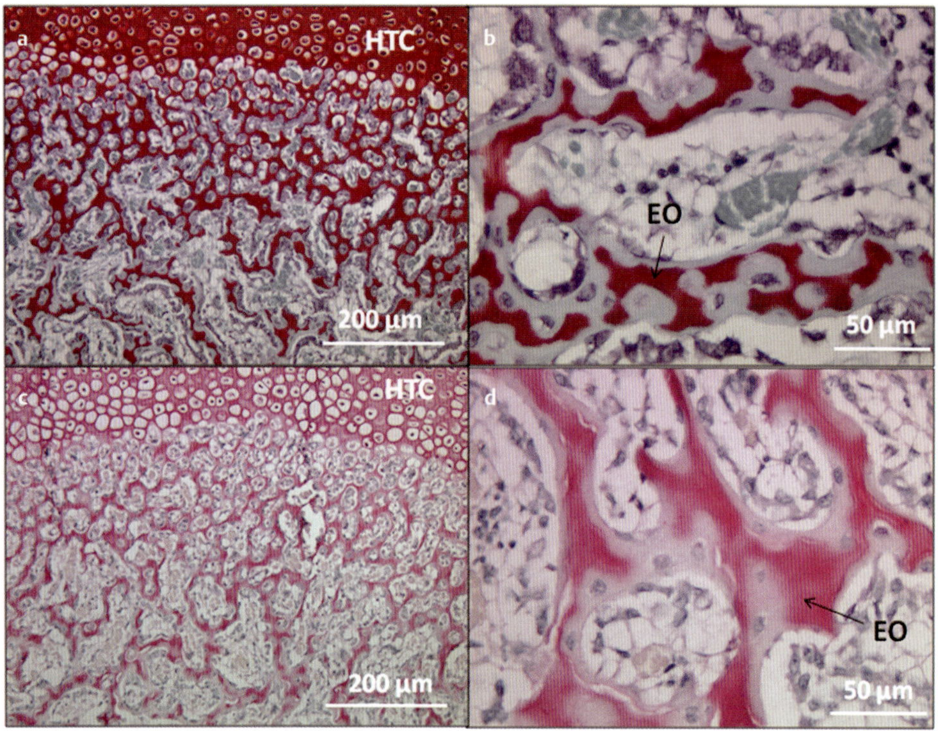

Fig. 5.3 Newly synthesized endochondral metaphyseal bone below the growth plate of a 4-month-old rabbit contains abundant sulfated glycosaminoglycans (GAG, red Safranin O stain, **a**, **b**) and collagen type II (pink immunostain, **c**, **d**) in addition to collagen type I (in this figure, the collagen type I matrix has been counterstained green by fast green in **a**, **b**, or blue by hematoxylin in **c**, **d**). In panel B, fast green counterstain also shows bone marrow and blood vessels. A 4-month-old rabbit knee femur end was fixed in formalin, decalcified in 0.5N HCl/0.1% glutaraldehyde, cut transversely in the trochlea, embedded in paraffin, and stained with Safranin O/Fast green/Iron hematoxylin or immunostained for collagen type II as previously described.[60,72,91] Examples of cartilage remnant present in the primary spongiosa formed by EO are indicated in (**b**) and (**d**). *Abbreviations:* EO, endochondral ossification; HTC, hypertrophic cartilage.

type X.[28] Mineral deposits form in the neonatal calcified layer of the articular cartilage in line with the collagen fibers.[34] Using fluorescent pulse labeling of the mineral phase, Oegema et al observed that the tidemark advances above the pulse-labeled mineralization front in 4-month-old rabbit patella at a rate of 8 μm/week, compared with 1 μm/week in the 7-month-old rabbit patella.[26] Creeping advancement of the tidemark is associated with thinning of the articular cartilage layer.[23]

Vascular channels are branched structures that supply the calcified cartilage of the articular layer[7,9,35] and have similarities with vascular channels in the vertebral endplate where bone abuts the cartilaginous nucleus pulposus.[36] The calcified cartilage zone is thus normally vascularized, whereas the nonmineralized cartilage above the tidemark is normally avascular. In a study by Bonde et al,[37] blood vessels were observed to only sporadically penetrate the tidemark into cartilage in normal patellar cadaveric subjects (less than one average blood vessel per normal patella from subjects 75 to 89 years old) compared with an average of nine tidemark-penetrating vessels per subject in osteoarthritis (OA) femoral condyle samples. Pathologic blood vessel passage beyond the tidemark is associated with occasional thrombosis, a thicker calcified cartilage layer, and tidemark duplication[37] (**Table 5.1**).

The calcified cartilage layer is semipermeable and permits passage of small molecules (< 500 Da) from the subchondral bone to

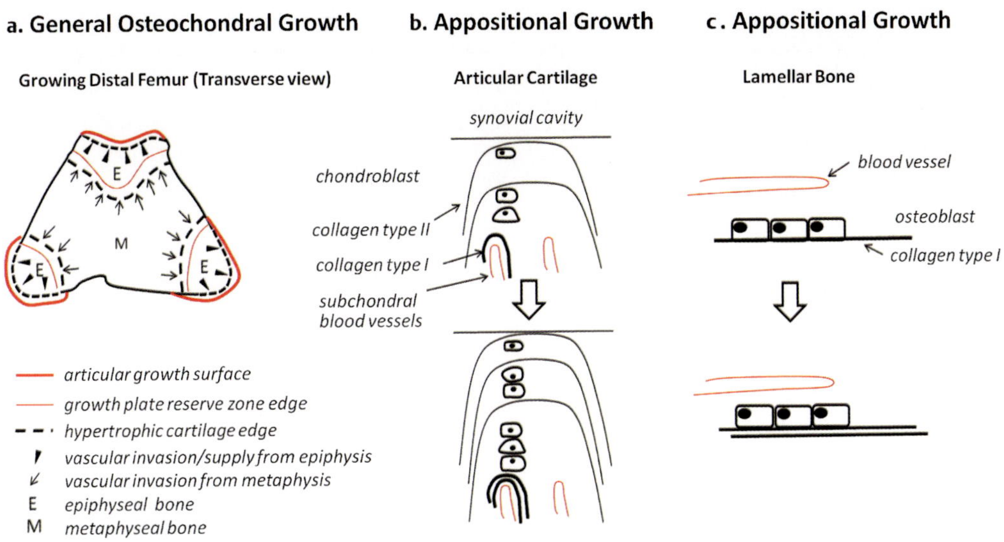

Fig. 5.4 Illustration of femur growth and appositional growth. (**a**) The distal femur grows (as illustrated in faithful tracings of a transverse section through the immature rabbit trochlea and proximal condyles) through vascular invasion, which drives endochondral ossification of the hypertrophic zone of the growth plates (*arrows, thick dashed lines*) while blood vessels from the epiphyseal bone supply the articular cartilage hypertrophic zone (*arrowheads*). During appositional growth of articular cartilage (**b**), proliferating chondroblasts deposit fibrillar collagen type II above a previously existing layer of collagen type II; the proliferating chondrocytes are situated above invading blood vessels frequently capped with collagen type I–positive mineralized bone. In intramembranous bone growth (**c**), osteoblasts and blood vessels are closely associated during inframembranous generation of lamellar bone. The mechanisms of endochondral bone growth are not illustrated in this diagram.

the articular cartilage layer.[38,39] Conversely, immersion of a mouse distal femur end in fluorescein allows full solute passage through the articular cartilage and selective fluorescein diffusion into chondrocytes in the calcified cartilage layer.[39] Calcified cartilage permeability was measured as fivefold less than noncalcified cartilage in mature horse metacarpal tissues.[38] It has been hypothesized that venous congestion in the synovium and subchondral bone could play a role in tidemark duplication.[37] Thickening of the calcified cartilage in OA could be expected to reduce the flow of small solutes from the vascularized subchondral bone to the deep zone chondrocytes.

Once formed, the tidemark and calcified cartilage layer persist as dynamic structures that can change and remodel over time. Below mature articular cartilage, the mineralization front is a relatively smooth and undulating plate-like surface, as illustrated in a micro-computed tomography (micro-CT) 3D image of the calcified cartilage and bone below the trochlear articular cartilage in a 30-month-old rabbit knee (**Fig. 5.5a**). The tidemark at this stage is a strong hematoxylin-stained line (**Fig. 5.5b**). Some remnant or newly duplicating tidemarks can be observed within the mature calcified cartilage layer (open arrowheads, **Fig. 5.5b**). In the skeletally immature 4-month-old rabbit, a more irregular mineralization front is observed at the articular cartilage–bone interface, which corresponds to the subchondral bone and a thin layer of calcified cartilage (**Fig. 5.5c**). At this stage a nascent tidemark can be visualized using a hematoxylin-eosin stain, and the vascular bone channels are just below, with close communication between the vasculature and calcified articular cartilage (**Fig. 5.5d**). By contrast, in the same 4-month-old rabbit trochlear specimen, the growth plate HTC contains a highly irregular and discontinuous mineralization front at the growing bone–cartilage interface with no visible tidemark (**Fig. 5.5e, f**). The mineral front at the base of the growth plate corresponds

Table 5.1 Published histomorphometric and stereological measures of articular cartilage, tidemark, calcified cartilage, bone plate thickness, and tidemark number or area in normal and OA human subjects and different animal species

	N/OA	Age	Site	Articular Cartilage (μm)	Tidemark (number or area)	Calcified Cartilage (μm)	Bone Plate
Lane and Bullough 1980[44]	NH	20–39	FH	—	1.2 ± 0.1	193 ± 28	—
Lane and Bullough 1980[44]	NH	40–59	FH	—	1.2 ± 0.1	141 ± 18	—
Lane and Bullough 1980[44]	NH	60–93	FH	—	1.8 ± 0.4	119 ± 24	—
Frisbie et al 2006[48]	NH	—	MFC	2,200	—	125	490
Frisbie et al 2006[48]	NE	—	MFC	2,000	—	210	375
Frisbie et al 2006[48]	NO	—	MFC	450	—	125	250
Frisbie et al 2006[48]	NR	—	MFC	200	—	100	250
Hunziker et al 2002[46]	NH	23–49	MFC	2,410	—	134	190
Wang et al 2009[43]	NH	20–45	MFC[a]	—	—	104 ± 21	—
Bonde et al 2005[37]	NH	65–85	PAT	—	2.5 cm^2 (1.8–3.9)[b]	—	—
Bonde et al 2005[37]	OA	47–86	MFC	—	7.7 cm^2 (2.4–13.3)[c]	—	—

[a]Weight-bearing area.
[b]Mean 0.5 (0 to 10) penetrating blood vessels in non-OA patellar tidemark.
[c]Mean 9 penetrating blood vessels (2 to 47) in OA tidemark area; 3 out of 21 vessels had thrombosis.
Abbreviations: —, not done; FH, femoral head; MFC, medial femoral condyle; NH, normal human; OA, osteoarthritic human; NE, normal equine; NO, normal ovine; NR, normal rabbit; PAT, patella.

with the vascular bone and newly deposited collagen type I (black arrowheads, **Fig. 5.5f**). In the growth plate hypertrophic zone, calcification of the collagen type II matrix is much delayed compared with the articular cartilage calcified layer. This is because after birth the mammalian joints require a suitable mechanically stable articular surface, while growth plates in the long bones are continually expanding, even beyond sexual maturity. Cartilage calcification is therefore occurring only at the end-stage of cartilage growth. After reaching skeletal maturity, growth plates are completely resorbed and replaced by collagen type I–positive mineralized bone.

To summarize, growth plates develop a relatively stable reserve zone–epiphyseal bone interface, with a purely collagen type II GAG-rich cartilage phase and a mixture of collagen type I and collagen type II in the newly forming primary spongiosa. Calcified cartilage becomes established at the edges of a "permanent" epiphyseal bone layer (i.e., proximal reserve zone and articular cartilage hypertrophic zone), and the tidemark serves as a barrier to vascular invasion and calcification of hyaline cartilage.

Structure and Mineral Content of the Mature Articular Cartilage–Bone Interface

Articular calcified cartilage is a mineralized layer in which extracellular matrix is chiefly composed of collagen type II, collagen type X, and GAG[19]; the layer also contains extracellular alkaline phosphatase.[31] Alkaline phosphatase can generate free phosphate from organophosphates such as β-glycerol phosphate, for incorporation into hydroxylapatite mineral (Ca-P).[40–42] In the calcified cartilage layer of normal human femoral condyles, chondrocytes are quiescent and present at

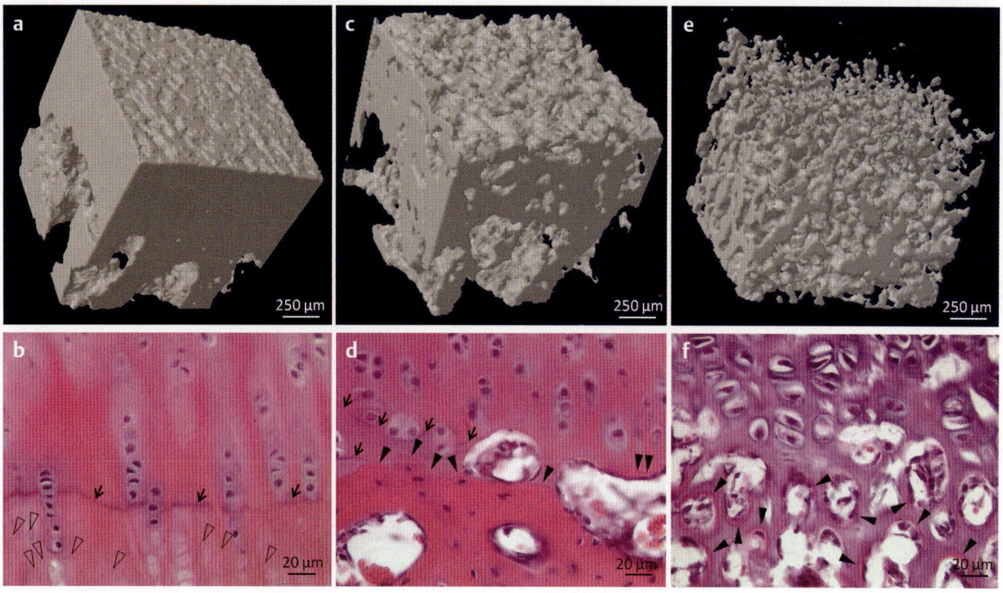

Fig. 5.5 The mineralization front at the cartilage–bone interface, in a skeletally aged 30-month-old rabbit trochlea articular cartilage-subchondral bone plate (**a**, **b**), and skeletally immature 4-month-old rabbit articular cartilage-epiphyseal bone interface (**c**, **d**) and hypertrophic growth plate–metaphyseal bone interface (**e**, **f**). (**A**), (**c**), and (**e**) show 3D reconstructions from a microcomputed tomography scan (SkyScan 1172 instrument, 9.8 µm/pixel resolution, NRecon and CTAn software) from the 30-month-old rabbit (**a**), and the 4-month-old rabbit trochlea shown in **Fig. 5.1c, e**. **B**, **d**, and **f** show the histological appearance of the cartilage–bone interface from the 30-month-old (**b**), and 4-month-old rabbit trochlea (**d**, **f**), hematoxylin & eosin stain. The small black arrows show the tidemark. Open arrowheads show vestigial or newly duplicating tidemarks (**b**) and the solid black arrowheads show areas of bone osteoid.

a much lower density compared with hyaline cartilage (average of 51 cells/mm^2 vs. 152 cells/mm^2).[43] The calcified cartilage layer is flanked by an undulating tidemark, and an even more irregular cement line adjacent to the bone. Wang et al[43] analyzed normal adult human bone (20- to 45-year-old cadaveric) by histomorphometry and stereology to show that hyaline cartilage is interlocked tightly in a "ravine-engomphosis" structure with the calcified cartilage zone, which is then attached in a "comb-anchor" to bone.[43] The surface roughness was determined to be 1.14 (tidemark) and 1.99 (cement line).[43] The more irregular cement line–bone interface is the end result of an inhomogeneous vascular invasion during development of the calcified cartilage layer (white arrowheads, **Fig. 5.1h**).

In normal human subjects, the mean calcified cartilage thickness is variable, from 20 µm to ~ 250 µm.[44–46] The calcified cartilage is tightly fused to the articular cartilage and subchondral bone plate composed of lamellar bone, along with punctate regions where the calcified cartilage is in direct contact with vascular channels.[7,47] In any group of individuals, the mean calcified cartilage thickness and mineral density will vary according to age, site in the joint, and mechanical loading (**Table 5.1**).[44,48,49] In a study of normal femoral head cadaveric specimens with no signs of OA by Lane and Bullough,[44] the calcified cartilage thickness in the femoral head varied from 79 µm to 243 µm, with a thicker calcified cartilage in less stressed areas of the hip joint. Müller-Gerbl et al[45] performed a similar study in normal cadaveric femur heads and found the calcified cartilage thickness varied from 20 µm to 230 µm, and that the ratio of calcified cartilage to total cartilage thickness was relatively constant. The calcified cartilage layer shows gradual thinning with age, along with tidemark duplication in subjects over 70 years old (see **Table 5.1**).[44]

Lane and Bullough concluded that the calcified layer is undergoing continual resorption and endochondral advancement over time.[44] These observations are consistent with the measured dwindling rate of advancement of the tidemark with age in rabbit patella.[26] In 31 normal human femoral condyles, an age-dependent loss in bone mass was measured in the subchondral bone plate.[50] The bone volume fraction (bone volume/total volume%) of the bone plate region was observed to decline from ~ 36% for subjects in their 20s to ~ 27% for those > 80 years old.[50] Bone loss was attributed to thinning of the subchondral trabeculae with age (as opposed to diminished trabecular number), and this occurred at a relatively steady rate (Trabecular Thickness = 141 μm − 0.63 × age).[50] By contrast, in OA, a pathological increase in the calcified cartilage layer thickness arises, along with abnormal tidemark duplication[37] (see Bonde et al, **Table 5.1**), and this is frequently accompanied by subchondral bone plate thickening and sclerosis.[51] The consequence or implication of tidemark duplication is not known, although Burr has proposed that microcracks at the bone–cartilage interface may be implicated in the etiology.[51]

The mineral component in the calcified cartilage layer is similar but distinct from that found in bone, and notably influenced by the uniform presence of GAG. In a study by Rey et al, pulverized calcifying cartilage from 2-month-old calves (collagen type II-positive and type I-negative) had a very low mineral content (2.8% by weight), with an immature, very poorly crystalline and low carbonate apatite mineral [calcium/(phosphate + carbonate)], compared with bone (normally ~ 0.20 carbonate/P).[41] The calcified cartilage mineral phase was also characterized by a large proportion of nonapatite "brushite-like" phosphate.[41] Unlike bone, this high nonmineral, labile phosphate content actually increases over time.[41] The low mineral content measured in immature calcified cartilage by Rey et al[41] is consistent with the quite irregular mineral surface of the epiphyseal growth plate shown in **Fig. 5.5c**. In this sample, the mineral surface most probably corresponds to mineralized collagen type I because the threshold level used in this three-dimensional reconstruction model would remove hypomineralized calcified cartilage from the image (i.e., areas between the black arrows and black arrowheads, **Fig. 5.5d**).

In situ elemental analyses of the skeletally mature normal human or OA human cartilage–bone interface has revealed the presence of calcium, phosphorus, potassium, sulfur, zinc, and strontium.[52,53] In mature osteochondral samples, the mineralized surface has a smoother texture and corresponds to the tidemark, with no visible difference by micro-CT between the calcified cartilage and osteoid immediately below (**Fig. 5.5a, b**).[54] In a normal human patella, mineralized cartilage showed a slightly but significantly higher calcium content than adjacent bone (25% vs. 23% w/v), and the mineral particles in bone and articular calcified cartilage were found to align with the direction of collagen organization.[55] Using two-dimensional nuclear magnetic resonance (2D-NMR) spectroscopy and X-ray diffraction, Duer et al[56] analyzed the mineral component of pulverized calcified cartilage samples from skeletally mature horse phalanx and distal radius. They concluded that the calcified cartilage mineral signal is similar to hydroxylapatite of bone, but with smaller peaks indicating small crystals or disorder in the mineral component. They also found evidence that, unlike bone, GAG present in calcified cartilage provides a more hydrated matrix, anionic side-chains (carboxylate and sulfate) for binding calcium in the mineral crystal surfaces, and hydroxyl groups to H-bond with surface water, mineral hydroxyl, and phosphate ions.[56] The spectra were also consistent with the presence of Gla residues (γ-carboxyglutamic acid) in the calcified cartilage layer; Gla-domain proteins are found in a variety of mineral-binding proteins such as osteocalcin.[27]

It is not well understood how the tidemark is formed, and knowledge of its precise composition is also limited. The tidemark is a 5-μm-thick structure that appears at the cartilage–calcified cartilage junction and can be visualized with hematoxylin, a blue dye that is intensified by metal ions. The tidemark could potentially arise simply by the accumulation and precipitation of chondrocyte-derived extracellular matrix species and ions at the calcified cartilage front, due to the sharp decrease in tissue permeability. The tidemark could serve to inhibit calcifying matrix vesicles released from the bone from penetrating into hyaline cartilage. Lectin staining has suggested that the tidemark contains variously branched alkali-resistant glycans with β-galactosyl or N-acetyl-lactosamine termini.[57] Zoeger et al[53] detected a specific accumulation of lead at the tidemark in normal cadaveric femoral head and patella. Oegema et al suggested that the superficial articular layer could be involved in a paracrine loop that controls deep zone chondrocyte hypertrophy and calcification, which could potentially explain thickening of the calcified cartilage in OA following loss of the superficial zone.[26] Alternatively, microcracks in the calcified layer could permit diffusion of bone-derived matrix vesicles farther into the deep zone, resulting in tidemark advancement.

Duplication of the tidemark in aging and OA is well documented.[26,44,51,58] In aging subjects, up to 5 duplicated tidemarks were observed in normal human subjects over 70 years old,[44] and as many as 10 tidemarks in primates over 20 years old.[58] A significant correlation was observed between increasing tidemark duplication, mineral density, and carbonate content in primates.[58] Repetitive knee microtrauma in a rabbit model during 9 weeks of loading was shown to lead to a mean 25% increase in the proximal tibial calcified cartilage layer thickness and tidemark duplication, with no change in mean articular cartilage thickness.[59] Multiple tidemarks were observed to form at the cartilage–bone interface in tissues surrounding an osteochondral defect in rabbit trochlea 6 months postoperative,[60] and in sheep above a metal implant placed in the subchondral bone.[59] Tidemark duplication could be related to uneven load-sharing following softening of a focal area of damaged cartilage.[60]

Cartilage–Bone Interface in Cartilage Repair

In articular cartilage lesions, the tidemark is either fully retained (Outerbridge grade I to III partial-thickness lesions) or missing to a variable extent (grade IV full-thickness lesions).[61] Most cartilage repair procedures start with debridement of the surface of the lesion to remove degenerated articular cartilage.[62] Depending on the repair approach, the debridement step may aim to retain the calcified cartilage layer for cell delivery[63–65] or to completely remove it, as during microfracture or marrow stimulation.[66–68] However, if present, the tidemark and calcified cartilage are technically very challenging to debride with precision. Light curettage usually leaves a thin layer of noncalcified deep zone articular cartilage, whereas shaving or vigorous curettage often removes a considerable amount of subchondral bone plate with the calcified cartilage.[54,64,65] Vascular channels containing erythrocytes terminate normally inside the calcified cartilage layer.[7] Therefore, in a joint with only one tidemark, debridement of the tidemark along with as little as 50 μm of the superficial mineralized layer is expected to generate some bleeding at the debrided surface, although bleeding from these tiny capillaries may not be macroscopically visible. Skeletally immature animals have a greater ease of debridement and different cell populations present in the epiphysis compared with the adult knee (**Figs. 5.1** and **5.5**). In addition, the epiphyseal blood vasculature in skeletally immature knees has active endothelial cell proliferation, whereas adult vasculature has postmitotic endothelia, and the subchondral bone no longer contains osteoclasts.[21] It is for these reasons that skeletally immature animals are improper cartilage repair models for adult knees.[69] These same cautionary notes hold for rats and mice, whose growth plates never close.

Scarce information is available on tidemark regeneration. One may reasonably wonder whether tidemark regeneration should be one of the goals of cartilage repair strategies. Frisbie et al showed that tidemark could regenerate in equine microfractured defects at 12 months postoperative only if the calcified cartilage layer were completely debrided.[70] In lesions that retained calcified cartilage and original tidemark, the new repair tissue had poor tissue integration with the base of the defect.[70] Sheep femoral condyle microfracture defects treated or not with a chitosan-based implant showed partial tidemark regeneration at 6 months postoperative, at the base of hyaline-like cartilage

(HyC) repair with a collagen type II-positive and collagen type I-negative deep zone fully integrated with bone.[71,72] In an equine case study, a cartilage defect treated by deep debridement and a composite implant regenerated a tidemark at 12 months postoperative in repair cartilage with a deep zone containing appropriate collagen fiber organization.[73] In a rabbit microdrill model of cartilage repair, a tidemark was observed at 6 months postoperative where the bone and cartilage tissue formed an integrated unit.[60]

Several groups have shown that chondrogenic foci will spontaneously form in drill or microfracture holes generated in skeletally mature knee cartilage defects.[74–77] Subchondral cartilage repair tissue contains cells with chondrocyte morphology that normally progress to hypertrophy, vascular invasion, and replacement by bone.[21,60,74–78] Chevrier et al[75] concluded that chondrogenic foci that appear near the top of the drill holes can mature to acquire a stratified structure with vascular invasion and endochondral resorption at the base after 2 months in a rabbit model, leaving an articular layer of hyaline repair cartilage.

In various models of marrow stimulation in the rabbit, microdrill holes that are created and left to bleed (as in clinical practice[79]) will spontaneously regenerate a fibrocartilage repair tissue that contains both collagen type I and collagen type II.[21,60,78] This type of spontaneous repair in a rabbit is illustrated in **Fig. 5.6**, at 2.5 months postoperative. In this rabbit, using a small arthrotomy, a 1.4-mm-diameter, 2-mm-deep microdrill hole was created in the distal femoral knee trochlea and allowed to bleed without further treatment. After 2.5 months postoperative, the drilled defect is still undergoing EO and repair. The drill hole has spontaneously regenerated fibrocartilage repair at the top of the drill hole, which is anchored

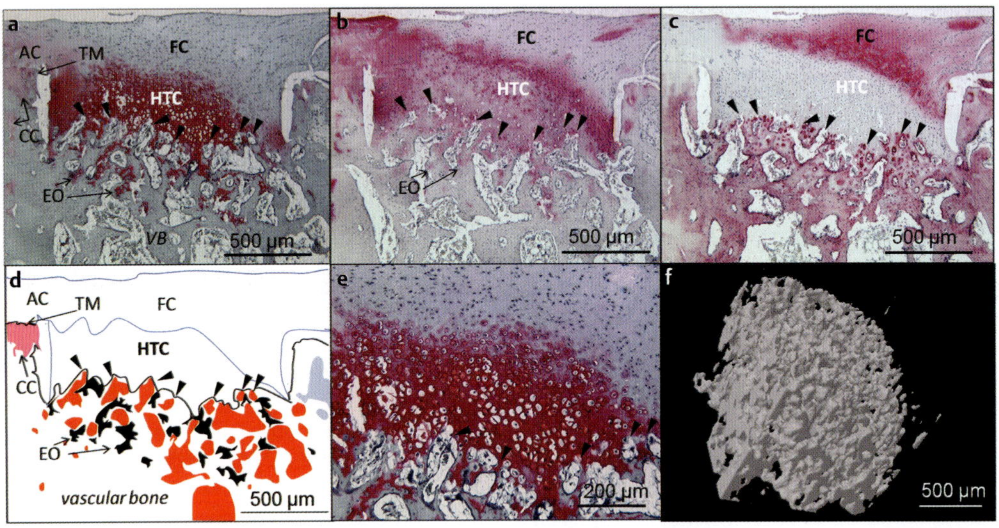

Fig. 5.6 Spontaneous repair of a 1.4-mm-diameter, 2-mm-deep osteochondral drill hole at 2.5 months postoperative in the knee trochlea of a skeletally aged (32 months) rabbit. Decalcified serial cryosections through the drill hole were stained for SafraninO/Fast green (**a**, **e**), or immunostained for collagen type II (**b**) or type I (**c**). The articular cartilage repair tissue is characterized as fibrocartilage or fibrous because it contains mainly collagen type I with little collagen type II and is depleted of GAG. (**D**) shows faithful tracings of structures from (**a**), including bone-associated GAG as a marker of endochondral ossification (EO, black), angiogenic bone marrow cavities (red), hypertrophic cartilage (HTC), and fibrocartilage (FC). Mineral formation below hypertrophic cartilage during endochondral ossification is shown in (**f**), by a reconstructed 3D image from a micro-CT scan (SkyScan 1172, 9.8 µm/pixel resolution, area corresponding to **e**) that was performed prior to decalcification. All protocols involving animals were approved by Institutional Ethics Committees. *Abbreviations:* AC, articular cartilage; FC, fibrocartilage; TM, tidemark; CC, calcified cartilage; EO, endochondral ossification; HTC, hypertrophic cartilage area; VB, vascular bone; arrowheads, bone mineralization front.

to HTC that sits below the tidemark within the flanking articular cartilage (**Fig. 5.6a–c**). Endochondral vascular invasion and mineralization are occurring at the base of the HTC (black arrowheads, **Fig. 5.6a–e**). Patches of GAG and collagen type II in the new repair bone trabeculae reveal the "tell-tale signs" of EO (**Fig. 5.6a–d**). From the histology drawing shown in **Fig. 5.6d**, we can appreciate that EO has been initiated at a previous time point in this defect because the patches of GAG, which reveal remnants of hyaline cartilage in the newly formed bone, occur in areas up to 400 μm below the blunted mineralization front (**Fig. 5.6d–f**). The morphology of the endochondral repair tissue at this point, including cryosectioning tears, resembles that of the growth plate, and no tidemark is visible.

In the same animal described above, a 1.4-mm-diameter, 2-mm-deep microdrill hole was created in the left knee trochlea and further treated by press-fitting a presolidified chitosan-blood implant into the hole.[80] Relative to the contralateral untreated drill hole, after 2.5 months of repair, a delayed and altered EO process is seen in the treated defect (**Fig. 5.7**). HyC repair tissue containing low levels of collagen type II and no collagen type I is observed above the mineralization front (**Fig. 5.7b, c**). The hyaline-like repair is overlaid with undifferentiated mesenchyme surrounded by collagen type I (**Fig. 5.7c**). In this implant-treated defect, the mineralization front has a more irregular appearance and consists in bony vascular invasion of hyaline tissue (**Fig. 5.7f**). One can appreciate that in this osteochondral defect the implant has delayed osteochondral ossification because the mineralized GAG is only beginning to form at the repair cartilage–bone interface (EO, **Fig. 5.7a, d, e**). Unlike the endochondral

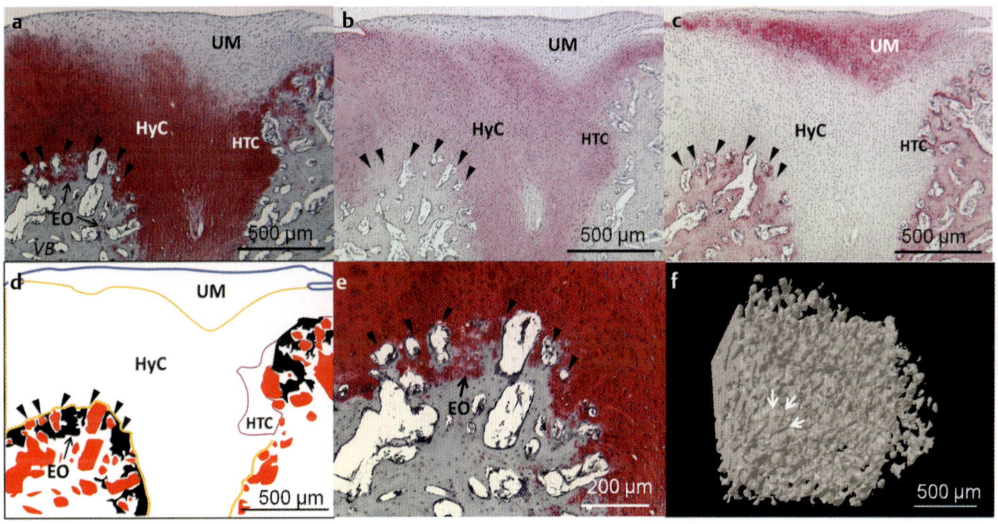

Fig. 5.7 Repair of a 1.4-mm-diameter, 2-mm-deep osteochondral drill hole at 2.5 months postsurgery in the knee trochlea of a skeletally mature (32 months) rabbit, where the drill hole was treated at surgery by press-fitting a presolidified chitosan-NaCl/autologous whole blood clot implant into the hole.[80] The defect was generated in the contralateral knee of the rabbit defect shown in **Fig. 5.6** under institutionally approved animal protocols. At 2.5 months postoperative, femur ends were fixed, micro-CT scanned (Skyscan 1172, 9.8 μm/pixel resolution), decalcified in EDTA, and cryosections stained for Safranin O/Fast green (**a**, **e**), immunostained for collagen type II (**b**), or collagen type I (**c**). (**D**) shows tracings of structures in (**a**), including bone-associated GAG as a marker of EO (black), angiogenic marrow cavities (red), HyC, area of HTC, and UM. Panel F shows a reconstructed 3D image from a micro-CT scan corresponding to the area shown in (**e**). Black arrowheads: mineralization front. The three white arrows in (**f**) show a bone-encased blood vessel very similar to a branched vascular invasion histology image previously published by Oegema et al.[26] *Abbreviations:* HyC, hyaline-like cartilage; HTC, hypertrophic cartilage area; EO, endochondral ossification; UM, undifferentiated mesenchyme; arrowheads, mineralization front.

bone formed during spontaneous repair, hypertrophic chondrocytes are scarcely present at the advancing interface of new bone and blood vessels (**Fig. 5.8**). Collagen type I of newly formed bone is being deposited from inside the invading bone marrow channels. Given that the chondrocytes present in the collagen type II repair matrix are not yet terminally differentiated to hypertrophic cells, the proximity of repair cells and invading blood vasculature can still drive cell proliferation and appositional growth of more collagen type II hyaline-like matrix. At one edge of the drill hole, the bone has regenerated to the native tidemark level, and a new tidemark can be observed (**Fig. 5.9**).

In some rabbit cartilage repair models involving complete debridement of the calcified cartilage layer, subchondral bone plate advancement beyond the native tidemark in flanking cartilage has been observed after 3 to 9 months of repair.[81,82] Bone plate advancement could be a consequence of delayed or failed tidemark regeneration during bone marrow-driven EO below hyaline-like repair tissue.

In human cartilage repair, the extent of tidemark formation in repair osteochondral biopsies has been added to a new histological scoring system generated by the International Cartilage Repair Society (ICRS II).[83] The score uses a visual analog scale (VAS) where the reader marks a line on a 10-mm scale that is then converted to a percentage between 0% (no tidemark) and 100%. In one randomized controlled clinical trial comparing characterized chondrocyte implantation (CCI) and microfracture (MFX), osteochondral repair biopsies were analyzed in a blinded fashion using the ICRS II scoring system. At 12 months postoperative, both CCI and MFX groups showed the same ~ 16-point mean clinical improvement from baseline in overall Knee injury and Osteoarthritis Outcome Score (KOOS)[84] that progressed to ~ 20 CCI versus ~ 15 MFX mean change from baseline KOOS at 5 years postoperative ($p = 0.116$).[85] Biopsies collected at 12 months postoperative from 48 out of 61 MFX-treated patients showed a mean ~ 18% tidemark formation along the biopsy width compared with a mean ~ 32% tidemark formation in biopsies from 38 out of 51 CCI-allocated

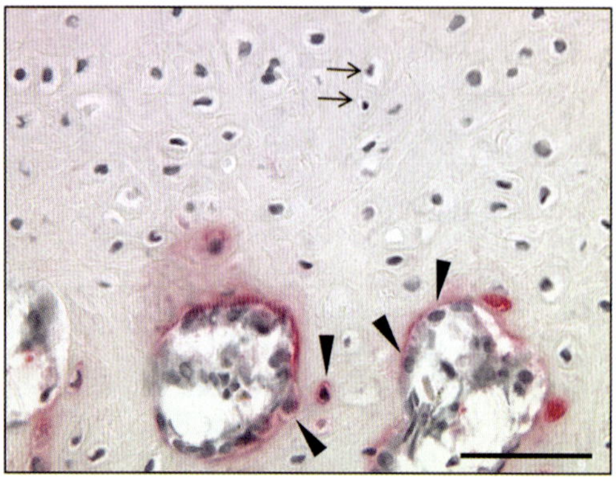

Fig. 5.8 Developing cartilage–bone interface in the subchondral area of an osteochondral defect treated with a presolidified chitosan-based implant at 2.5 months postoperative (from the same histology image shown in **Fig. 5.7c**). The section was immunostained for collagen type I (red stain) with iron hematoxylin counterstain. Arrows show round and crescent-shaped chondrocyte cells in hyaline-like repair cartilage, and arrowheads show collagen type I-expressing cells that cap the invading blood vessel bone marrow cavities with new osteoid, similar to that described by Gilmore and Palfrey in neonatal human lateral femoral articular cartilage.[7] Both round and cuboidal cells express collagen type I, as is seen in growth plate endochondral bone. Scale bar: 50 μm.

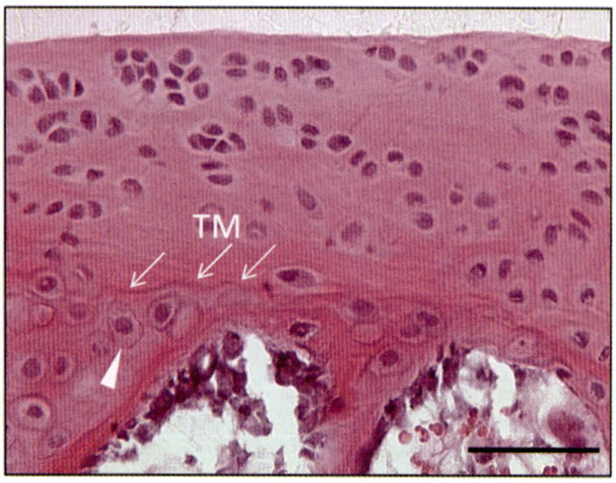

Fig. 5.9 Evidence of tidemark formation at 2.5 months postoperative at the edge of the repairing rabbit trochlear osteochondral hole at the level of the tidemark in an osteochondral drill hole treated with presolidified chitosan-blood implant. Hematoxylin-eosin stained EDTA-decalcified cryosection. The arrows show the new tidemark (TM) and the arrowhead shows a terminally differentiated hypertrophic chondrocyte inside the newly forming calcified cartilage layer. The image was taken from a field near the upper right corner of **Fig. 5.7a**. Scale bar: 50 μm.

patients ($p = 0.036$).[84] Given that cell therapy aims to retain the calcified cartilage layer at surgery,[86] significantly greater tidemark present in CCI biopsies could be partly due to a lighter debridement of the initial tidemark in the CCI lesions.

Another randomized controlled clinical trial compared MFX to MFX and chitosan–glycerol phosphate/blood implant (BST-CarGel), at 12 months postoperative.[87,88] Blinded ICRS II scoring of osteochondral biopsies revealed more tidemark present in nine biopsies from MFX-treated defects compared with 12 implant-treated defects.[88] Some biopsies from this study showed zonal collagen organization resembling native articular cartilage,[89] and a hyaline-like deep zone containing collagen type II and no collagen type I.[87,88] One MFX biopsy collected at 12 months postoperatively consisted in a collagen type I+/collagen type II+ fibrocartilage repair with an irregular bone interface, and 0% tidemark formation.[72]

To summarize, tidemark has been observed in some human cartilage repair osteochondral biopsies 12 months following bone marrow stimulation or cell therapy. The presence of a tidemark could arise through hyaline cartilage regeneration via EO, or by incomplete debridement and persistence of the native tidemark in the treated lesion. Finally, despite great care, it is also possible that some biopsies with a complete tidemark may have been taken from outside the area of the initial lesion.[72]

Cartilage repair is a complex process that takes place over a long period of time. The notion of cartilage repair as an isolated event should be discarded for the more comprehensive view of osteochondral repair, given the extensive cross-talk between cartilage repair tissues, bone, and blood vessels in the developing interface. New calcified cartilage layer/tidemark can be regenerated in a pure type II collagen matrix containing GAG integrated to endochondral bone near the articular surface. Residual cartilage and calcified cartilage can block cell migration and vascular invasion during marrow-derived cartilage regeneration; therefore, new tools or methods that permit the surgeon to verify the presence of residual cartilage and calcified cartilage at the debridement step would help control this important variable. To better evaluate the progression and success of different cartilage repair therapies, patient-reported outcomes[90] need to be correlated with repair tissue architecture.[72] A better understanding of cartilage repair tissue maturity will be reached with new histological methods that can distinguish between native and regenerated tidemark, standardized measures of tidemark and calcified cartilage formation, and further research on the mechanisms of cartilage calcification.

◆ Summary Points and Clinical Relevance

1. All cartilage–bone interfaces develop from an initially cartilaginous structure that undergoes coordinated invasion by blood vessels and osteoblasts. Formation of a tidemark anatomically stabilizes the cartilage–bone interface and arrests cartilage calcification and blood vessel invasion. Vascularization of the calcified cartilage layer and subchondral bone plate is an important feature of a healthy cartilage–bone interface.
2. The cartilage–bone interface is a mineralized blood vessel boundary where collagen type II is integrated with collagen type I.
3. Animals that have permanently open growth plates (mice and rats) and skeletally immature animals with open growth plates (rabbits less than 7 months old, and large animals less than ~ 2 years old) are improper cartilage repair models for establishing the efficacy of therapies intended for use in adult human knees.
4. Chronic medications (i.e., steroids), drugs (i.e., smoking), or surgical procedures that produce chronic ischemia in the epiphyseal bone may contribute to articular cartilage degeneration and/or suppress cartilage regeneration. Conversely, treatments that stimulate revascularization of subchondral bone damaged by drilling or microfracture have the potential to drive epiphyseal endochondral repair.
5. With increasing age, microtrauma, and advanced OA, the calcified cartilage layer either thins out or thickens and becomes more mineralized (**Fig. 5.5a** vs. **5.5c,**

Table 5.1). Therefore, during clinical surgical procedures that aim to debride the calcified cartilage layer, the potential thickness and extent of mineralization of the calcified cartilage layer should be taken into account.

6. Calcified cartilage and osteoid in the adult subchondral bone have a similar mineral level, which means that during debridement procedures the tidemark can be fully removed only by cleanly and carefully scraping off a specific mineralized depth from the entire lesional surface.
7. The tidemark and calcified cartilage layer are technically very challenging to debride with precision. Light curettage usually leaves a thin layer of noncalcified deep zone articular cartilage, whereas shaving or vigorous curettage often removes a considerable amount of subchondral bone plate with the calcified cartilage. In a joint with only one tidemark, debridement of the tidemark along with as little as 50 μm of the superficial mineralized layer is expected to generate bleeding, which may or may not be immediately visible.
8. Bone plate advancement could be a consequence of delayed or failed tidemark regeneration during bone marrow-driven EO below hyaline-like repair tissue.
9. When evaluating outcomes of cartilage repair procedures, it is important to realize that the presence of a tidemark could arise through true hyaline cartilage regeneration or by incomplete debridement and persistence of the native tidemark.

Acknowledgments

Funding for this work was from the Canadian Institutes of Health Research (CIHR, 185810-BME), Natural Sciences and Engineering Research Council of Canada (NSERC, STGP 365025), the Canada Foundation for Innovation (CFI), and Fonds de Recherche Santé du Québec (FRSQ, Groupe de Recherche en Sciences et Technologies Biomédicales, GRSTB). Salary support for C.D.H. was from an FRSQ bourse Chercheur National/National Researcher Career Award. We thank Geneviève Picard and Jun Sun for excellent technical contributions, and Julie Tremblay for quality assurance.

References

1. Sasano Y, Mizoguchi I, Kagayama M, Shum L, Bringas P Jr, Slavkin HC. Distribution of type I collagen, type II collagen and PNA binding glycoconjugates during chondrogenesis of three distinct embryonic cartilages. Anat Embryol (Berl) 1992;186(3):205–213
2. Craig FM, Bentley G, Archer CW. The spatial and temporal pattern of collagens I and II and keratan sulphate in the developing chick metatarsophalangeal joint. Development 1987;99(3):383–391
3. Olsen BR, Reginato AM, Wang W. Bone development. Annu Rev Cell Dev Biol 2000;16(10810706):191–220
4. Sandberg M, Vuorio E. Localization of types I, II, and III collagen mRNAs in developing human skeletal tissues by in situ hybridization. J Cell Biol 1987;104(4):1077–1084
5. Shapiro F. Bone development and its relation to fracture repair. The role of mesenchymal osteoblasts and surface osteoblasts. Eur Cell Mater 2008;15:53–76
6. Hunziker EB. The elusive path to cartilage regeneration. Adv Mater (Deerfield Beach Fla) 2009; 21(32–33):3419–3424
7. Gilmore RS, Palfrey AJ. A histological study of human femoral condylar articular cartilage. J Anat 1987;155:77–85
8. Leboy PS, Vaias L, Uschmann B, Golub E, Adams SL, Pacifici M. Ascorbic acid induces alkaline phosphatase, type X collagen, and calcium deposition in cultured chick chondrocytes. J Biol Chem 1989;264(29):17281–17286
9. Clark JM. The structure of vascular channels in the subchondral plate. J Anat 1990;171:105–115
10. Haines RW. The histology of epiphyseal union in mammals. J Anat 1975;120(Pt 1):1–25
11. D'Angelo M, Yan Z, Nooreyazdan M, et al. MMP-13 is induced during chondrocyte hypertrophy. J Cell Biochem 2000;77(4):678–693
12. Sakiyama H, Inaba N, Toyoguchi T, et al. Immunolocalization of complement C1s and matrix metalloproteinase 9 (92kDa gelatinase/type IV collagenase) in the primary ossification center of the human femur. Cell Tissue Res 1994;277(2):239–245
13. Stickens D, Behonick DJ, Ortega N, et al. Altered endochondral bone development in matrix metalloproteinase 13-deficient mice. Development 2004;131(23):5883–5895
14. Little CB, Meeker CT, Hembry RM, et al. Matrix metalloproteinases are not essential for aggrecan turnover during normal skeletal growth and development. Mol Cell Biol 2005;25(8):3388–3399
15. Vu TH, Shipley JM, Bergers G, et al. MMP-9/gelatinase B is a key regulator of growth plate angiogenesis and apoptosis of hypertrophic chondrocytes. Cell 1998;93(3):411–422
16. Reich A, Jaffe N, Tong A, et al. Weight loading young chicks inhibits bone elongation and promotes growth plate ossification and vascularization. J Appl Physiol 2005;98(6):2381–2389
17. Farnum CE, Lee AO, O'Hara K, Wilsman NJ. Effect of short-term fasting on bone elongation rates: an analysis of catch-up growth in young male rats. Pediatr Res 2003;53(1):33–41
18. Kim HK, Su P-H, Qiu Y-S. Histopathologic changes in growth-plate cartilage following ischemic

necrosis of the capital femoral epiphysis. An experimental investigation in immature pigs. J Bone Joint Surg Am 2001;83-A(5):688–697

19. Alini M, Matsui Y, Dodge GR, Poole AR. The extracellular matrix of cartilage in the growth plate before and during calcification: changes in composition and degradation of type II collagen. Calcif Tissue Int 1992;50(4):327–335

20. Cackowski FC, Anderson JL, Patrene KD, et al. Osteoclasts are important for bone angiogenesis. Blood 2010;115(1):140–149

21. Chen G, Sun J, Lascau-Coman V, Chevrier A, Marchand C, Hoemann CD. Acute osteoclast activity following subchondral drilling is promoted by chitosan and associated with improved cartilage tissue integration. Cartilage 2011;2:173–185

22. Hashimoto S, Ochs RL, Rosen F, et al. Chondrocyte-derived apoptotic bodies and calcification of articular cartilage. Proc Natl Acad Sci U S A 1998;95(6): 3094–3099

23. Hunziker EB, Kapfinger E, Geiss J. The structural architecture of adult mammalian articular cartilage evolves by a synchronized process of tissue resorption and neoformation during postnatal development. Osteoarthritis Cartilage 2007;15(4):403–413

24. Namba RS, Meuli M, Sullivan KM, Le AX, Adzick NS. Spontaneous repair of superficial defects in articular cartilage in a fetal lamb model. J Bone Joint Surg Am 1998;80(1):4–10

25. Hayes AJ, MacPherson S, Morrison H, Dowthwaite G, Archer CW. The development of articular cartilage: evidence for an appositional growth mechanism. Anat Embryol (Berl) 2001;203(6):469–479

26. Oegema TR Jr, Carpenter RJ, Hofmeister F, Thompson RC Jr. The interaction of the zone of calcified cartilage and subchondral bone in osteoarthritis. Microsc Res Tech 1997;37(4):324–332

27. Heinegård D, Oldberg A. Structure and biology of cartilage and bone matrix noncollagenous macromolecules. FASEB J 1989;3(9):2042–2051

28. Gannon JM, Walker G, Fischer M, Carpenter R, Thompson RC Jr, Oegema TR Jr. Localization of type X collagen in canine growth plate and adult canine articular cartilage. J Orthop Res 1991;9(4):485–494

29. Jiang J, Leong NL, Mung JC, Hidaka C, Lu HH. Interaction between zonal populations of articular chondrocytes suppresses chondrocyte mineralization and this process is mediated by PTHrP. Osteoarthritis Cartilage 2008;16(1):70–82

30. Sundaramurthy S, Mao JJ. Modulation of endochondral development of the distal femoral condyle by mechanical loading. J Orthop Res 2006;24(2):229–241

31. Stephens M, Kwan APL, Bayliss MT, Archer CW. Human articular surface chondrocytes initiate alkaline phosphatase and type X collagen synthesis in suspension culture. J Cell Sci 1992;103(Pt 4): 1111–1116

32. Ali SY, Sajdera SW, Anderson HC. Isolation and characterization of calcifying matrix vesicles from epiphyseal cartilage. Proc Natl Acad Sci USA 1970;67(3):1513–1520

33. Shum L, Nuckolls G. The life cycle of chondrocytes in the developing skeleton. Arthritis Res 2002;4(2):94–106

34. Robinson RA, Cameron DA. Electron microscopy of the primary spongiosa of the metaphysis at the distal end of the femur in the newborn infant. J Bone Joint Surg Am 1958;40-A(3):687–697

35. Boyde A, Firth EC. Articular calcified cartilage canals in the third metacarpal bone of 2-year-old thoroughbred racehorses. J Anat 2004;205(6): 491–500

36. Kobayashi S, Baba H, Takeno K, et al. Fine structure of cartilage canal and vascular buds in the rabbit vertebral endplate. Laboratory investigation. J Neurosurg Spine 2008;9(1):96–103

37. Bonde HV, Talman MLM, Kofoed H. The area of the tidemark in osteoarthritis—a three-dimensional stereological study in 21 patients. APMIS 2005; 113(5):349–352

38. Arkill KP, Winlove CP. Solute transport in the deep and calcified zones of articular cartilage. Osteoarthritis Cartilage 2008;16(6):708–714

39. Pan J, Zhou XZ, Li W, Novotny JE, Doty SB, Wang LY. In situ measurement of transport between subchondral bone and articular cartilage. J Orthop Res 2009;27(10):1347–1352

40. Bellows CG, Aubin JE, Heersche JNM. Initiation and progression of mineralization of bone nodules formed in vitro: the role of alkaline phosphatase and organic phosphate. Bone Miner 1991;14(1): 27–40

41. Rey C, Beshah K, Griffin R, Glimcher MJ. Structural studies of the mineral phase of calcifying cartilage. J Bone Miner Res 1991;6(5):515–525

42. Hoemann CD, El-Gabalawy H, McKee MD. In vitro osteogenesis assays: influence of the primary cell source on alkaline phosphatase activity and mineralization. Pathol Biol (Paris) 2009;57(4):318–323

43. Wang FY, Ying Z, Duan XJ, et al. Histomorphometric analysis of adult articular calcified cartilage zone. J Struct Biol 2009;168(3):359–365

44. Lane LB, Bullough PG. Age-related changes in the thickness of the calcified zone and the number of tidemarks in adult human articular cartilage. J Bone Joint Surg Br 1980;62(3):372–375

45. Müller-Gerbl M, Schulte E, Putz R. The thickness of the calcified layer of articular cartilage: a function of the load supported? J Anat 1987;154:103–111

46. Hunziker EB, Quinn TM, Häuselmann HJ. Quantitative structural organization of normal adult human articular cartilage. Osteoarthritis Cartilage 2002;10(7):564–572

47. Lyons TJ, McClure SF, Stoddart RW, McClure J. The normal human chondro-osseous junctional region: evidence for contact of uncalcified cartilage with subchondral bone and marrow spaces. BMC Musculoskelet Disord 2006;7:52

48. Frisbie DD, Cross MW, McIlwraith CW. A comparative study of articular cartilage thickness in the stifle of animal species used in human pre-clinical studies compared to articular cartilage thickness in the human knee. Vet Comp Orthop Traumatol 2006;19(3):142–146

49. Madry H, van Dijk CN, Mueller-Gerbl M. The basic science of the subchondral bone. Knee Surg Sports Traumatol Arthrosc 2010;18(4):419–433

50. Koszyca B, Fazzalari NL, Vernon-Roberts B. Calcified cartilage, subchondral and cancellous bone morphometry within the knee of normal subjects. Knee 1996;3:15–22

51. Burr DB, Schaffler MB. The involvement of subchondral mineralized tissues in osteoarthrosis: quantitative microscopic evidence. Microsc Res Tech 1997;37(4):343–357

52. Kaabar W, Iaklouk A, Bunk O, Baily M, Farquharson MJ, Bradley D. Compositional and structural studies of the bone-cartilage interface using PIXE and SAXS techniques. Nuclear Instruments & Methods in Physics Research Section Accelerators Spectrometers Detectors and Associated Equipment 2010;619(1–3):78–82

53. Zoeger N, Roschger P, Hofstaetter JG, et al. Lead accumulation in tidemark of articular cartilage. Osteoarthritis Cartilage 2006;14(9):906–913

54. Marchand C, Chen HM, Buschmann MD, Hoemann CD. Standardized three-dimensional volumes of interest with adapted surfaces for more precise subchondral bone analyses by micro-computed tomography. Tissue Eng Part C Methods 2011;17(4):475–484

55. Zizak I, Roschger P, Paris O, et al. Characteristics of mineral particles in the human bone/cartilage interface. J Struct Biol 2003;141(3):208–217

56. Duer MJ, Friscić T, Murray RC, Reid DG, Wise ER. The mineral phase of calcified cartilage: its molecular structure and interface with the organic matrix. Biophys J 2009;96(8):3372–3378

57. Lyons TJ, Stoddart RW, McClure SF, McClure J. The tidemark of the chondro-osseous junction of the normal human knee joint. J Mol Histol 2005;36(3):207–215

58. Miller LM, Novatt JT, Hamerman D, Carlson CS. Alterations in mineral composition observed in osteoarthritic joints of cynomolgus monkeys. Bone 2004;35(2):498–506

59. Burr DB. Anatomy and physiology of the mineralized tissues: role in the pathogenesis of osteoarthrosis. Osteoarthritis Cartilage 2004;12(Suppl A): S20–S30

60. Marchand C, Chen G, Tran-Khanh N, et al. Microdrilled cartilage defects treated with thrombin-solidified chitosan/blood implant regenerate a more hyaline, stable, and structurally integrated osteochondral unit compared to drilled controls. Tissue Eng Part A 2012;18(5–6):508–519

61. Cameron ML, Briggs KK, Steadman JR. Reproducibility and reliability of the outerbridge classification for grading chondral lesions of the knee arthroscopically. Am J Sports Med 2003;31(1): 83–86

62. Mithoefer K, McAdams TR, Scopp JM, Mandelbaum BR. Emerging options for treatment of articular cartilage injury in the athlete. Clin Sports Med 2009;28(1):25–40

63. Peterson L, Minas T, Brittberg M, Nilsson A, Sjögren-Jansson E, Lindahl A, Sjögren-Jansson E. Two- to 9-year outcome after autologous chondrocyte transplantation of the knee. Clin Orthop Relat Res 2000;374(374):212–234

64. Drobnic M, Radosavljevic D, Cör A, Brittberg M, Strazar K. Debridement of cartilage lesions before autologous chondrocyte implantation by open or transarthroscopic techniques: a comparative study using post-mortem materials. J Bone Joint Surg Br 2010;92(4):602–608

65. Mika J, Clanton TO, Pretzel D, Schneider G, Ambrose CG, Kinne RW. Surgical preparation for articular cartilage regeneration without penetration of the subchondral bone plate: in vitro and in vivo studies in humans and sheep. Am J Sports Med 2011;39(3):624–631

66. Steadman JR, Rodkey WG, Singleton SB, Briggs KK. Microfracture technique for full-thickness chondral defects: technique and clinical results. Oper Tech Orthop 1997;7(4):300–304

67. Mithoefer K, Williams RJ III, Warren RF, et al. The microfracture technique for the treatment of articular cartilage lesions in the knee. A prospective cohort study. J Bone Joint Surg Am 2005;87(9):1911–1920

68. Knutsen G, Engebretsen L, Ludvigsen TC, et al. Autologous chondrocyte implantation compared with microfracture in the knee. A randomized trial. J Bone Joint Surg Am 2004;86-A(3):455–464

69. Hurtig MB, Buschmann MD, Fortier LA, et al. Preclinical Studies for Cartilage Repair: Recommendations from the International Cartilage Repair Society. Cartilage. 2011;2(2):137–152

70. Frisbie DD, Morisset S, Ho CP, Rodkey WG, Steadman JR, McIlwraith CW. Effects of calcified cartilage on healing of chondral defects treated with microfracture in horses. Am J Sports Med 2006;34(11):1824–1831

71. Hoemann CD, Hurtig M, Rossomacha E, et al. Chitosan-glycerol phosphate/blood implants improve hyaline cartilage repair in ovine microfracture defects. J Bone Joint Surg Am 2005;87(12): 2671–2686

72. Hoemann CD, Kandel R, Roberts S, et al. International Cartilage Repair Society (ICRS) Recommended Guidelines for Histological Endpoints for Cartilage Repair Studies in Animal Models and Clinical Trials. Cartilage 2011;2(2):153–172

73. Kon E, Mutini A, Arcangeli E, et al. Novel nanostructured scaffold for osteochondral regeneration: pilot study in horses. J Tissue Eng Regen Med 2010;4(4):300–308

74. Shapiro F, Koide S, Glimcher MJ. Cell origin and differentiation in the repair of full-thickness defects of articular cartilage. J Bone Joint Surg Am 1993;75(4):532–553

75. Chevrier A, Hoemann CD, Sun J, Buschmann MD. Temporal and spatial modulation of chondrogenic foci in subchondral microdrill holes by chitosan-glycerol phosphate/blood implants. Osteoarthritis Cartilage 2011;19(1):136–144

76. Hoemann CD, Chen G, Marchand C, et al. Scaffold-guided subchondral bone repair: implication of neutrophils and alternatively activated arginase-1 + macrophages. Am J Sports Med 2010; 38(9):1845–1856

77. Chevrier A, Hoemann CD, Sun J, Buschmann MD. Chitosan-glycerol phosphate/blood implants increase cell recruitment, transient vascularization and subchondral bone remodeling in drilled cartilage defects. Osteoarthritis Cartilage 2007; 15(3):316–327

78. Chen H, Hoemann CD, Sun J, et al. Depth of subchondral perforation influences the outcome of bone marrow stimulation cartilage repair. J Orthop Res 2011;29(8):1178–1184

79. Insall JN. Intra-articular surgery for degenerative arthritis of the knee. A report of the work of the late K. H. Pridie. J Bone Joint Surg Br 1967; 49(2):211–228

80. Guzmàn-Morales J, Lafantaisie-Favreau C, Sun J, Rivard G, Hoemann CD. Analysis of the mid-term effects of chitosan-NaCl/blood pre-solidified implants in an in vivo osteochondral repair model. Paper presented at: International Cartilage Repair Society, 25 Sept, 2010; Barcelona, Spain.

81. Qiu YS, Shahgaldi BF, Revell WJ, Heatley FW. Observations of subchondral plate advancement during osteochondral repair: a histomorphometric and mechanical study in the rabbit femoral condyle. Osteoarthritis Cartilage 2003;11(11):810–820

82. Chen HM, Chevrier A, Hoemann CD, Sun J, Ouyang W, Buschmann MD. Characterization of subchondral bone repair for marrow-stimulated chondral defects and its relationship to articular cartilage resurfacing. Am J Sports Med 2011;39(8):1731–1740

83. Mainil-Varlet P, Van Damme B, Nesic D, Knutsen G, Kandel R, Roberts S. A new histology scoring system for the assessment of the quality of human cartilage repair: ICRS II. Am J Sports Med 2010;38(5):880–890

84. Saris DB, Vanlauwe J, Victor J, et al. Characterized chondrocyte implantation results in better structural repair when treating symptomatic cartilage defects of the knee in a randomized controlled trial versus microfracture. Am J Sports Med 2008;36(2):235–246

85. Vanlauwe J, Saris DBF, Victor J, Almqvist KF, Bellemans J, Luyten FP; TIG/ACT/01/2000&EXT Study Group. Five-year outcome of characterized chondrocyte implantation versus microfracture for symptomatic cartilage defects of the knee: early treatment matters. Am J Sports Med 2011;39(12):2566–2574

86. Brittberg M, Lindahl A, Nilsson A, Ohlsson C, Isaksson O, Peterson L. Treatment of deep cartilage defects in the knee with autologous chondrocyte transplantation. N Engl J Med 1994;331(14):889–895

87. Hoemann CD, Tran-Khan N, Méthot S, et al. Correlation of tissue histomorphometry with ICRS histology scores in biopsies obtained from a randomized controlled clinical trial comparing BST-CarGel™ versus microfracture. Paper presented at: International Cartilage Repair Society, 27 Sept, 2010; Barcelona, Spain

88. Méthot S, Hoemann CD, Rossomacha E, et al. ICRS Histology Scores of Biopsies from an Interim Analysis of a Randomized Controlled Clinical Trial Show Significant Improvement in Tissue Quality at 13 Months for BST-CarGel versus Microfracture. Paper presented at: International Cartilage Repair Society, 27 Sept, 2010; Barcelona, Spain

89. Changoor A, Nelea M, Méthot S, et al. Structural characteristics of the collagen network in human normal, degraded and repair articular cartilages observed in polarized light and scanning electron microscopies. Osteoarthritis Cartilage 2011;19(12):1458–1468

90. Roos EM, Engelhart L, Ranstam J, et al. ICRS Recommendation Document: Patient-Reported Outcome Instruments for Use in Patients with Articular Cartilage Defects. Cartilage 2011;2(2):122–136

91. Lascau-Coman V, Buschmann MD, Hoemann CD. Rapid EDTA Microwave Decalcification of Rabbit Osteochondral Samples Preserves Enzyme Activity and Antigen Epitopes. J Bone Miner Res 2008; 23:S408–S408

92. Chevrier A, Rossomacha E, Buschmann MD, Hoemann CD. Optimization of histoprocessing methods to detect glycosaminoglycan, collagen type II, and collagen type I in decalcified rabbit osteochondral sections. J Histotechnol 2005;28(3):165–175

6

Science and Animal Models of Marrow Stimulation for Cartilage Repair

Lisa A. Fortier, Brian J. Cole, and C. Wayne McIlwraith

Microfracture perforation of the subchondral bone for cartilage repair was originally described by Steadman in 1994.[1] Microfracture involves penetration of the subchondral bone plate with an arthroscopic awl to allow bone marrow contents to fill the defect and form a "superclot."[2] In humans[1,3,4] and nonhuman primates,[5] microfracture results in increased tissue volume and improved patient comfort and function for an average of 2 to 3 years. There are other described methods of bone marrow stimulation such as drilling and abrasion, but less research and clinical data are available to critically evaluate the efficacy of these techniques.

◆ The Superclot

In theory, enhanced cartilage repair following microfracture is the result of the superclot thought to be laden with bone marrow-derived mesenchymal stem cells (MSCs) and growth factors.[6] Although there have been several in vitro and in vivo animal studies aimed at understanding how microfracture repair tissue remodels over time, it has never been well documented that the superclot contains MSCs or growth factors. In a small study of 11 human patients with femoral condylar defects, superclot from microfracture was compared with bone marrow aspirate from the iliac crest and concentrated by centrifugation.[7] The two cellular populations were different with respect to cell surface markers. Neither cell type carried CD34 or CD45 marker expression, suggesting that there were no hematopoietic cells in either bone marrow aspirate concentrate or microfracture superclot. This result might suggest that neither cell source is derived from the bone marrow, but it must be interpreted with great caution because both cell sources were cultured for at least two passages and the cells were treated with trypsin before flow cytometry analysis, both of which have been documented to alter cell surface protein expression on stem cells.[8,9] In a similar study in which cells were derived from subchondral corticospongious bone and cultured over time, the cells retained their multilineage potential to undergo trilineage differentiation into cartilage, adipose, and bone phenotypes.[10] Interestingly, MSC-based cartilage studies continue to focus predominantly on the ability of the cells to differentiate into and form neocartilage despite the growing evidence that MSCs function at least in part to modulate the local environment through a paracrine effect and recruitment of other progenitor cells and immunomodulation.[11,12]

Understanding the source and type of cells that populate microfracture defects is critically important. There are a number of studies that have evaluated the various effects of drugs, growth factors, devices, scaffolds, gene therapy, and rehabilitation on microfracture. Some of these modifying factors are being promoted and even marketed predicated on the concept that they enhance chemotaxis, adherence, and/or proliferation of bone marrow-derived MSCs.[13–19] These cited studies represent only a few of the many studies investigating the use of scaffolds, devices, and drugs, in vitro and in rabbit, canine, ovine, laprine, or equine animal models for augmentation of microfracture to enhance articular cartilage repair. This intense level of investigation into scaffold/device-augmented microfracture and its potential recruitment of MSCs lies in the thought that these technologies could improve the clinical results of microfracture alone and the relative ease and marketability of such technologies when compared with cultured or manipulated stem cell articular cartilage grafts.

If the cell population of the subchondral bone is truly different from that of bone marrow aspirated from a bone marrow space, then perhaps the results of in vitro studies done on bone marrow aspirate or metaphyseal-derived MSCs are not directly applicable to microfracture, in which the cell is likely derived from the subchondral bone plate in the area 2 to 4 mm underlying the calcified cartilage layer.[13,15] During the process of maturation, the cell population in a superclot might be composed of cells derived from the bone marrow, subchondral bone, surrounding host cartilage, synovium, synovial fluid, or a combination thereof. Studies are routinely performed in vitro, and using bone marrow-derived MSCs to investigate a method to improve microfracture and the results can change clinical practice. For example, a recent study showed that chondrogenic differentiation of bone marrow-derived MSCs is impaired by rheumatoid arthritis synovial fluid as compared with synovial fluid from patients with osteoarthritis or normal patients.[20] Another study suggested that age in males, but not in females, negatively affects their ability to undergo chondrogenic differentiation.[21] The potential clinical ramifications of this study, where clinicians might presume failure of microfracture in patients with rheumatoid arthritis or in older males, underscore the need for a more refined understanding of the basic biology of microfracture.

Animal Model Studies

Animal model studies provide insight into temporal changes following microfracture (**Fig. 6.1**). Early animal model studies on microfracture repair were done in the horse.[22,23] The horse model was also used to validate the subjective clinical impression that removal of the calcified cartilage layer was important to optimize volume and attachment of repair tissue.[24] Further equine studies indicated that the volume of repair tissue did not change between 4 and 12 months postmicrofracture in direct weight-bearing sites (distal medial femoral condyle and distal radiocarpal bones), which at a minimum suggests that the repair tissue did not deteriorate by 12 months postoperatively.[22] Histologic assessment revealed that there was more type II collagen present at 12 months than at 4 months, suggesting continued chondrogenic maturation of repair tissue to 12 months, but the aggrecan content remained far below normal.

To provide information in a physiologic and anatomic environment more closely related to the human, similar studies were performed in cynomolgus macaques.[5] In this study, repair tissue was studied at 6 and 12 weeks postmicrofracture and indicated that the repair tissue underwent progressive chondrogenic remodeling during this time period based on postmortem gross and histologic assessments. It is interesting to note that progressive maturation of microfracture repair tissue is not appreciated using arthroscopy with validated categorical scoring systems,[25] which makes it difficult for a surgeon to make decisions regarding success based on arthroscopic observation only.[26] Noninvasive dGEMRIC and T2 mapping has been used to evaluate repair tissue following microfracture at 24 and 48 weeks postoperatively in a goat model.[27] The achieved objective of the study was to validate dGEMRIC and T2 mapping as surrogate markers of biochemical and histologic integrity of repair tissue. In addition, the study was the first to demonstrate

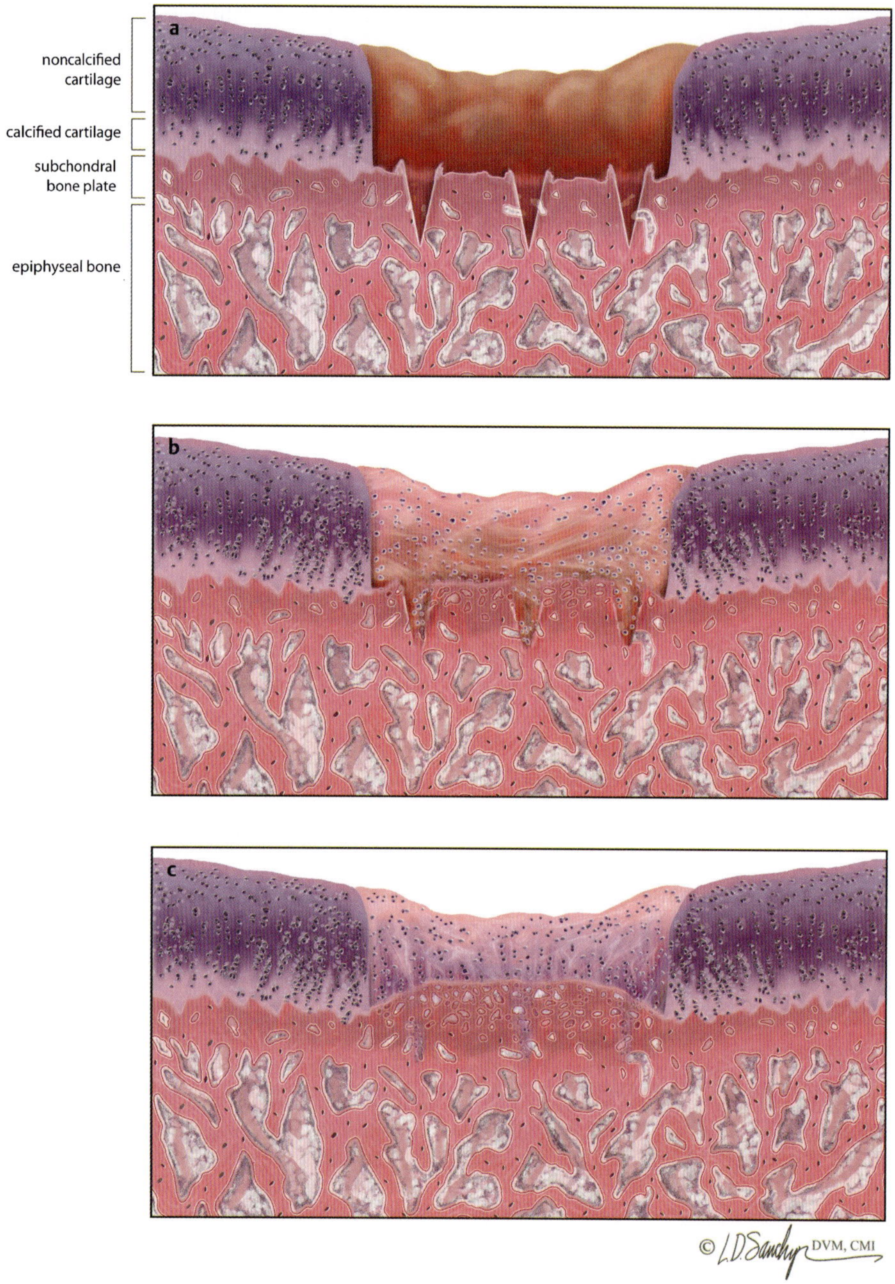

Fig. 6.1 Schematic representation of microfracture maturation over time. (**a**) At time 0, the cartilage defect is debrided to include removal of calcified cartilage. Microfracture is performed to a depth of 2 to 4 mm to penetrate the subchondral plate, thereby allowing bone marrow to gain access to the cartilage defect and form a "superclot." (**b**) At 4 to 6 months postmicrofracture, there is progressive chondrogenic remodeling of the fibrocartilage repair tissue filling the defect. The repair tissue is hypercellular. Proteoglycan (purple in base of repair tissue) and type II collagen content (not depicted) progressively increase but remain low compared with normal tissue. Chondrocyte cloning is evident in the adjacent host cartilage. The microfracture holes progressively heal during this time period. (**c**) At 12 months postmicrofracture, the repair tissue has improved cellular organization and proteoglycan content but not type II collagen. The microfracture holes are healed, and in some instances the subchondral bone is sclerotic and/or extends into the cartilage defect forming a "central osteophyte."

increased glycosaminoglycan and total collagen content between 24 and 48 weeks postmicrofracture measured with both ΔR1 (1/s) and high-performance liquid chromatography. Combined, these results suggest that microfracture continues to mature for the first 12 months after surgery, but the lack of normal matrix molecules translates to tissue with inferior biomechanical properties compared with normal cartilage, which renders the repair tissue prone to injury and deterioration. Based on animal model studies, it is unclear what biochemical or mechanical changes happen beyond 12 months and when, why, or how microfracture repair tissue fails or not. In unpublished data by L.A.F, 2-year data are being analyzed in the horse. Clinically, it may have less to do with the breakdown of microfracture repair tissue than with ability of the repair tissue to "shield" the subchondral bone from load that is theoretically associated with the manifestation of symptoms. If this theory is correct, then methods to enhance or retain proteoglycan content in the repair tissue would increase the compressive stiffness of the repair tissue and should improve long-term results.

Central Osteophyte Formation and Subchondral Bone Sclerosis

Microfracture has long been thought of as a "can't hurt" or "burn no bridges" type of procedure. However, in more recent years, there is heightened awareness and concern about the formation of central/intralesional osteophytes, which are protrusions of subchondral bone extending above the level of the adjacent, normal subchondral plate (**Fig. 6.2**).[28] Formation of central osteophytes is not specifically investigated a priori or mentioned in most animal studies despite being quite obvious in figures contained in published articles, irrespective of the animal model studied. Figures presented in articles can be too high in magnification or focused on the repair-host tissue interface to appreciate central osteophyte formation. It should be noted that central osteophyte formation has been observed in microfracture defects in the horse

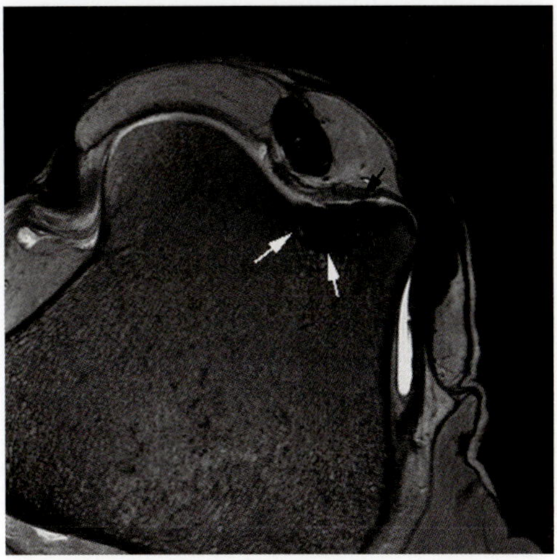

Fig. 6.2 Fast-spin echo magnetic resonance image (MRI) of a microfracture-treated defect on the lateral trochlear ridge of the femur, 12 months after surgery. Subchondral bone sclerosis (*white arrows*) and protrusion of the subchondral plate into the cartilage defect (*black arrow*) are evident.

model in both the distal femur (**Fig. 6.3**)[22,29] and lateral trochlear ridge (**Fig. 6.2**)[30] in ovine,[14] and in nonhuman primates.[5] Central osteophyte formation clearly does not occur in every case of microfracture, and there are too few instances in the animal model studies for robust observations into causality.

Subchondral bone sclerosis has also been noted following microfracture in horses when the repair tissue was assessed with radiographs or magnetic resonance imaging (MRI).[26,30] Most animal studies evaluate repair with histology and not with radiographs or MRI, making assessment of subchondral bone sclerosis difficult and subjective. Sclerosis of the subchondral bone has been postulated as an initiating event in the development of osteoarthritis.[31–33] It should be restated that the animal model studies are limited to 1-year duration so the long-term presence or consequences of this subchondral bone sclerosis on the microfracture repair tissue or clinical outcome of the patient are not evident.

Microfracture by definition is fracturing of the subchondral bone, and the results of subchondral bone sclerosis or central osteophyte formation might be anticipated knowing the natural course of healing following microfracture of cancellous subchondral bone.

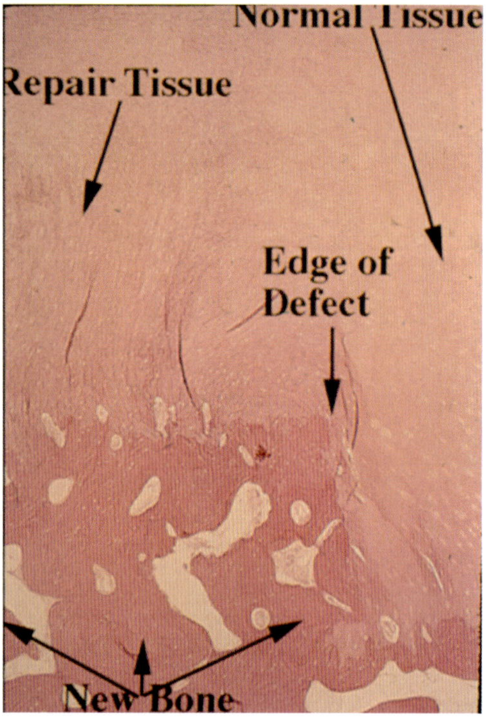

Fig. 6.3 Histologic appearance of a microfracture-treated defect on the medial femoral condyle, 12 months after surgery. The fibrocartilage is well adhered to the surrounding normal cartilage tissue and to the underlying, protruding new subchondral bone. (Reproduced with permission from Frisbie, et al. Vet Surg 1999;28(4):242–255.)

Trabecular microfractures of the femoral head, spine, patella, and acetabulum have been studied since the 1960s.[34] These naturally occurring microfractures heal with woven bone microcallus. It is reasonable to presume that penetration of the subchondral plate with a microfracture awl to gain access to bone marrow elements stimulates a similar bone repair response. What circumstances lead to an overexuberant reaction with resultant central osteophyte formation is not clear. Bone repair/regeneration is complex and is influenced by many factors including age, mechanical and cellular environments, bone mineral content, and genetics.[34,35] There are also differences in the response of cells to mechanical loading, and this too might influence cells in the superclot to differentiate down osteogenic or chondrogenic lineages.[33,36] The ability of progenitor cells to differentiate into osteogenic or chondrogenic cell lines should be remembered and investigated simultaneously when developing technologies for augmentation of microfracture.

Subchondral Cystic Formation

In animal models when the medial femoral condyle is used as the treatment site, violation of the subchondral bone plate can result in formation of subchondral bone cysts.[37–39] In preparation of a cartilage bed for microfracture, overexuberant debridement of the calcified cartilage layer to include removal of the subchondral bed can lead to subchondral cyst formation.[39] Precise attention to the technical aspects of microfracture and the use of skeletally mature animals (tidemark is fully formed and the calcified cartilage layer is visible) are crucial for successful modeling. Radiolucent "cyst-like" areas in the medial femoral condyle have been observed following microfracture, but there was no evidence of a cyst on histologic analysis.[22] Although MRI was not performed, the authors were of the opinion that the radiolucency represented bone edema.

Microfracture Compared with Microdrilling

In a rabbit study comparing microdrilling with microfracture at a depth of 2 mm, microcomputed tomography imaging performed 1 day postoperatively indicated that microfracture led to more compaction of bone in the holes than did microdrilling.[40] The authors concluded that this impaction of bone might impede the ability of bone marrow to reach the articular defect and thereby might negatively affect repair. Bleeding in only one of four microfracture holes was observed intraoperatively, but all defects were filled with a blood clot. The lack of bleeding from the microfracture holes has not been reported, nor is it consistent with the clinical experiences of the authors in humans or horses. Thus, it is likely a flaw of the rabbit as an animal model or, more likely, as the authors suggested, a result of the type of homemade microfracture awl specifically created for the study, which had a collar to limit the depth of penetration to 2 mm. The collar likely restricted movement of bone from the microfracture holes, creating impaction fractures in the subchondral bone. However, impaction of subchondral bone surrounding the microfracture hole is seen using standard arthroscopic microfracture awls without a collar (**Fig. 6.4**; **Videos 6.1, 6.2**). Microdrilling might be as effective as microfracture but obviously requires more surgical instrumentation such as a drill compared with a handheld awl to generate a superclot.

In summary, basic science and animal model studies indicate that microfracture results in improved repair tissue that continues to mature and becomes more cartilaginous for at least 1 year after surgery. The superclot clearly remodels but does remain quite inferior to normal articular cartilage in matrix molecule composition and therefore biomechanical function. Numerous studies have been performed to augment microfracture even though we don't fully understand the fundamental biology of microfracture and therefore how to improve upon current results. A potential detriment to the use of microfracture is the formation of central/intralesional osteophytes, which are unpredictable and have been associated with persistent or recurrent pain in human studies. Microfracture remains a commonly performed and investigated cartilage repair procedure because it is easy to do and requires minimal equipment, and clinical results in human patients are encouraging.

Video 6.1

Continued impaction of bone from the depth to the surface of the microfracture hole.
Online content including video sequences viewable at www.thieme-connect.com/ejournals/html/jks/doi/10.1055/s-0032-1310389

Video 6.2

Impacted bone surrounding the entire circumference of the microfracture hole.
Online content including video sequences viewable at www.thieme-connect.com/ejournals/html/jks/doi/10.1055/s-0032-1310389

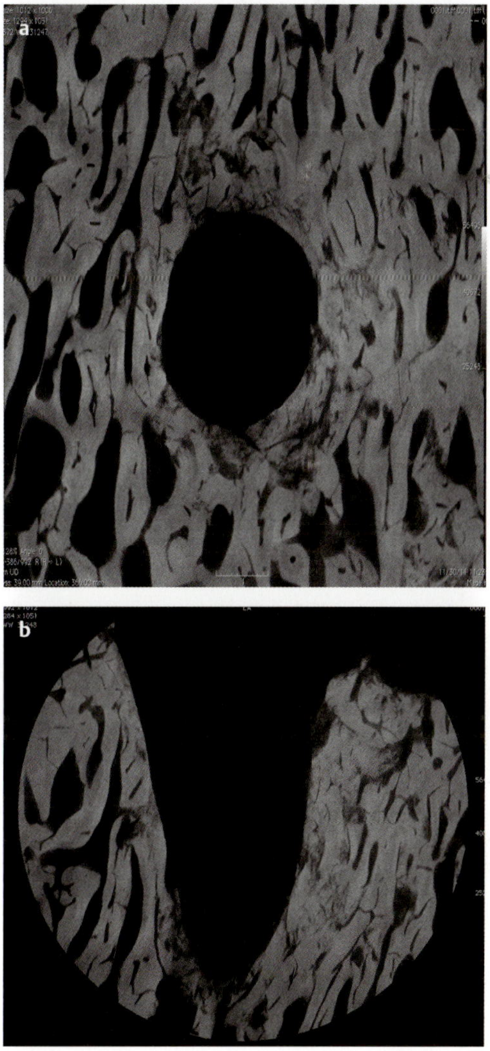

Fig. 6.4 Microcomputed tomography of normal equine lateral trochlear ridge subjected to microfracture. (**a**) Axial view of a microfracture hole demonstrating impaction of surrounding subchondral bone. (**b**) Sagittal view of a microfracture hole demonstrating impaction of surrounding subchondral bone. Minimum-intensity projection of micro-CT data acquired with 4-μm *x-y-z* voxel size.

Acknowledgment

The authors thank Ms. Paula Sharp for her assistance in the generation of this chapter and Mark Riccio for generation of the micro-CT images.

References

1. Rodrigo JJ, Steadman JR, Silliman JF, Fulstone AH. Improvement of full-thickness chondral defect healing in the human knee after debridement and microfracture using continuous passive motion. Am J Knee Surg 1994;7:109–116
2. Steadman JR, Rodkey WG, Singleton SB, Briggs KK. Microfracture technique for full-thickness chondral defects: technique and clinical results. Oper Tech Orthop 1997;7:300–304
3. Knutsen G, Engebretsen L, Ludvigsen TC, et al. Autologous chondrocyte implantation compared with microfracture in the knee. A randomized trial. J Bone Joint Surg Am 2004;86-A(3):455–464
4. Blevins FT, Steadman JR, Rodrigo JJ, Silliman J. Treatment of articular cartilage defects in athletes: an analysis of functional outcome and lesion appearance. Orthopedics 1998;21(7):761–767, discussion 767–768
5. Gill TJ, McCulloch PC, Glasson SS, Blanchet T, Morris EA. Chondral defect repair after the microfracture procedure: a nonhuman primate model. Am J Sports Med 2005;33(5):680–685
6. Shapiro F, Koide S, Glimcher MJ. Cell origin and differentiation in the repair of full-thickness defects of articular cartilage. J Bone Joint Surg Am 1993;75(4):532–553
7. de Girolamo L, Bertolini G, Cervellin M, Sozzi G, Volpi P. Treatment of chondral defects of the knee with one step matrix-assisted technique enhanced by autologous concentrated bone marrow: in vitro characterisation of mesenchymal stem cells from iliac crest and subchondral bone. Injury 2010;41(11):1172–1177
8. Hackett CH, Flaminio MJ, Fortier LA. Analysis of CD14 expression levels in putative mesenchymal progenitor cells isolated from equine bone marrow. Stem Cells Dev 2011;20(4):721–735
9. Radcliffe CH, Flaminio MJ, Fortier LA. Temporal analysis of equine bone marrow aspirate during establishment of putative mesenchymal progenitor cell populations. Stem Cells Dev 2010;19 (2):269–282
10. Neumann K, Dehne T, Endres M, et al. Chondrogenic differentiation capacity of human mesenchymal progenitor cells derived from subchondral cortico-spongious bone. J Orthop Res 2008;26 (11):1449–1456
11. Bunnell BA, Betancourt AM, Sullivan DE. New concepts on the immune modulation mediated by mesenchymal stem cells. Stem Cell Res Ther 2010;1(5):34
12. Prockop DJ. Repair of tissues by adult stem/progenitor cells (MSCs): controversies, myths, and changing paradigms. Mol Ther 2009;17(6):939–946
13. Endres M, Neumann K, Häupl T, et al. Synovial fluid recruits human mesenchymal progenitors from subchondral spongious bone marrow. J Orthop Res 2007;25(10):1299–1307
14. Erggelet C, Neumann K, Endres M, Haberstroh K, Sittinger M, Kaps C. Regeneration of ovine articular cartilage defects by cell-free polymer-based implants. Biomaterials 2007;28(36):5570–5580
15. Endres M, Andreas K, Kalwitz G, et al. Chemokine profile of synovial fluid from normal, osteoarthritis and rheumatoid arthritis patients: CCL25, CXCL10 and XCL1 recruit human subchondral mesenchymal progenitor cells. Osteoarthritis Cartilage 2010;18(11):1458–1466
16. Kang SW, Bada LP, Kang CS, et al. Articular cartilage regeneration with microfracture and hyaluronic acid. Biotechnol Lett 2008;30(3):435–439
17. Kramer J, Böhrnsen F, Lindner U, Behrens P, Schlenke P, Rohwedel J. In vivo matrix-guided human mesenchymal stem cells. Cell Mol Life Sci 2006;63(5):616–626
18. Lee CH, Cook JL, Mendelson A, Moioli EK, Yao H, Mao JJ. Regeneration of the articular surface of the rabbit synovial joint by cell homing: a proof of concept study. Lancet 2010;376(9739):440– 448
19. Kuo AC, Rodrigo JJ, Reddi AH, Curtiss S, Grotkopp E, Chiu M. Microfracture and bone morphogenetic protein 7 (BMP-7) synergistically stimulate articular cartilage repair. Osteoarthritis Cartilage 2006;14(11):1126–1135
20. Krüger JP, Endres M, Neumann K, Häupl T, Erggelet C, Kaps C. Chondrogenic differentiation of human subchondral progenitor cells is impaired by rheumatoid arthritis synovial fluid. J Orthop Res 2010;28(6):819–827
21. Payne KA, Didiano DM, Chu CR. Donor sex and age influence the chondrogenic potential of human femoral bone marrow stem cells. Osteoarthritis Cartilage 2010;18(5):705–713
22. Frisbie DD, Trotter GW, Powers BE, et al. Arthroscopic subchondral bone plate microfracture technique augments healing of large chondral defects in the radial carpal bone and medial femoral condyle of horses. Vet Surg 1999;28(4):242–255
23. Frisbie DD, Oxford JT, Southwood L, et al. Early events in cartilage repair after subchondral bone microfracture. Clin Orthop Relat Res 2003(407): 215–227
24. Frisbie DD, Morisset S, Ho CP, Rodkey WG, Steadman JR, McIlwraith CW. Effects of calcified cartilage on healing of chondral defects treated with microfracture in horses. Am J Sports Med 2006; 34(11):1824–1831
25. van den Borne MP, Raijmakers NJ, Vanlauwe J, et al; International Cartilage Repair Society. International Cartilage Repair Society (ICRS) and Oswestry macroscopic cartilage evaluation scores validated for use in Autologous Chondrocyte Implantation (ACI) and microfracture. Osteoarthritis Cartilage 2007;15(12):1397–1402
26. Fortier LA, McCarrel TM, Bradica G, et al. Evaluation of a biphasic graft for osteochondral repair in an equine model. Transactions of the 9th World Congress of the International Cartilage Repair Society. CD-ROM, 2010
27. Watanabe A, Boesch C, Anderson SE, Brehm W, Mainil Varlet P. Ability of dGEMRIC and T2 mapping to evaluate cartilage repair after microfracture: a goat study. Osteoarthritis Cartilage 2009;17 (10):1341–1349

28. Gomoll AH, Madry H, Knutsen G, et al. The subchondral bone in articular cartilage repair: current problems in the surgical management. Knee Surg Sports Traumatol Arthrosc 2010;18(4):434–447
29. McIlwraith CW, Frisbie DD, Rodkey WG, et al. Evaluation of intra- articular mesenchymal stem cells to augment healing of microfractured chondral defects. Arthroscopy 2011;27(11):1552–1561
30. Fortier LA, Potter HG, Rickey EJ, et al. Concentrated bone marrow aspirate improves full-thickness cartilage repair compared with microfracture in the equine model. J Bone Joint Surg Am 2010; 92(10):1927–1937
31. Brandt KD, Mazzuca SA. Experience with a placebo-controlled randomized clinical trial of a disease-modifying drug for osteoarthritis: the doxycycline trial. Rheum Dis Clin North Am 2006; 32(1):217–234, xi–xii
32. Burr DB, Radin EL. Microfractures and microcracks in subchondral bone: are they relevant to osteoarthrosis? Rheum Dis Clin North Am 2003; 29(4):675–685
33. Sanchez C, Pesesse L, Gabay O, et al. Regulation of subchondral bone osteoblast metabolism by cyclic compression. Arthritis Rheum 2012;64(4): 1193–1203
34. Fazzalari NL. Trabecular microfracture. Calcif Tissue Int 1993;53 (Suppl 01):S143–S146, discussion S146–S147
35. Dimitriou R, Jones E, McGonagle D, Giannoudis PV. Bone regeneration: current concepts and future directions. BMC Med 2011;9:66
36. Potier E, Noailly J, Ito K. Directing bone marrow-derived stromal cell function with mechanics. J Biomech 2010;43(5):807–817
37. Kold SE, Hickman J, Melsen F. An experimental study of the healing process of equine chondral and osteochondral defects. Equine Vet J 1986; 18(1):18–24
38. Ray CS, Baxter GM, McIlwraith CW, et al. Development of subchondral cystic lesions after articular cartilage and subchondral bone damage in young horses. Eq Vet J 1996;28(3):225–232
39. McIlwraith CW, Fortier LA, Frisbie DD, Nixon AJ. Equine models of articular cartilage repair. Cartilage 2011;2(4):317–326
40. Chen H, Sun J, Hoemann CD, et al. Drilling and microfracture lead to different bone structure and necrosis during bone-marrow stimulation for cartilage repair. J Orthop Res 2009;27(11): 1432–1438

7

Microfracture and Augments

Andreas H. Gomoll

Microfracture was originally developed by Steadman in the 1980s in response to perceived limitations of the then commonly used marrow stimulation techniques (MSTs), abrasion arthroplasty and subchondral (Pridie) drilling—namely, destabilization of the subchondral plate and heat necrosis, respectively. Microfracture was quickly adopted for the treatment of cartilage defects, first in the knee, followed by the ankle, shoulder, elbow, and hip. The technique aims to induce the formation of a reparative tissue by the creation of channels in the subchondral plate, allowing the migration of mesenchymal stem cells (MSCs) from the subchondral marrow space into the defect. Here, they differentiate and produce a fibrocartilaginous tissue to fill the defect. The underlying biology is discussed in greater detail by Dr. Fortier and co-authors in chapter 6.

This chapter will review the indications, clinical application, rehabilitation, and outcomes of the standard microfracture procedure. In addition, it will present an overview of new technologies currently under development that aim to augment microfracture through the use of biomaterials and growth factors in hopes of improving outcomes and broadening the indications.

The Microfracture Procedure

Indications

The following indications are based on findings from multiple studies discussed further in the Results section. Microfracture is primarily indicated for the treatment of full-thickness articular cartilage defects without significant bone loss (Outerbridge Grade 3 and 4; International Cartilage Research Society [ICRS] Grade 3) measuring less than 2 to 4 cm^2 on the femoral condyles. Articular comorbidities such as malalignment and meniscal deficiency do not represent a contraindication provided they are corrected in a staged or concomitant fashion. Elevated body mass index (BMI) over 30 kg/m^2, defect size larger than 2 to 4 cm^2, defect location in the patellofemoral compartment or on the tibial plateau, and age older than 40 years are associated with worse outcomes.

Technique

Microfracture is generally performed as an all-arthroscopic procedure utilizing standard anteromedial and anterolateral portals. Rarely, accessory portals may become

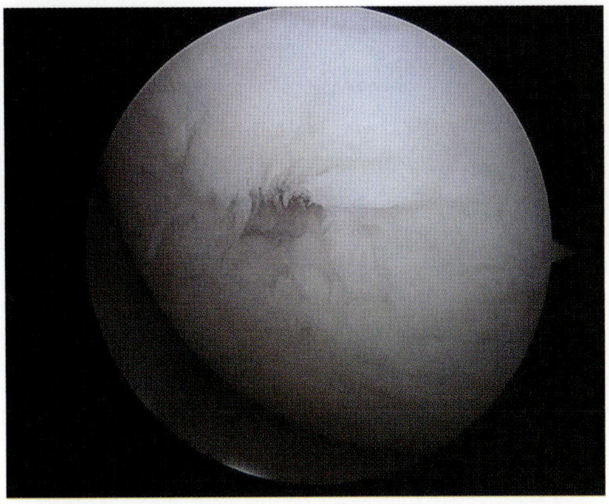

Fig. 7.1 Arthroscopic view of a femoral condyle cartilage defect.

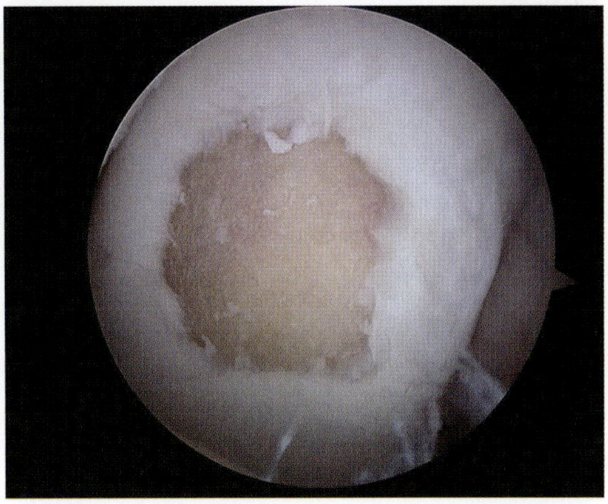

Fig. 7.2 Same defect after debridement of degenerated tissue and the layer of calcified cartilage with creation of stable shoulders.

necessary for optimal access. When performed together with other intra-articular procedures such as anterior cruciate ligament reconstruction or meniscal repair, microfracture should be performed last to preserve the developing blood clot that could otherwise be irrigated away by the arthroscopic fluid. The use of a tourniquet is optional.

Once the entire joint has been carefully evaluated and any articular comorbidities have been addressed, the cartilage defect (**Fig. 7.1**) is prepared. First, all degenerated cartilage is removed with a sharp curet, including any areas of surrounding cartilage that is delaminated. Vertical shoulders of stable cartilage are thus created. Next, the layer of calcified cartilage is removed with the curet; however, avoid excessive force that can injure the subchondral plate (**Fig. 7.2**). Generally, a motorized shaver can assist in removing larger flaps but is inadequate to appropriately prepare the defect by itself. Microfracture awls are available with different angled tips; depending on the location of the defect, the awl providing perpendicular alignment of the tip to the defect surface should be chosen. If the angle of placement is too oblique, furrows are created rather than holes, with increased damage to the subchondral plate. Microfracture holes are now created, starting at the periphery to improve edge integration (**Fig. 7.3**). The holes should be placed ~ 3 to 4 mm apart to prevent holes from becoming confluent and destabilizing the subchondral plate. At the end, the tourniquet should be deflated or the pump pressure lowered to observe fat droplets and bone marrow from each hole; otherwise, individual holes can be revisited and deepened with the awl. The use of intra-articular drains should be avoided because removal of the intra-articular hematoma would be counterproductive to the formation of the desired marrow clot.

Rehabilitation

The postoperative rehabilitation is a critical and inherent component of the microfracture procedure, and its contribution to

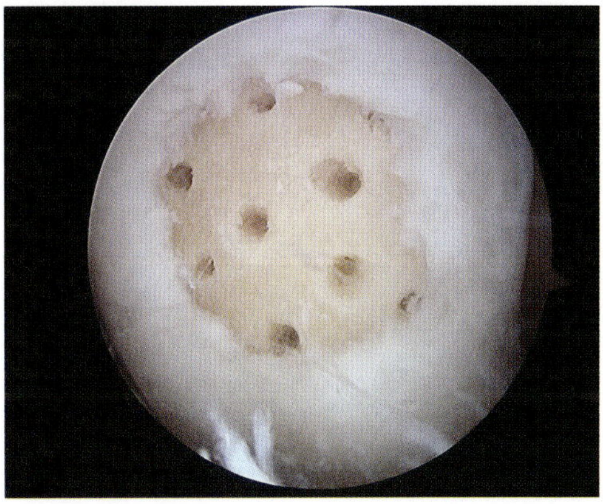

Fig. 7.3 Same defect after microfracture.

the overall success cannot be overemphasized.[1–3] Weightbearing restrictions are tailored to the individual patient: small and well-shouldered defects (< 1 cm^2) require less protection than larger defects with compromised shoulders. Generally, patients are kept touch-down weightbearing on two crutches for 4 to 8 weeks. A continuous passive motion machine is started on postoperative day 1 and continued for 6 weeks, 6 to 8 hours per day, increasing range of motion (ROM) as tolerated. Quadriceps isometrics and straight-leg raises can be started immediately, adding resistance bands and mini-squats at 2 months. More aggressive weight training is delayed until 4 months postoperatively. Impact sports, especially those involving cutting or pivoting, should not be resumed until 6 to 9 months after the procedure, and only once swelling has resolved and adequate muscle strength and proprioception have returned.

Results

Several authors have reported on the outcome of microfracture using both case series and randomized controlled trials (RCTs). Steadman's group has the largest experience with microfracture and has published extensively on this technique in various subpopulations. They demonstrated symptom improvement in 80% of cases in a minimum 7-year follow-up study of patients younger than 45 years without concomitant intra-articular comorbidities; at 3 years postoperatively 16% rated themselves unchanged and 4% considered their symptoms worse than preoperatively.[4] A more challenging group of patients with degenerative defects reported improvements in pain and function; 13 of 81 (16%) patients required repeat surgery for lysis of adhesions, and 5 (6%) patients underwent repeat microfracture or revision to arthroplasty at an average of 23 months postoperatively.[5] In their experience, results were not affected by lesion size or location.[4] Other groups have reported good and excellent results in 60 to 80% of patients[4,6–9] but have recommended more narrow indications, most reporting worse outcomes in defects larger than 2 to 4 cm^2.[7,9–12] The treatment of patellofemoral lesions with microfracture has been associated with worse outcomes than the treatment of femoral condyles.[11] Microfracture in patients older than 35 to 40 years resulted in worse outcomes than in younger patients.[4,9,10,12,13] The influence of defect chronicity appears controversial, with Steadman et al reporting no effect, while Mithoefer et al demonstrated better outcomes with lesions less than 1 year old.[4,6] Finally, BMI over 25 to 30 kg/m^2 appears negatively correlated with outcomes.[6,10]

When comparing microfracture to other procedures, Knutsen et al demonstrated overall comparable results to autologous chondrocyte implantation (ACI) for various lesions sizes in an RCT,[12] while Saris et al showed better histological and functional outcomes with ACI.[14,15] Coleman et al reported a trial of microfracture versus ACI, showing 44 versus 22% increased Cincinnati scores, respectively. Magnetic resonance imaging (MRI) scores were better for ACI; however, this did not correlate with functional outcomes.[16] Basad et al specifically focused on size-related outcomes in an RCT of microfracture versus ACI in defects larger than 4 cm^2, which demonstrated better results for ACI.[17] Kon et al presented results from a cohort study comparing microfracture to ACI with comparable results at 2 years, but worse results for microfracture at 5 years.[8] Gudas et al compared microfracture to osteochondral autograft transfer (OAT) in two RCTs. One demonstrated better arthroscopic, histologic, and MRI appearance, and higher return to play with OAT than microfracture (93 vs. 52%, respectively) in athletes.[9] Gudas et al's second RCT randomized patients with osteochondritis dissecans lesions to OAT versus microfracture, showing better outcome with OAT at 4 years (83 vs. 63%, respectively).[18]

Several studies used MRI to evaluate the quality of the repair tissue at follow-up, reporting good and excellent fill in approximately half of patients or fewer.[6,9,19,20] Poor fill on MRI correlated with worsening symptoms after an initial period of improvement.[6] When serial MRIs were performed, quality improved in the early postoperative period up to 2 years.[6]

Complications

Significant surgical complications are rare with this arthroscopic procedure. Depending on lesion size and location, patients can experience catching until the defect has filled with repair tissue. Over the mid to long term, microfracture has been shown to result in the formation of intralesional osteophytes, subchondral sclerosis, and cysts in up to 50% of patients (**Fig. 7.4**).[6,9,20] The potential influence of these subchondral changes on subsequent revision surgery with ACI has been reported in several studies. Some investigators reported no negative influence of prior microfracture in subanalyses of studies designed for general outcomes after ACI.[21,22] Conversely, two publications specifically tailored to investigate this question reported failure rates of ACI after prior marrow stimulation that were up to three times the failure rates seen in ACI in not previously treated defects.[23,24]

Augmentation Techniques

Although microfracture provides good short-term outcomes for many patients, it results in a fibrocartilaginous, rather than hyaline-like, repair tissue.[25] Mid- to long-term studies have demonstrated gradual decreases in functional outcomes after 24 to 36 months, potentially due to tissue degradation over time.[8,11] Increased interest in augmenting the body's own reparative response to improve quality and functional outcomes has led to the development of various biomaterials for implantation in conjunction with MSTs. Several of these treatment approaches will be discussed briefly below.

Drilling

Pridie developed subchondral drilling as treatment for cartilage defects in the 1950s and reported patient satisfaction of 77%; 64% of the knees were rated as good.[26] Utilizing an open arthrotomy and postoperative immobilization to perform the drilling, however, resulted in a high number of patients with stiffness, and heat necrosis of the subchondral bone was a concern. The rise of arthroscopic instrumentation facilitated the development of microfracture, addressing both the need for an open approach as well as any concern for heat necrosis.

More recently, several studies have pointed to certain benefits of drilling over microfracture, leading to a possible renaissance of this procedure. Since the current drilling technique is performed arthroscopically in an aqueous environment, heat necrosis is of lesser or no concern. Animal models have demonstrated that drilling with actual drill bits (rather than smooth K-wires) results in better marrow clot formation. This has been explained by the deeper channels created with drilling rather than microfracture, which can access more of the subchondral marrow space. Also, while microfracture

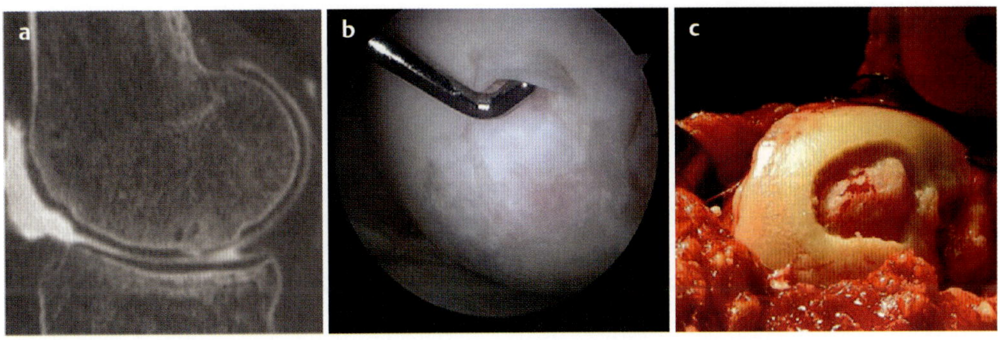

Fig. 7.4 Intralesional osteophyte of the medial femoral condyle after failed microfracture depicted by computed tomography arthrogram (**a**), arthroscopy (**b**), and surgical image (**c**) during revision with ACI.

compacts the bone around the hole and potentially seals off the marrow space, drilling removes bone and allows for more blood flow into the defect. A rabbit model demonstrated more complete fill and higher-quality tissue after subchondral drilling than microfracture.[27,28] The same group also demonstrated better repair of the subchondral bone with deeper drilling rather than the more shallow microfracture, with less incidence of cystic and sclerotic abnormalities.[28,29] Even though many of the following techniques have been tested with microfracture, drilling could be substituted, and additional studies should confirm its benefits.

Biomaterials

Microfracture relies on the formation of a blood clot containing MSCs from the bone marrow to fill the defect. The reparative tissue slowly matures with time but is vulnerable in the early phase, requiring protected weightbearing and ROM restrictions. Many groups are investigating modifications to the original microfracture technique, utilizing biomaterials to augment the mechanical stability of the early clot, hoping to retain more of the early, MSC-rich blood in the defect. Furthermore, bioactive materials could improve cell proliferation, differentiation, and matrix production, guiding the tissue toward a more hyaline-like histological appearance with potentially better long-term stability. The former function uses the biomaterial as a scaffold to maintain cells in situ until the new repair tissue has achieved adequate mechanical stability; thereafter, the scaffold has fulfilled its purpose and should resorb on its own. Several materials are being investigated for this role, including the autologous fibrin clot forming in standard microfracture, exogenous fibrin glue, alginate or agarose, collagen, hyaluronic acid, chitosan, and artificial polymers, such a polylactic and polyglycolic acid and their modifications. The second function as a bioactive substrate is far more complex and requires careful investigation to ensure correct differentiation signals—that is, chondrogenic rather than osteogenic differentiation of the MSCs. The following section will review preclinical and clinical data on biomaterial augmentation of marrow stimulation (some studies use drilling rather than microfracture to generate the marrow clot).

Autologous Matrix-Induced Chondrogenesis (AMIC; Chondro-Gide, Geistlich Biomaterials, Switzerland)

AMIC is a commonly used augmentation technique in Europe that utilizes Chondro-Gide, a type I/III collagen membrane to stabilize the clot in a marrow-stimulated defect. As such, the procedure is performed in an open fashion to size and secure the membrane with either sutures or fibrin glue. Originally introduced by Behrens, various authors have reported on mid-term outcomes of this technique, generally showing improved pain and function.[30–33] One group added platelet-rich plasma to AMIC for patellar defects in a five-patient pilot project, demonstrating good clinical outcomes but formation of intralesional osteophytes in three of five patients.[31] Due to reports on the benefits of drilling versus microfracture, Behrens most recently reported changing his technique to subchondral drilling for marrow stimulation.[27,28,30]

BST-CarGel (Piramal Healthcare Inc., Laval, Quebec, Canada)

This technique uses chitosan mixed with autologous blood to form a gel that is then implanted in a marrow-stimulated cartilage defect.[34] Chitosan is a polysaccharide primarily composed of polyglucosamine; it is thrombogenic, self-adhering, and completely resorbable. It has been the focus of extensive research for multiple tissue engineering applications, including cartilage repair. In preclinical animal models, chitosan improved the histological quality of the repair tissue when compared with marrow-stimulated control defects.[35–37] Preliminary data from an RCT were recently presented: 81 patients were randomized to microfracture with and without BST-CarGel, of whom 41 had completed 1-year follow-up. Evaluation demonstrated no clinical differences at this early time point, but better MRI and histological appearance of the augmented group.[38]

Chondrotissue (Bio Tissue AG, Freiburg, Germany)

Chondrotissue is a nonwoven polyglycolic acid fleece infused with hyaluronic acid that is implanted into cartilage defects after microfracture. Implantation of the scaffold in an ovine model demonstrated the formation of cartilaginous repair tissue.[39,40] Reports on the clinical outcomes are limited at this point with only a case report available.[41]

Gelrin C (Regentis Biomaterials, Or-Akiva, Israel)

This scaffold is a biodegradable photopolymerized hydrogel of polyethylene glycol (PEG) diacrylate bound to fibrinogen. The scaffold is injected into the previously microfractured defect as a gel that polymerizes in situ and completely degrades within 6 to 12 months. The major degradation products are polyethylene glycolated peptides, amino acids, and PEG, and have been shown to be nontoxic.[42] A clinical trial is currently under way.

Hyaluronan (HA)

The use of viscosupplementation for the treatment of osteoarthritis is widespread in clinical practice, although not without controversy.[43] Its application for microfracture has been explored by several groups, demonstrating that postoperative injections of HA had anti-inflammatory effects and improved repair tissue quality.[44–46]

Bioactive (Growth) Factors

Bioactive factors are proteins that can promote and inhibit cell differentiation, proliferation, and matrix production in a complex interaction. Marrow stimulation in its original form relies on MSCs to differentiate into the correct (chondrogenic) cell lineage, rather than into an osteogenic phenotype. Bioactive factors have been added to marrow-stimulated defects in hopes of improving the quality of the developing reparative tissue.[47,48] The delivery of bioactive factors remains challenging. Direct application of the growth factor in liquid form has been explored, but residence time in the defect is limited. Other groups have investigated repeated postoperative injections, viral vectors, and matrix-bound proteins with varying degrees of success.

Gelse et al investigated the use of thrombospondin-1 and osteogenic protein-1 (OP-1) after microfracture in a minipig model. Microfracture alone produced an inferior fibrocartilaginous tissue. The addition of OP-1 stimulated chondrogenesis but also induced enchondral ossification, which in turn was negated by treatment with thrombospondin-1. The authors concluded that the combination of the two factors has the potential to improve tissue quality after microfracture and reduce ossification.[49] Klinger et al found similar inhibitory effects on enchondral ossification in microfractured defects with delivery of chondromodulin-I through viral vectors. Additional chondrogenic and antiangiogenic effects were observed in this minipig model.[50] Sellers, Yang, Zhang, and Kuo and their colleagues described improved tissue quality in microfractured defects in a rabbit model treated with BMP-2, -4, and -7, respectively.[51–54] Morisset et al treated microfractured defects in horses with injections of adenoviral vectors carrying the genes of interleukin-1 receptor antagonist protein and insulin-like growth factor-1, showing improved defect healing in comparison with saline injection controls.[55] Feeley et al reported on the negative influence of postoperative parathyroid hormone treatment on cartilage formation in a rabbit microfracture model and recommended against its use.[56]

Stem Cells

There is considerable interest in the application of stem cells for cartilage repair. MSTs, such as microfracture, attempt to recruit stem cells from the underlying marrow cavity. These techniques have been criticized due to the low concentration of MSCs in the subchondral bone marrow, and also due to the resultant damage to the subchondral plate from the perforations. Techniques are being investigated that obtain stem cells from other areas such as the iliac crest or subcutaneous fat, concentrate the cells, and then introduce them either intraoperatively into the defect or postoperatively with injections. A comprehensive review of this topic is beyond the scope of this chapter, but several authors have investigated their use in conjunction with MSTs, applied either during surgery or as injections in the postoperative period.

Saw et al reported on five patients undergoing second-look arthroscopy after previous drilling followed by 5 weekly injections of peripheral blood progenitor cells and hyaluronic acid. The patients were part of a larger group of 180 undergoing this treatment. He reported articular cartilage regeneration and histology demonstrating features of hyaline cartilage.[57] Gobbi et al reported on the use of concentrated bone marrow aspirate without marrow stimulation for the repair of large cartilage defects in 15 patients. The marrow clot was implanted under a type I/III collagen matrix. At 2-year follow-up, patients reported significant improvements in pain and functional scores.[58]

Conclusion

Microfracture is a widely accepted treatment option for full-thickness articular cartilage defects. Its ready availability, low cost, and minimal invasiveness make it attractive as a first-line treatment option. Initially hailed as a "nonbridge burning" procedure, it is now being recognized as altering the subchondral bone through sclerosis of the subchondral plate, formation of subchondral cysts, and intralesional osteophytes in over one-third of patients. Following strict indications, however, microfracture continues to present a useful treatment option for the repair of cartilage defects smaller than 2 to 4 cm^2, primarily in the femoral condyles in younger patients. Broadening of these indications and improved long-term outcomes might be achieved through modification of the standard microfracture technique through the use of biomaterials and bioactive factors.

References

1. Irrgang JJ, Pezzullo D. Rehabilitation following surgical procedures to address articular cartilage lesions in the knee. J Orthop Sports Phys Ther 1998;28(4):232–240
2. Steadman JR, Rodkey WG, Rodrigo JJ. Microfracture: surgical technique and rehabilitation to treat chondral defects. Clin Orthop Relat Res 2001;(391, Suppl):S362–S369
3. Hurst JM, Steadman JR, O'Brien L, Rodkey WG, Briggs KK. Rehabilitation following microfracture for chondral injury in the knee. Clin Sports Med 2010;29(2):257–265, viii
4. Steadman JR, Briggs KK, Rodrigo JJ, Kocher MS, Gill TJ, Rodkey WG. Outcomes of microfracture for traumatic chondral defects of the knee: average 11-year follow-up. Arthroscopy 2003;19(5):477–484
5. Miller BS, Steadman JR, Briggs KK, Rodrigo JJ, Rodkey WG. Patient satisfaction and outcome after microfracture of the degenerative knee. J Knee Surg 2004;17(1):13–17
6. Mithoefer K, Williams RJ III, Warren RF, et al. The microfracture technique for the treatment of articular cartilage lesions in the knee. A prospective cohort study. J Bone Joint Surg Am 2005;87(9):1911–1920
7. Mithoefer K, Williams RJ III, Warren RF, Wickiewicz TL, Marx RG. High-impact athletics after knee articular cartilage repair: a prospective evaluation of the microfracture technique. Am J Sports Med 2006;34(9):1413–1418
8. Kon E, Gobbi A, Filardo G, Delcogliano M, Zaffagnini S, Marcacci M. Arthroscopic second-generation autologous chondrocyte implantation compared with microfracture for chondral lesions of the knee: prospective nonrandomized study at 5 years. Am J Sports Med 2009;37(1):33–41
9. Gudas R, Kalesinskas RJ, Kimtys V, et al. A prospective randomized clinical study of mosaic osteochondral autologous transplantation versus microfracture for the treatment of osteochondral defects in the knee joint in young athletes. Arthroscopy 2005;21(9):1066–1075
10. Asik M, Ciftci F, Sen C, Erdil M, Atalar A. The microfracture technique for the treatment of full-thickness articular cartilage lesions of the knee: midterm results. Arthroscopy 2008;24(11):1214–1220
11. Kreuz PC, Steinwachs MR, Erggelet C, et al. Results after microfracture of full-thickness chondral defects in different compartments in the knee. Osteoarthritis Cartilage 2006;14(11):1119–1125
12. Knutsen G, Engebretsen L, Ludvigsen TC, et al. Autologous chondrocyte implantation compared with microfracture in the knee. A randomized trial. J Bone Joint Surg Am 2004;86-A(3):455–464
13. Kreuz PC, Erggelet C, Steinwachs MR, et al. Is microfracture of chondral defects in the knee associated with different results in patients aged 40 years or younger? Arthroscopy 2006;22(11):1180–1186
14. Saris DB, Vanlauwe J, Victor J, et al; TIG/ACT/01/2000&EXT Study Group. Treatment of symptomatic cartilage defects of the knee: characterized chondrocyte implantation results in better clinical outcome at 36 months in a randomized trial compared to microfracture. Am J Sports Med 2009;37(Suppl 1):10S–19S
15. Saris DB, Vanlauwe J, Victor J, et al. Characterized chondrocyte implantation results in better structural repair when treating symptomatic cartilage defects of the knee in a randomized controlled trial versus microfracture. Am J Sports Med 2008;36(2):235–246

16. Coleman SH, Malizia R, Macgillivray J, Warren RF. Treatment of isolated articular cartilage lesions of the medial femoral condyle. A clinical nad MR comparison of autologous chondrocyte implantation vs. microfracture. Ortop Traumatol Rehabil 2001;3(2):224–226
17. Basad E, Ishaque B, Bachmann G, Stürz H, Steinmeyer J. Matrix-induced autologous chondrocyte implantation versus microfracture in the treatment of cartilage defects of the knee: a 2-year randomised study. Knee Surg Sports Traumatol Arthrosc 2010;18(4):519–527
18. Gudas R, Simonaityte R, Cekanauskas E, Tamosiūnas R. A prospective, randomized clinical study of osteochondral autologous transplantation versus microfracture for the treatment of osteochondritis dissecans in the knee joint in children. J Pediatr Orthop 2009;29(7):741–748
19. Von Keudell A, Atzwanger J, Forstner R, Resch H, Hoffelner T, Mayer M. Radiological evaluation of cartilage after microfracture treatment: a long-term follow-up study. Eur J Radiol 2011; June 16 (Epub ahead of print)
20. Brown WE, Potter HG, Marx RG, Wickiewicz TL, Warren RF. Magnetic resonance imaging appearance of cartilage repair in the knee. Clin Orthop Relat Res 2004;(422):214–223
21. Zaslav K, Cole B, Brewster R, et al; STAR Study Principal Investigators. A prospective study of autologous chondrocyte implantation in patients with failed prior treatment for articular cartilage defect of the knee: results of the Study of the Treatment of Articular Repair (STAR) clinical trial. Am J Sports Med 2009;37(1):42–55
22. McNickle AG, L'Heureux DR, Yanke AB, Cole BJ. Outcomes of autologous chondrocyte implantation in a diverse patient population. Am J Sports Med 2009;37(7):1344–1350
23. Minas T, Gomoll AH, Rosenberger R, Royce RO, Bryant T. Increased failure rate of autologous chondrocyte implantation after previous treatment with marrow stimulation techniques. Am J Sports Med 2009;37(5):902–908
24. Pestka JM, Bode G, Salzmann G, Südkamp NP, Niemeyer P. Clinical outcome of autologous chondrocyte implantation for failed microfracture treatment of full-thickness cartilage defects of the knee joint. Am J Sports Med 2011; November 5 (Epub ahead of print)
25. Hunziker EB. Articular cartilage repair: basic science and clinical progress. A review of the current status and prospects. Osteoarthritis Cartilage 2002;10(6):432–463
26. Insall JN. Intra-articular surgery for degenerative arthritis of the knee. A report of the work of the late K. H. Pridie. J Bone Joint Surg Br 1967;49(2):211–228
27. Chen H, Hoemann CD, Sun J, et al. Depth of subchondral perforation influences the outcome of bone marrow stimulation cartilage repair. J Orthop Res 2011;29(8):1178–1184
28. Chen H, Sun J, Hoemann CD, et al. Drilling and microfracture lead to different bone structure and necrosis during bone-marrow stimulation for cartilage repair. J Orthop Res 2009;27(11): 1432–1438
29. Chen H, Chevrier A, Hoemann CD, Sun J, Ouyang W, Buschmann MD. Characterization of subchondral bone repair for marrow-stimulated chondral defects and its relationship to articular cartilage resurfacing. Am J Sports Med 2011;39(8):1731–1740
30. Benthien JP, Behrens P. The treatment of chondral and osteochondral defects of the knee with autologous matrix-induced chondrogenesis (AMIC): method description and recent developments. Knee Surg Sports Traumatol Arthrosc 2011;19(8): 1316–1319
31. Dhollander AA, De Neve F, Almqvist KF, et al. Autologous matrix-induced chondrogenesis combined with platelet-rich plasma gel: technical description and a five pilot patients report. Knee Surg Sports Traumatol Arthrosc 2011;19(4):536–542
32. Benthien JP, Behrens P. Autologous matrix-induced chondrogenesis (AMIC). A one-step procedure for retropatellar articular resurfacing. Acta Orthop Belg 2010;76(2):260–263
33. Gille J, Schuseil E, Wimmer J, Gellissen J, Schulz AP, Behrens P. Mid-term results of Autologous Matrix-Induced Chondrogenesis for treatment of focal cartilage defects in the knee. Knee Surg Sports Traumatol Arthrosc 2010;18(11):1456–1464
34. Kumar MN, Muzzarelli RA, Muzzarelli C, Sashiwa H, Domb AJ. Chitosan chemistry and pharmaceutical perspectives. Chem Rev 2004;104(12):6017–6084
35. Chevrier A, Hoemann CD, Sun J, Buschmann MD. Chitosan-glycerol phosphate/blood implants increase cell recruitment, transient vascularization and subchondral bone remodeling in drilled cartilage defects. Osteoarthritis Cartilage 2007; 15(3):316–327
36. Hoemann CD, Sun J, McKee MD, et al. Chitosan-glycerol phosphate/blood implants elicit hyaline cartilage repair integrated with porous subchondral bone in microdrilled rabbit defects. Osteoarthritis Cartilage 2007;15(1):78–89
37. Hoemann CD, Hurtig M, Rossomacha E, et al. Chitosan-glycerol phosphate/blood implants improve hyaline cartilage repair in ovine microfracture defects. J Bone Joint Surg Am 2005;87(12): 2671–2686
38. Stanish WD, Restrepo A, Macdonald PB, et al. An International Randomized Clinical Trial Evaluating BST-CarGel: A New Gel Implant for Cartilage Repair. Presented at: ISAKOS, 8th Biennial Congress; Rio de Janeiro, Brazil; 2011: #149
39. Erggelet C, Neumann K, Endres M, Haberstroh K, Sittinger M, Kaps C. Regeneration of ovine articular cartilage defects by cell- free polymer-based implants. Biomaterials 2007;28(36):5570– 5580
40. Erggelet C, Endres M, Neumann K, et al. Formation of cartilage repair tissue in articular cartilage defects pretreated with microfracture and covered with cell-free polymer-based implants. J Orthop Res 2009;27(10):1353–1360
41. Patrascu JM, Freymann U, Kaps C, Poenaru DV. Repair of a post-traumatic cartilage defect with a cell-free polymer-based cartilage implant: a follow-up at two years by MRI and histological review. J Bone Joint Surg Br 2010;92(8):1160–1163
42. Gonen-Wadmany M, Oss-Ronen L, Seliktar D. Protein-polymer conjugates for forming photopolymerizable biomimetic hydrogels for tissue engineering. Biomaterials 2007;28(26):3876–3886
43. Bellamy N, Campbell J, Robinson V, Gee T, Bourne R, Wells G. Viscosupplementation for the treatment of osteoarthritis of the knee. Cochrane Database Syst Rev 2006;(2):CD005321

44. Strauss E, Schachter A, Frenkel S, Rosen J. The efficacy of intra-articular hyaluronan injection after the microfracture technique for the treatment of articular cartilage lesions. Am J Sports Med 2009;37(4):720–726
45. Legović D, Zorihić S, Gulan G, et al. Microfracture technique in combination with intraarticular hyaluronic acid injection in articular cartilage defect regeneration in rabbit model. Coll Antropol 2009;33(2):619–623
46. Kang SW, Bada LP, Kang CS, et al. Articular cartilage regeneration with microfracture and hyaluronic acid. Biotechnol Lett 2008;30(3):435–439
47. Angel MJ, Sgaglione NA, Grande DA. Clinical applications of bioactive factors in sports medicine: current concepts and future trends. Sports Med Arthrosc 2006;14(3):138–145
48. Frenkel SR, Di Cesare PE. Scaffolds for articular cartilage repair. Ann Biomed Eng 2004;32(1):26–34
49. Gelse K, Klinger P, Koch M, et al. Thrombospondin-1 prevents excessive ossification in cartilage repair tissue induced by osteogenic protein-1. Tissue Eng Part A 2011;17(15–16):2101–2112
50. Klinger P, Surmann-Schmitt C, Brem M, et al. Chondromodulin 1 stabilizes the chondrocyte phenotype and inhibits endochondral ossification of porcine cartilage repair tissue. Arthritis Rheum 2011;63(9):2721–2731
51. Kuo AC, Rodrigo JJ, Reddi AH, Curtiss S, Grotkopp E, Chiu M. Microfracture and bone morphogenetic protein 7 (BMP-7) synergistically stimulate articular cartilage repair. Osteoarthritis Cartilage 2006;14(11):1126–1135
52. Sellers RS, Zhang R, Glasson SS, et al. Repair of articular cartilage defects one year after treatment with recombinant human bone morphogenetic protein-2 (rhBMP-2). J Bone Joint Surg Am 2000;82(2):151–160
53. Zhang X, Zheng Z, Liu P, et al. The synergistic effects of microfracture, perforated decalcified cortical bone matrix and adenovirus-bone morphogenetic protein-4 in cartilage defect repair. Biomaterials 2008;29(35):4616–4629
54. Yang HS, La WG, Bhang SH, et al. Hyaline cartilage regeneration by combined therapy of microfracture and long-term bone morphogenetic protein-2 delivery. Tissue Eng Part A 2011;17(13–14):1809–1818
55. Morisset S, Frisbie DD, Robbins PD, Nixon AJ, McIlwraith CW. IL-1ra/IGF-1 gene therapy modulates repair of microfractured chondral defects. Clin Orthop Relat Res 2007;462:221–228
56. Feeley BT, Doty SB, Devcic Z, Warren RF, Lane JM. Deleterious effects of intermittent recombinant parathyroid hormone on cartilage formation in a rabbit microfracture model: a preliminary study. HSS J 2010;6(1):79–84
57. Saw K-Y, Anz A, Merican S, et al. Articular cartilage regeneration with autologous peripheral blood progenitor cells and hyaluronic acid after arthroscopic subchondral drilling: a report of 5 cases with histology. Arthroscopy 2011;27(4):493–506
58. Gobbi A, Karnatzikos G, Scotti C, Mahajan V, Mazzucco L, Grigolo B. One-Step Cartilage Repair with Bone Marrow Aspirate Concentrated Cells and Collagen Matrix in Full-Thickness Knee Cartilage Lesions Results at 2-Year Follow-up. Cartilage 2011;2(3):286–299

8

ACI and MACI

Elizaveta Kon, Giuseppe Filardo, Alessandro Di Martino, and Maurilio Marcacci

The ultrastructure of articular cartilage is unique: chondrocytes are sparsely distributed within the surrounding matrix, maintaining minimal cell-to-cell contact. The interaction between cells, collagen framework, aggrecan, and fluid constitutes the complex ultrastructure of hyaline cartilage, making its replacement or reproduction difficult.[1]

New, ambitious regenerative procedures are emerging as potential therapeutic options for the treatment of chondral lesions, aiming to re-create a hyaline-like tissue, thus restoring a biologically and biomechanically valid articular surface with durable clinical results.

Autologous chondrocyte implantation (ACI) was introduced in 1987 in Sweden, and in 1994 Brittberg et al[2] published the first clinical report showing satisfactory results for isolated femoral condyle lesions. Since then, several studies have followed, documenting both the production of a hyaline-like articular surface and good results in the majority of the patients at medium-long follow-up. Treatment indications have been broadened,[3–5] and this cell-based technique has gained increasing interest worldwide.[6–10] The development of bioengineering technology further improved this regenerative treatment approach—essentially, transplanting biodegradable molecules that are used as temporary scaffolds for the growth of living cells.[11] Matrix-assisted ACI (MACI) techniques were introduced in the clinical practice one decade ago, showing good clinical results while at the same time overcoming most of the concerns related to the first-generation ACI.[12] The use of cell-loaded scaffolds to regenerate a cartilage-like tissue presents advantages from both the biological and the surgical point of view, thus aiming to further optimize this regenerative surgical procedure.

Surgical Technique

The surgical technique of both ACI and MACI consists of two steps. The first one is an arthroscopic procedure in which a biopsy of healthy cartilage is harvested from a non–weight-bearing site on the articular surface (usually intercondylar notch) for autologous chondrocyte cell culture, and the second step consists of implanting the expanded chondrocytes (**Fig. 8.1**).

The ACI procedure involves the implantation of a liquid cell culture, thus requiring the use of a flap to avoid leakage of chondrocytes from the defect area. An autologous periosteal patch has been traditionally used for its biological activity, but recently the use of a collagen xenograft membrane is becoming more popular. Through a parapatellar arthrotomy, the flap is sutured to the defect rim, and fibrin glue or sealant is applied to create a watertight seal before the cultured cells are injected.[10]

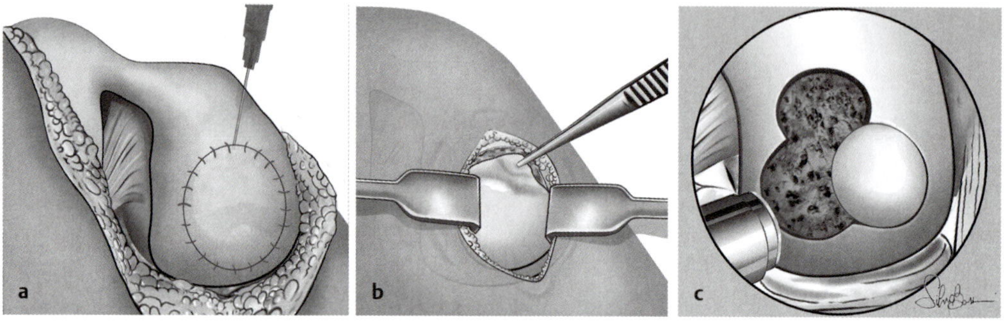

Fig. 8.1 Schematic representation. (**a**) ACI, open approach; (**b**) MACI, mini-open approach; (**c**) MACI, arthroscopic approach. *Abbreviations:* ACI, autologous chondrocyte implantation; MACI, matrix-assisted ACI.

The bioengineered MACI technology simplified the second step, which differs depending on the scaffold used. A mini-open approach can be used to prepare the lesion site, debriding the defect area down to the subchondral bone. Afterward, using a foil template reflecting the size and geometry of the defect, the chondrocyte-loaded matrix is cut to size and fitted into the defect with the cell-loaded surface facing the subchondral bone. In the case of an arthroscopic approach, dedicated instruments are used:[13] a circular area with regular margins for graft implantation is prepared with a specially designed cannulated low-profile drill. The delivery device is then filled with the bioengineered tissue, which is transported and positioned in the prepared area.

Depending on the adhesive characteristics of the grafts, no fibrin glue or sutures are needed, but for some biomaterials fibrin glue is used to fix the implant, and a transosseous fixation technique has been proposed to ensure secure fixation of the graft even in defects without stable shoulders.[12,14]

Rehabilitation Protocol

A similar rehabilitation protocol is used for both treatment approaches.

In the early stage (0 to 6 weeks), the rehabilitation strategies are focused on controlling pain, effusion, loss of motion, and muscle atrophy, and on protecting the transplant by preventing weight bearing for ~ 4 weeks. Continued passive motion is usually applied intensively until 90 degrees of flexion is attained, to avoid joint adherence and favor chondral nutrition and regenerative processes. Controlled mobilization exercises with reduced range of motion, early isometric and isotonic exercises, and controlled mechanical compression are performed. In the fourth week progressive touch-down weight bearing with crutches is allowed and usually advanced to full weight bearing within 6 to 8 weeks after surgery. Gait training in a swimming pool can be prescribed to facilitate the recovery of normal gait phases. Subsequently, active functional training can be started if there are no symptoms of overloading, such as pain, effusion, and tenderness. Proprioceptive, strength and endurance exercises, and aerobic training are then introduced, aiming to return to a correct running pathway. The remainder of the rehabilitation is dedicated to the return to previous sport activity, which is usually allowed no earlier than 1 year after surgery. However, time needed to recover may vary markedly depending on the procedure used. The bioengineered tissue significantly reduces the inherent fragility of the culture implant during the early postoperative stage and makes an accelerated patient recovery possible. The arthroscopic approach results in lower surgical morbidity and may enable a further acceleration of the functional recovery.[15,16]

Results

ACI

Since its conception 20 years ago, satisfactory clinical and radiographic (magnetic resonance imaging [MRI]) outcomes have been reported consistently at medium- to long-term follow-up for the ACI procedure.[3–10]

After the preliminary promising results,[2] the indication of this treatment has been broadened, and good patient-reported outcomes have been reported also for more challenging lesions. Browne et al[8] documented good results in large defects and in patients who previously failed prior cartilage repair. Minas et al[4] treated patients affected by early osteoarthritis successfully. Rosenberger et al[17] analyzed ACI treatment in older patients and found outcomes comparable to those reported in the literature for younger patients if all articular comorbidities were recognized and treated concomitantly. Farr[18] and Pascual-Garrido et al[19] showed that even more complex patellofemoral lesions can successfully be treated as long as corrective osteotomies are being performed to unload the repair tissue. The ACI procedure has also been modified to expand the treatment to deep osteochondral lesions. This "sandwich technique" procedure shows good results for the treatment of osteochondritis dissecans (OCD) at medium-term follow-up.[5]

Recently, Peterson et al[7] investigated the 20-year outcomes of the ACI procedure with periosteum in 224 patients. The subjective scores documented a significant improvement compared with the preoperative values. Seventy-four percent of the patients reported that they were better or stable, and 92% were satisfied with the operation and would undergo the ACI again. Further analysis was performed to determine factors that could influence the final outcome. The authors found that the age at the time of the operation and the size of the lesion did not correlate with the results, and interestingly the presence of meniscal injuries before ACI or history of bone marrow procedures before the implantation did not affect the outcome in this series either. This is in contrast to a report by Minas et al, who showed a three times higher rate of failures for defects that had prior treatments affecting the subchondral bone.[20] Better results have been obtained in cases of isolated femoral condyle lesions and OCD, whereas patients with multiple lesions undergo a progressive decline, with the bipolar lesions having an inferior outcome at 20 years.

These studies demonstrated that patients report good and excellent clinical and functional outcomes after ACI at long-term follow-up. With regard to imaging, encouraging data have also been documented.[21] Even though intralesional osteophytes, subchondral cysts, and bone marrow edema were common, the defect area was restored in most patients, with the quality of the repair tissue being similar to the surrounding normal cartilage, thus confirming the good long-term outcome offered by this procedure.

However, these good results have to be weighed against several limitations related to the complexity and morbidity of the surgical procedure, which requires a large open surgical approach and thus entails a high risk of joint stiffness and arthrofibrosis. The periosteal patch is believed to have biological properties, but it also requires a second incision and causes a high rate of hypertrophy with a high reoperation rate.[9,22–24]

A much lower complication rate has been shown by several authors using a type I/III collagen membrane in place of periosteum.[25,26] A further improvement of the procedure has been introduced with the development of a new technology: characterized chondrocyte implantation (CCI), which aims to improve the results of articular regeneration through the use of a selected cell population (approximately 5% deselection of inferior whole culture populations of chondrocytes).[27,28]

However, despite the substantial development undergone by the procedure since its introduction, some problems remain unsolved. One of those factors is the concern about the maintenance of the chondrocytic phenotype during the prolonged monolayer culture, which is a critical factor. In fact, it is known that chondrocytes in two-dimensional cell cultures alter their phenotype and dedifferentiate to fibroblast cells that no longer possess the capacity to produce type II collagen or proteoglycans,[6] and it is still unclear whether transplanted cells re-express their phenotype after transplantation. Another important concern is whether chondrocytes will be homogeneously distributed in the three-dimensional (3D) space of the defect when used in liquid cell suspension. Even with meticulous technique, there is the risk of chondrocyte leakage.

MACI

MACI has been developed with the attempt to address most of the concerns related to the cell culture and the surgical technique observed with the first-generation approach. Tissue engineering technology allows for the use of cell-loaded scaffolds. These are 3D structures developed to provide a support for cell adhesion, proliferation and production of matrix, and in the end the formation of a cartilage-like tissue.[29] Maintenance of a chondrocyte differentiated phenotype, lost during liquid culture, has been documented with the use of 3D scaffolds.[30] In addition, the MACI technique can be done arthroscopically or through minimal exposure, thus avoiding the large open approach necessary for the traditional ACI procedure.[13,14]

Several different second-generation tissue-engineered products have been introduced. Most of the currently available products are either collagen- or hyaluronan-based matrices. While there are many studies reporting good short-term success, long-term evaluations of these procedures are not available to date.[11,31–34]

Behrens et al[32] treated localized cartilage defects using a cell-seeded collagen matrix. They obtained good clinical outcome in 8 out of 11 patients at 5 years after transplantation. Ebert et al[33] used a different chondrocyte-seeded collagen membrane and found clinical and functional improvements and a high patient satisfaction rate. They also reported good MRI outcomes in 41 patients at 5 years' follow-up. Kreuz et al[34] used a bioresorbable two-component gel-polymer scaffold for the treatment of mild degenerative and focal osteoarthritic defects of the knee, showing that the good clinical outcome was stable over the course of a period of 4 years. Ferruzzi et al[35] analyzed the results obtained with a 3D hyaluronic acid scaffold, reporting ~ 50 patients affected by OCD and traumatic lesions. They reported good and excellent clinical outcomes at a minimum 5 years' follow-up. This study also reported MRI findings suggesting that the cartilage repair tissue was well integrated in 93% of the patients. Nehrer et al[36] confirmed good results over time with this scaffold. Gobbi et al[37] treated more challenging patellofemoral defects and reported a decline in clinical outcome at 5 years. Interestingly, even though the data showed a small decline, clinical and histological scores were still good, suggesting that autologous chondrocytes seeded on a hyaluronan-based scaffold can be considered a viable treatment option even for these lesions.

A nonrandomized prospective cohort study documented good and excellent results over time in the MACI group. Interestingly, over the same time frame, the microfractures control group showed a decline in patient-reported outcomes.[31] These results were confirmed in a study on highly demanding patients, soccer players; despite similar success in returning to competitive sport, microfractures allowed a faster recovery but presented a clinical deterioration over time, whereas arthroscopic second-generation ACI offered more durable clinical results.[38] A clinical improvement was found using MACI even in older patients.[16] Patients over 40 years old were treated with chondrocytes seeded on a collagen-based membrane or a hyaluronic-based scaffold. Despite the higher failure rate in this population and inferior results with respect to those previously found for younger patients, a significant improvement was documented. Results were consistent comparing the two treatment groups, with the only difference of a faster recovery in the group in which an arthroscopic approach was used. More recently, stable results were reported up to 7 years of follow-up,[39] and a good outcome was documented, using a modified technique with an associated bone grafting, also for the treatment of OCD, at 6 years of follow-up.[40]

Many other scaffolds have been recently proposed in preclinical studies; only a few have been introduced in clinical practice, though, presenting promising results at shorter follow-up.[41,42]

◆ Discussion

Despite thousands of treated patients and many published studies suggesting good and durable clinical results of this regenerative surgical treatment approach, there is no agreement to date about the effective superiority of one of these techniques over another, and both indications and results are still being discussed controversially.[10]

Studies comparing ACI with mosaicplasty are inconclusive; whereas the outcome is similar or even worse in small to medium-size lesions,[43,44] it seems that the regenerative (ACI) approach may offer better results for bigger lesions.[45] The comparison between ACI and bone marrow stimulation procedures also reported contradictory findings. Fu et al observed that patients who received ACI obtained higher levels of knee function and greater relief from pain and swelling at 3 years compared with debridement.[46]

Visna et al[47] evaluated an original method of chondrograft preparation based on cultivated autologous chondrocytes in a 3D carrier—fibrin glue, showing a better short-term outcome when compared with abrasive techniques, and Basad et al[48] compared MACI with microfractures, reporting MACI superiority at 2 years.

By contrast, Knutsen et al[49,50] showed that there was no difference in clinical outcomes between ACI and microfractures at up to 5 years' follow-up, and no differences were detected in the macroscopic or histological results at 2 years' and radiographic findings at 5 years' follow-up. However, they also demonstrated that none of the failures presented with high-quality repair cartilage, and that patients with ACI tended to have a more hyaline-like appearance of the repair tissue in the biopsies that were taken. The studies of Saris et al[27,28] confirmed that the repair tissue quality influences risk of failure and outcome over time. In fact, whereas the superior histomorphometric and histological score observed by Saris et al in the CCI group did not correlate with the short-term outcome, the quality of the repair tissue significantly influenced the later follow-up; at 3 years, CCI offered a further improvement with better clinical results compared with microfractures, whose results reached a plateau after 18 months.

These most recent findings suggest the higher healing potential offered by regenerative procedures, but the role of many variables that may influence the final outcome still need to be clarified to optimize results, and further systematic long-term evaluation is necessary to confirm the promising preliminary results obtained. The modern regenerative procedures can replace the injured articular surface with a hyaline-like tissue, but the properties of the healthy cartilage tissue are still unmatched by any available substitute. Additionally the current techniques are still cell culture-based; thus, they still pose the problem of cost-effectiveness and two-step surgery.

Even though different approaches have been studied to avoid the problems related to the ex vivo chondrocyte culture and expansion in a scaffold, cell-free implants[51–53] sufficiently "intelligent" to introduce the appropriate cues to induce orderly and durable tissue regeneration are under investigation in numerous animal studies and are gaining increasing interest in the clinical practice. These new and emerging technologies represent a new field of regenerative treatments that will be exciting to follow in the future.

References

1. Buckwalter JA, Mankin HJ. Articular cartilage: tissue design and chondrocyte-matrix interactions. Instr Course Lect 1998;47:477–486
2. Brittberg M, Lindahl A, Nilsson A, Ohlsson C, Isaksson O, Peterson L. Treatment of deep cartilage defects in the knee with autologous chondrocyte transplantation. N Engl J Med 1994; 331(14):889–895
3. Vasiliadis HS, Lindahl A, Georgoulis AD, Peterson L. Malalignment and cartilage lesions in the patellofemoral joint treated with autologous chondrocyte implantation. Knee Surg Sports Traumatol Arthrosc 2011;19(3):452–457
4. Minas T, Gomoll AH, Solhpour S, Rosenberger R, Probst C, Bryant T. Autologous chondrocyte implantation for joint preservation in patients with early osteoarthritis. Clin Orthop Relat Res 2010; 468(1):147–157
5. Peterson L, Minas T, Brittberg M, Lindahl A. Treatment of osteochondritis dissecans of the knee with autologous chondrocyte transplantation: results at two to ten years. J Bone Joint Surg Am 2003;85-A(Suppl 2):17–24
6. Peterson L, Brittberg M, Kiviranta I, Akerlund EL, Lindahl A. Autologous chondrocyte transplantation. Biomechanics and long-term durability. Am J Sports Med 2002;30(1):2–12
7. Peterson L, Vasiliadis HS, Brittberg M, Lindahl A. Autologous chondrocyte implantation: a long-term follow-up. Am J Sports Med 2010;38(6): 1117–1124
8. Browne JE, Anderson AF, Arciero R, et al. Clinical outcome of autologous chondrocyte implantation at 5 years in US subjects. Clin Orthop Relat Res 2005;(436):237–245
9. Micheli LJ, Browne JE, Erggelet C, et al. Autologous chondrocyte implantation of the knee: multicenter

experience and minimum 3-year follow-up. Clin J Sport Med 2001;11(4):223–228

10. Vasiliadis HS, Wasiak J. Autologous chondrocyte implantation for full thickness articular cartilage defects of the knee. Cochrane Database Syst Rev 2010;(10):CD003323
11. Kon E, Delcogliano M, Filardo G, Montaperto C, Marcacci M. Second generation issues in cartilage repair. Sports Med Arthrosc 2008;16(4):221–229
12. Kon E, Verdonk P, Condello V, et al. Matrix-assisted autologous chondrocyte transplantation for the repair of cartilage defects of the knee: systematic clinical data review and study quality analysis. Am J Sports Med 2009;37(1, Suppl 1):156S–166S
13. Marcacci M, Zaffagnini S, Kon E, Visani A, Iacono F, Loreti I. Arthroscopic autologous chondrocyte transplantation: technical note. Knee Surg Sports Traumatol Arthrosc 2002;10(3):154–159
14. Erggelet C, Sittinger M, Lahm A. The arthroscopic implantation of autologous chondrocytes for the treatment of full-thickness cartilage defects of the knee joint. Arthroscopy 2003;19(1):108–110
15. Della Villa S, Kon E, Filardo G, et al. Does intensive rehabilitation permit early return to sport without compromising the clinical outcome after arthroscopic autologous chondrocyte implantation in highly competitive athletes? Am J Sports Med 2010;38(1):68–77
16. Kon E, Filardo G, Condello V, et al. Second-generation autologous chondrocyte implantation: results in patients older than 40 years. Am J Sports Med 2011;39(8):1668–1675
17. Rosenberger RE, Gomoll AH, Bryant T, Minas T. Repair of large chondral defects of the knee with autologous chondrocyte implantation in patients 45 years or older. Am J Sports Med 2008; 36(12):2336–2344
18. Farr J. Autologous chondrocyte implantation improves patellofemoral cartilage treatment outcomes. Clin Orthop Relat Res 2007;463:187–194
19. Pascual-Garrido C, Slabaugh MA, L'Heureux DR, Friel NA, Cole BJ. Recommendations and treatment outcomes for patellofemoral articular cartilage defects with autologous chondrocyte implantation: prospective evaluation at average 4-year follow-up. Am J Sports Med 2009;37(1, Suppl 1):33S–41S
20. Minas T, Gomoll AH, Rosenberger R, Royce RO, Bryant T. Increased failure rate of autologous chondrocyte implantation after previous treatment with marrow stimulation techniques. Am J Sports Med 2009;37(5):902–908
21. Vasiliadis HS, Danielson B, Ljungberg M, McKeon B, Lindahl A, Peterson L. Autologous chondrocyte implantation in cartilage lesions of the knee: long-term evaluation with magnetic resonance imaging and delayed gadolinium-enhanced magnetic resonance imaging technique. Am J Sports Med 2010;38(5):943–949
22. Peterson L, Minas T, Brittberg M, Nilsson A, Sjögren-Jansson E, Lindahl A. Two- to 9-year outcome after autologous chondrocyte transplantation of the knee. Clin Orthop Relat Res 2000;(374): 212–234
23. Minas T. Autologous chondrocyte implantation for focal chondral defects of the knee. Clin Orthop Relat Res 2001;(391, Suppl):S349–S361
24. Henderson I, Francisco R, Oakes B, Cameron J. Autologous chondrocyte implantation for treatment of focal chondral defects of the knee—a clinical, arthroscopic, MRI and histologic evaluation at 2 years. Knee 2005;12(3):209–216
25. Gomoll AH, Probst C, Farr J, Cole BJ, Minas T. Use of a type I/III bilayer collagen membrane decreases reoperation rates for symptomatic hypertrophy after autologous chondrocyte implantation. Am J Sports Med 2009;37(1, Suppl 1):20S–23S
26. Gooding CR, Bartlett W, Bentley G, Skinner JA, Carrington R, Flanagan A. A prospective, randomised study comparing two techniques of autologous chondrocyte implantation for osteochondral defects in the knee: Periosteum covered versus type I/III collagen covered. Knee 2006;13(3):203–210
27. Saris DB, Vanlauwe J, Victor J, et al. Characterized chondrocyte implantation results in better structural repair when treating symptomatic cartilage defects of the knee in a randomized controlled trial versus microfracture. Am J Sports Med 2008;36(2):235–246
28. Saris DB, Vanlauwe J, Victor J, et al; TIG/ACT/01/2000&EXT Study Group. Treatment of symptomatic cartilage defects of the knee: characterized chondrocyte implantation results in better clinical outcome at 36 months in a randomized trial compared to microfracture. Am J Sports Med 2009;37(1, Suppl 1):10S–19S
29. Freed LE, Marquis JC, Nohria A, Emmanual J, Mikos AG, Langer R. Neocartilage formation in vitro and in vivo using cells cultured on synthetic biodegradable polymers. J Biomed Mater Res 1993;27(1):11–23
30. Grigolo B, Lisignoli G, Piacentini A, et al. Evidence for redifferentiation of human chondrocytes grown on a hyaluronan-based biomaterial (HYAff 11): molecular, immunohistochemical and ultrastructural analysis. Biomaterials 2002;23(4):1187–1195
31. Kon E, Gobbi A, Filardo G, Delcogliano M, Zaffagnini S, Marcacci M. Arthroscopic second-generation autologous chondrocyte implantation compared with microfracture for chondral lesions of the knee: prospective nonrandomized study at 5 years. Am J Sports Med 2009;37(1):33–41
32. Behrens P, Bitter T, Kurz B, Russlies M. Matrix-associated autologous chondrocyte transplantation/implantation (MACT/MACI)—5-year follow-up. Knee 2006;13(3):194–202
33. Ebert JR, Robertson WB, Woodhouse J, et al. Clinical and magnetic resonance imaging-based outcomes to 5 years after matrix-induced autologous chondrocyte implantation to address articular cartilage defects in the knee. Am J Sports Med 2011;39(4):753–763
34. Kreuz PC, Müller S, Ossendorf C, Kaps C, Erggelet C. Treatment of focal degenerative cartilage defects with polymer-based autologous chondrocyte grafts: four-year clinical results. Arthritis Res Ther 2009;11(2):R33
35. Ferruzzi A, Buda R, Faldini C, et al. Autologous chondrocyte implantation in the knee joint: open compared with arthroscopic technique. Comparison at a minimum follow-up of five years. J Bone Joint Surg Am 2008;90(4, Suppl 4):90–101
36. Nehrer S, Dorotka R, Domayer S, Stelzeneder D, Kotz R. Treatment of full-thickness chondral defects with hyalograft C in the knee: a prospective clinical case series with 2 to 7 years' follow-up. Am J Sports Med 2009;37(Suppl 1):81S–87S

37. Gobbi A, Kon E, Berruto M, et al. Patellofemoral full-thickness chondral defects treated with second-generation autologous chondrocyte implantation: results at 5 years' follow-up. Am J Sports Med 2009;37(6):1083–1092
38. Kon E, Filardo G, Berruto M, et al. Articular cartilage treatment in high-level male soccer players: a prospective comparative study of arthroscopic second-generation autologous chondrocyte implantation versus microfracture. Am J Sports Med 2011;39(12):2549–2557
39. Filardo G, Kon E, Di Martino A, Iacono F, Marcacci M. Arthroscopic second-generation autologous chondrocyte implantation: a prospective 7-year follow-up study. Am J Sports Med 2011;39(10):2153–2160
40. Filardo G, Kon E, Berruto M, et al. Arthroscopic second generation autologous chondrocytes implantation associated to bone grafting for the treatment of knee osteochondritis dissecans: results at 6 years. Knee 2011; November 22 (Epub ahead of print)
41. Selmi TA, Verdonk P, Chambat P, et al. Autologous chondrocyte implantation in a novel alginate-agarose hydrogel: outcome at two years. J Bone Joint Surg Br 2008;90(5):597–604
42. Nehrer S, Chiari C, Domayer S, Barkay H, Yayon A. Results of chondrocyte implantation with a fibrin-hyaluronan matrix: a preliminary study. Clin Orthop Relat Res 2008;466(8):1849–1855
43. Dozin B, Malpeli M, Cancedda R, et al. Comparative evaluation of autologous chondrocyte implantation and mosaicplasty: a multicentered randomized clinical trial. Clin J Sport Med 2005;15(4):220–226
44. Horas U, Pelinkovic D, Herr G, Aigner T, Schnettler R. Autologous chondrocyte implantation and osteochondral cylinder transplantation in cartilage repair of the knee joint. A prospective, comparative trial. J Bone Joint Surg Am 2003;85-A(2):185–192
45. Bentley G, Biant LC, Carrington RW, et al. A prospective, randomised comparison of autologous chondrocyte implantation versus mosaicplasty for osteochondral defects in the knee. J Bone Joint Surg Br 2003;85(2):223–230
46. Fu FH, Zurakowski D, Browne JE, et al. Autologous chondrocyte implantation versus debridement for treatment of full-thickness chondral defects of the knee: an observational cohort study with 3-year follow-up. Am J Sports Med 2005;33(11):1658–1666
47. Visna P, Pasa L, Cizmár I, Hart R, Hoch J. Treatment of deep cartilage defects of the knee using autologous chondrograft transplantation and by abrasive techniques—a randomized controlled study. Acta Chir Belg 2004;104(6):709–714
48. Basad E, Ishaque B, Bachmann G, Stürz H, Steinmeyer J. Matrix-induced autologous chondrocyte implantation versus microfracture in the treatment of cartilage defects of the knee: a 2-year randomised study. Knee Surg Sports Traumatol Arthrosc 2010;18(4):519–527
49. Knutsen G, Engebretsen L, Ludvigsen TC, et al. Autologous chondrocyte implantation compared with microfracture in the knee. A randomized trial. J Bone Joint Surg Am 2004;86-A(3):455–464
50. Knutsen G, Drogset JO, Engebretsen L, et al. A randomized trial comparing autologous chondrocyte implantation with microfracture. Findings at five years. J Bone Joint Surg Am 2007; 89(10):2105–2112
51. Gille J, Schuseil E, Wimmer J, Gellissen J, Schulz AP, Behrens P. Mid-term results of autologous matrix-induced chondrogenesis for treatment of focal cartilage defects in the knee. Knee Surg Sports Traumatol Arthrosc 2010;18(11):1456–1464
52. Bedi A, Foo LF, Williams RJ, et al. The maturation of synthetic scaffolds for osteochondral donor sites of the knee: an MRI and T2-mapping analysis. Cartilage 2010;1(1):20–28
53. Kon E, Delcogliano M, Filardo G, Busacca M, Di Martino A, Marcacci M. Novel nano-composite multilayered biomaterial for osteochondral regeneration: a pilot clinical trial. Am J Sports Med 2011;39(6):1180–1190

9

Particulated Articular Cartilage: CAIS and DeNovo NT

Jack Farr, Brian J. Cole, Seth Sherman, and Vasili Karas

Articular cartilage lesions are a common cause of knee symptoms.[1,2] The ultimate goal of surgical intervention is to restore the patient's comfort and function while the secondary goal is to prevent or delay osteoarthritis.[3–5] As with other tissues, articular cartilage form follows function, and recent studies suggest that improved clinical results correlate with better cartilage restoration constructs.[6] Current surgical treatment options for symptomatic cartilage lesions include debridement/lavage, marrow stimulation, osteochondral autograft implantation, fresh osteochondral allograft implantation, and autologous chondrocyte implantation (ACI).[7–12] More recently, minced cartilage autograft (Cartilage Autograft Implantation System [CAIS]; DePuy/Mitek, Raynham, MA) and particulated juvenile cartilage allograft (DeNovo Natural Tissue [NT]; ISTO, St. Louis, MO) options have been reported.[3,13]

One repair strategy is to use a bioactive component (i.e., cells or growth factors) that drives the biological process and a matrix (biomaterial that serves as a carrier or scaffold) that provides architectural support and facilitates the integration of the repaired tissue with the contiguous tissue.[3] Current treatment options have unique advantages and disadvantages. Autograft osteochondral plugs provide a living osteochondral unit but are limited to smaller lesions ranging from 1 to 2.5 cm^2.[14,15] Marrow stimulation is easy to perform but may also be limited with regard to the extent of durable hyaline-like cartilage formation, lesion size, and long-term sustained clinical gains.[15,16] ACI was the first cultured chondrocyte-based therapy, but it has variable long-term benefits when compared with microfracture and is technically tedious. Today, it is indicated for second-line treatment, especially for those patients with larger chondral defects.[8,17,18] Similar to ACI, other cultured chondrocyte techniques (e.g., ChondroCelect; DePuy/Mitek, Raynham, MA) have promising mid-term results.[6,16] Second- and third-generation cultured chondrocyte techniques culture the chondrocytes on a matrix, which improves the technical aspects, yet the results are similar to first generation and still require two-staged surgical procedures for harvest and implantation. Until recently, allograft treatment options have been limited to osteochondral grafts, as graft incorporation to host tissue was possible only at the bone level. The biologic requirement of transplant bone remodeling/incorporation to host bone at the basilar bone layer remains a challenge, and availability is limited.[1,2] In light of these limitations, ongoing research continues to search for cartilage restoration techniques that form durable tissue, are technically easier for the surgeon to perform, and are less disruptive to patients' lives during the recovery phase.

The concept that cartilage could be transplanted without its underlying bony component and heal would have been considered heretical even a few years ago by most cartilage surgeons. However, the potential safety and efficacy of both CAIS and DeNovo NT are challenging this paradigm. As in many aspects of science, the key to advances is "seeing" what has been there all along. While the phenomenon of hyaline cartilage repair using particulated articular cartilage is relatively new to the English language literature, a thorough literature review reveals a published report by Albrecht et al in the German literature dating back to 1983.[19] Their work showed that cartilage autograft implantation without bone can lead to cartilage defect healing if the cartilage is cut into small pieces. Most scientists in the English-speaking world were unaware of this article until recently, after US-based scientists noted the production of new chondrocyte and matrix formation adjacent to minced cartilage fragments. Researchers Lu and Binette began a series of experiments to investigate these findings. What followed was a rapid progression from in vitro experiments to the mouse, goat, and finally the horse model.[20,21] All these studies together demonstrated that autograft cartilage, when mechanically minced into cubes of 1 to 2 mm, could affect cartilage repair.[19,21] In essence, chondrocytes in the cartilage pieces could "escape" from the extracellular matrix, migrate, multiply, and form a new hyaline-like cartilage tissue matrix that would integrate with the surrounding host tissue. In addition, unlike cultured chondrocytes that take on a spindle-shaped morphology during culture, the chondrocytes from the minced cartilage retained the standard chondrocyte spheroid shape.[21]

These preclinical data were compelling enough for the Food and Drug Administration (FDA) to approve a proof of concept and safety pilot study of what the sponsor referred to as CAIS.[3] The clinical outcomes are now published at 2 years, and an extension follow-up study is complete to 4 years postoperative with publication to follow. Based on a parallel European study, the technique received a CE (Conformité Européenne) mark and is available through a limited release in Europe. In the United States, the FDA has approved a pivotal study of the technique, which began recruiting patients in 2010 and will enroll over 300 patients for a randomized prospective comparison of CAIS to MFX (microfracture) (i.e., CAIS is not available for general use in the United States; use is limited to study patients).

In another laboratory, Yao noted these early preclinical reports and decided to evaluate similar studies using particulated juvenile cartilage allograft (DeNovo NT; distributed by Zimmer, Warsaw, IN) in place of autograft.[1,2] This alternative approach was based on two factors: (1) allograft allows conceptually no limit to the amount of harvested tissue and (2) juvenile cartilage has the potential of more robust cellular activity than older cartilage tissue.[13,22–25] Yao demonstrated that new extracellular matrix can be formed from juvenile cartilage cubes in an explant culture study.[1] In addition, he demonstrated that particulated juvenile articular cartilage xenografts healed chondral defects on the trochlea of the horse knee joint.[1] Given the momentum from these positive results, the FDA now considers DeNovo NT as a minimally manipulated human tissue allograft, regulated as a 361 HCT/P product similar to fresh osteochondral allograft and bone-tendon-bone allograft. It is available for use in clinical applications without an investigational device exemption (IDE) study and, to date, over 2,200 patients have received this product.[2] During this same market release, the sponsor supported a prospective study of 25 patients in a multicenter study with preliminary results reported that complement a case report in the literature.[2,26]

◆ Indications/Contraindications

The indications for CAIS (DePuy/Mitek) and DeNovo NT (ISTO) are evolving. In general, they mirror the selection criteria for other cell-based cartilage procedures. On the basis of limited clinical trials, these products are indicated for treatment of symptomatic articular cartilage defects in patients from age 18 to 55. Prior to treatment, same-day arthroscopic evaluation should confirm a cartilage lesion that is at least International Cartilage Repair Society (ICRS) grade 3 or higher. After peripheral cartilage debridement, lesion size should range from 1 to 5 cm^2. As with the treatment of all cartilage defects, careful attention must be paid to meniscal status and to restoring or maintaining knee alignment and stability. Potential contraindications to CAIS (DePuy/Mitek) or DeNovo NT (ISTO) include bipolar lesion > ICRS grade 2, significant underlying subchondral bony edema, or osteochondritis dissecans lesion with > 6 mm subchondral bone loss, as the two last scenarios may require an osteochondral allograft or alternative techniques.

◆ Surgical Technique

CAIS

Standard arthroscopic portals are established and the lesion(s) are evaluated to confirm size, location, and appropriateness for treatment. If CAIS (DePuy/Mitek) is indicated, hyaline cartilage is then harvested arthroscopically from a low load-bearing surface (i.e., lateral wall of the intercondylar notch or trochlear margin with an amount similar to that harvested for ACI, roughly 200 mg) using a unique device that minces the cartilage into 1- to 2-mm pieces. After harvest, the device (DePuy/Mitek, Raynham, MA) uniformly disperses the minced cartilage onto a biodegradable scaffold. (The CAIS scaffold implant consists of an absorbable copolymer foam of 35% polycaprolactone and 65% polyglycolic acid, reinforced with a polydioxanone [PDO mesh] [Advanced Technologies and Regenerative Medicine, Raynham, MA].) The polymer foam is designed to keep the tissue fragments in place and serves as a three-dimensional scaffold for cartilage matrix generation. The reinforcing PDO mesh enables the foam to have adequate mechanical strength during implant handling. The fragments are then secured to this scaffold using a commercially available fibrin sealant (Tisseel, Baxter, IL). A mini-arthrotomy is performed, and the defect is identified and prepared similar to the technique used for ACI, whereby vertical lesion walls are created and the damaged cartilage is removed to the level of the subchondral bone using a ring curet. If bleeding is noted, hemostasis is achieved using epinephrine-soaked sponges and/or punctuated amounts of fibrin glue. An arthroscopic ruler is used to measure width, length, and depth of the prepared lesion. Subsequently, a template sizes the area of the lesion. Sterile paper or foil is used to make a template of the cartilage defect and used to cut the minced cartilage/scaffold construct to the appropriate size. The trimmed CAIS scaffold (DePuy/Mitek) implant is transferred to the defect with the cartilage fragments facing the subchondral bone and affixed with two or more biodegradable staple anchors (prototype, Advanced Technologies and Regenerative Medicine), which consist of PDO straps and tip (Advanced Technologies and Regenerative Medicine).

DeNovo NT

After confirmatory arthroscopy, a limited medial or lateral arthrotomy is performed to fully visualize the lesion(s) as shown in **Fig. 9.1a** The defect is outlined with a scalpel to create a shoulder (vertical peripheral wall) of normal or nearly normal host articular cartilage. The cartilage within the outlined area is removed carefully with a curet to the vertical wall of the host cartilage shoulder and the base of the defect (**Fig. 9.1b**). The base is cleared of all cartilage tissue, including the calcified layer, without entering into the subchondral bone. No marrow stimulation procedure is performed. Hemostasis, without a tourniquet, is achieved with epinephrine-soaked cottonoids and fibrin glue. After measuring the defect dimensions and recording the visual findings with photographs, a thin aluminum sterile foil is pressed into the defect to create a three-dimensional mold, as a complete replica of the defect (**Fig. 9.1c**). Once formed, the foil mold is removed from the defect and placed on the back table of the operating room. Using the measured defect dimensions, the defect surface area was calculated. One package of DeNovo NT graft (ISTO) is used for each 2.5-cm^2 defect. Larger defects require proportionally more packages of DeNovo NT graft (ISTO).

The DeNovo NT graft (ISTO), in a specially formulated nutrient preservation medium, is shipped in an aseptic temperature-controlled packaging (**Fig. 9.1d**). The medium is aspirated

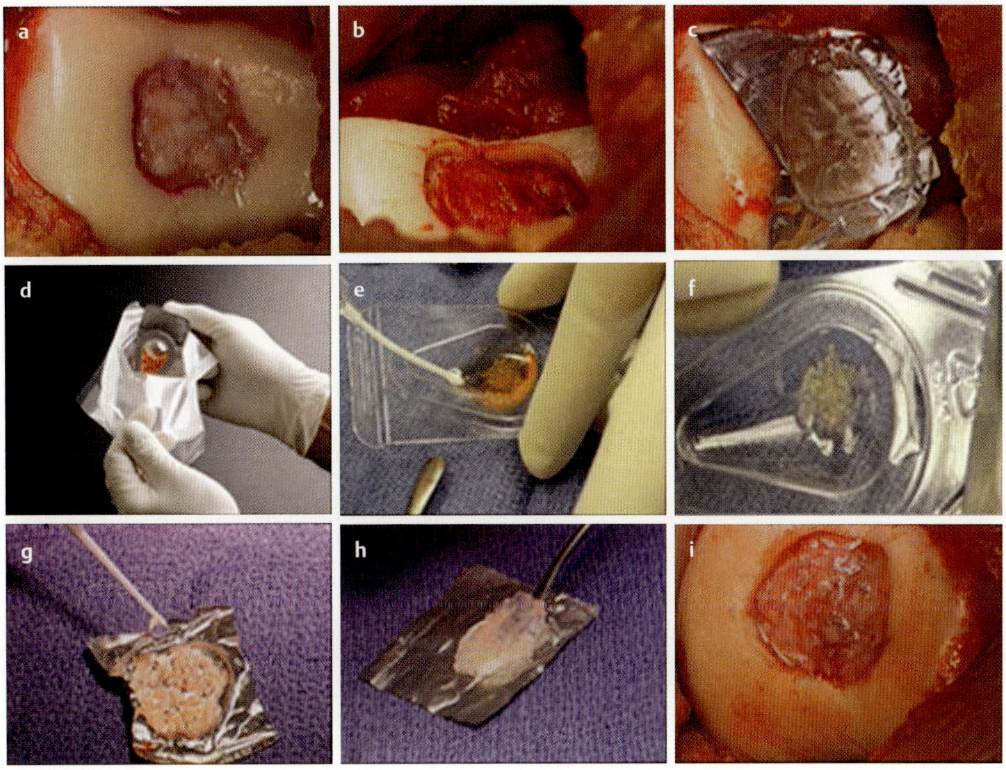

Fig. 9.1 Surgical technique for DeNovo NT. (**a**) The defect is identified, and a curet is used to clear the base of the defect (**b**). (**c**) A thin aluminum sterile foil is pressed into the defect to create a three-dimensional mold. The DeNovo NT is shipped in a temperature-controlled package (**d**). The nutrient-preserving medium (**e**, **f**) is aspirated. The cells are then transferred to the foil mold approximately 1 mm apart (**g**). Fibrin glue is then added and the cells are allowed to cure (**g**). At this time the construct is lifted in one piece (**h**). Fibrin glue is added to the base of the defect and the cartilage construct is then placed in the defect and sealed with fibrin glue and allowed to cure for 10 minutes. The cartilage construct is recessed relative to the surrounding cartilage (**i**). (Reprinted with permission from Farr J, Yao JQ. Chondral defect repair with particulated juvenile cartilage allograft. Cartilage 2011;2(4):346–353.)

(**Fig. 9.1e**) and the particulated cartilage pieces are transferred to the foil mold and distributed ~ 1 to 2 mm apart (potentially less separation depending upon the ratio between the implanted tissue volume and the surface area of the defect) (**Fig. 9.1f**). Fibrin glue is then added to the cartilage pieces until the foil mold is filled to within ~ 1 mm of its full depth (**Fig. 9.1g**). The glue is allowed to cure (typically 3 to 10 minutes). At that point, the fibrin glue/cartilage tissue construct is gently separated and then lifted from the foil in one piece (**Fig. 9.1h**). Fresh fibrin glue is applied at the base of the patient's cartilage lesion and the fibrin glue/particulated cartilage construct is pressed into the defect and the glue allowed to cure (**Fig. 9.1i**). As an alternative to the Zimmer/ISTO technique, some surgeons are directly applying the particulated cartilage into the defect and gluing it in situ.[26] It is imperative that the fibrin glue–cartilage tissue construct is thinner (average 1 mm) than the surrounding cartilage shoulders (average 2 to 3 mm), to minimize the potential for shear or direct compressive load.

Rehabilitation Protocol

In general, the rehabilitation program focuses initially on protection of the cartilage repair process and then progresses toward controlled loading, increased range of motion, and progressive muscle strengthening.[3] Patients receive a different rehabilitation protocol depending on whether they had a lesion in the patellofemoral compartment or the tibiofemoral compartment. Immediately after surgery, all patients receive a hinged knee brace locked in extension. Patients with a lesion on the femoral condyle are made non-weightbearing for the first 2 weeks and are advanced to partial weight bearing with an unlocked brace from weeks 2 through 6. Patients with a trochlear lesion are allowed to bear weight as tolerated immediately with the brace locked in extension. Regardless of lesion location, the brace is removed each day for continuous passive motion during the first 4 weeks, which is progressively increased (as tolerated) during the subsequent 3 weeks. Muscle strength is maintained using isometric quadriceps sets, straight-leg raises, and isometric contraction of the hamstrings, hip abductors, and hip adductors. When tolerated, patients use a stationary bike without resistance to maintain passive range of motion. Patients return to low-load activity levels at week 6 to 8 and progress in activity as strength and comfort permitted.

Clinical Results

CAIS

There is only one study in the clinical literature reporting outcomes of single-stage CAIS (DePuy/Mitek) for symptomatic knee cartilage defects.[3] The goal of this FDA-approved study was to establish the safety of CAIS (DePuy/Mitek) and to test whether CAIS (DePuy/Mitek) improves quality of life by using standardized outcomes assessment tools. A total of 29 patients was randomized with the intent to treat with either MFX or CAIS. Patients were followed at predetermined time points for 2 years using several standardized outcomes assessment tools (Short Form-36 [SF-36], International Knee Documentation Committee [IKDC], Knee Injury and Osteoarthritis Outcome Score [KOOS]). Magnetic resonance imaging (MRI) was performed at baseline, 3 weeks, and 6, 12, and 24 months.

Lesion size and ICRS grade were similar in both groups. General outcome measures (e.g., physical component score of the SF-36) indicated an overall improvement in both groups, and no differences in the number of adverse effects were noted in comparisons between the CAIS (DePuy/Mitek) and MFX groups. The IKDC score of the CAIS (DePuy/Mitek) group was significantly higher compared with the MFX group at both 12 and 24 months. Select subdomains (4⁄5) in the KOOS instrument were significantly different at 12 and 18 months, and all subdomains (Symptoms and Stiffness, Pain, Activities of Daily Living, Sports and Recreation, Knee-Related Quality of Life) were significantly increased at 24 months in CAIS (DePuy/Mitek) versus MFX. These significant improvements were maintained at 24 months in both IKDC and KOOS.

Qualitative analysis of the imaging data did not note differences between the two groups in fill of the graft bed, tissue integration, or presence of subchondral cysts. Patients treated with MFX had a significantly higher incidence of intralesional osteophyte formation (54 and 70% of total number of lesions treated) at 6 and 12 months when compared with CAIS (DePuy/Mitek) (8 and 25% of total number of lesions treated).

DeNovo NT

To date, there are only two clinical studies on the use of DeNovo NT (ISTO) for symptomatic cartilage lesions in the knee that are reported in the literature.[2,26]

The first is a case report on the use of particulated juvenile cartilage tissue for a symptomatic full-thickness patella cartilage defect.[26] At 2-year follow-up, the patient experienced substantial clinical improvement in both pain and function when evaluated with both the IKDC subjective evaluation and the KOOS outcome measures. MRI at final follow-up demonstrated fill of the defect with repair tissue, and nearly complete resolution of preoperative bony edema. **Figure 9.2** shows a preoperative MRI from a patient implanted with DeNovo NT (ISTO) and at 21 months postoperatively (**Fig. 9.3**).

The second is an early interim report of patients who are a part of an ongoing multicenter, prospective, single-arm study of 25 subjects.[2] This study is designed to

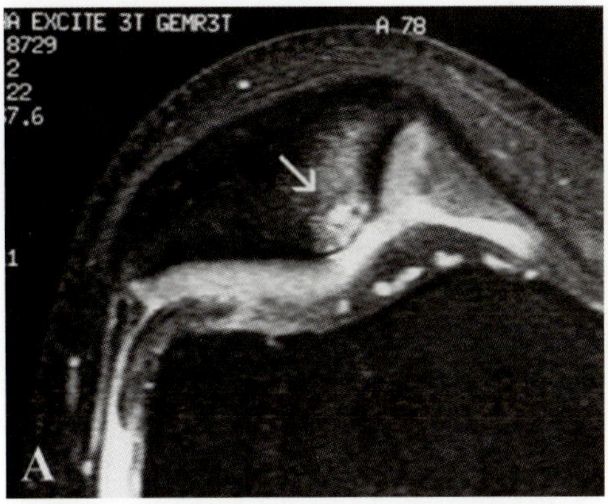

Fig. 9.2 Preoperative magnetic resonance imaging demonstrating a full-thickness chondral defect of the patella with underlying bone edema and early subchondral cyst formation (indicated by the arrow). (Reprinted with permission from Bonner KF, Daner WD, Yao JQ. 2-year postoperative evaluation of a patient with a symptomatic full-thickness patellar cartilage defect repaired with particulated juvenile cartilage tissue. J Knee Surg 2010;23:109–113.)

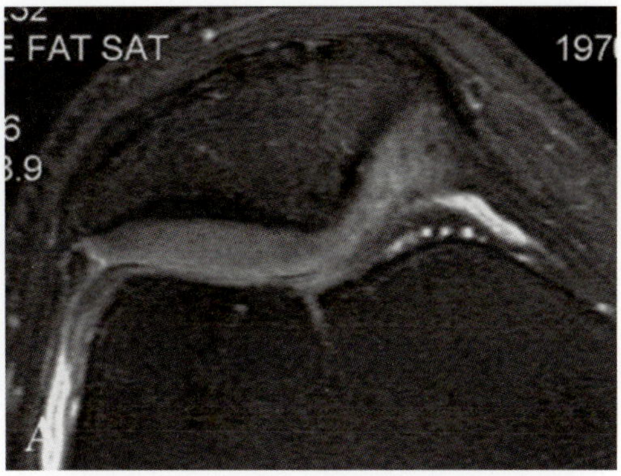

Fig. 9.3 Postoperative magnetic resonance imaging at 21 months reveals near-resolution of the bone edema and repair tissue within the previous defect site. (Reprinted with permission from Bonner KF, Daner WD, Yao JQ. 2-year postoperative evaluation of a patient with a symptomatic full-thickness patellar cartilage defect repaired with particulated juvenile cartilage tissue. J Knee Surg 2010;23:109–113.)

evaluate clinical outcomes such as IKDC, KOOS, and visual analog scale (VAS) scores, as well as extent and quality of repair with MRI and optional biopsies. To date, 25 patients with one or two chondral lesions on the femoral condyle or trochlea have been enrolled at three study sites. Four patients have completed 24 months of follow-up and their outcomes have been recently reported in *Cartilage*.[2] Detailed results of all 25 patients will be reported once they have all reached the 2-year postoperative follow-up milestone.

Of the four patients with 2-year follow-up, three had nontraumatic cartilage lesions and one had a traumatic cartilage injury. The average age was 43 ± 5.4 years and body mass index was 27 ± 5.8 lb/in^2. The average lesion size was 2.71 ± 1.2 cm^2. Two patients had isolated trochlear lesions, one had an isolated condylar lesion, and one had focal lesions of both the femoral condyle and the trochlea. KOOS, IKDC, and VAS scores demonstrate clear improvements in all scores across the 24-month follow-up period. Most of these improvements, especially in KOOS and VAS, were achieved at the 12-month mark and maintained throughout the study period. A representative patient MRI taken preoperatively at 12 months and at 24 months is shown in **Fig. 9.4**. This demonstrates good defect fill at 24 months postoperative.

Discussion

CAIS (DePuy/Mitek) and DeNovo NT (ISTO) are somewhat similar surgical procedures that involve the single-stage implantation of minced articular cartilage using either autograft or juvenile allograft, respectively. Several in vitro and in vivo models demonstrate the unique ability of both particulated autograft and juvenile allograft chondrocytes to escape from the extracellular matrix, migrate, and form new hyaline-like cartilage tissue that integrates with the surrounding host cartilage.[19–21] In short-term clinical studies, both procedures appear to be safe, feasible, and effective, with improvements in subjective patient scores and with MRI evidence of good defect fill.[2,3,26]

There are several potential advantages to these techniques. Both CAIS (DePuy/Mitek) and DeNovo NT (ISTO) do not require the violation of the subchondral bone, as is necessary for marrow stimulation procedures.

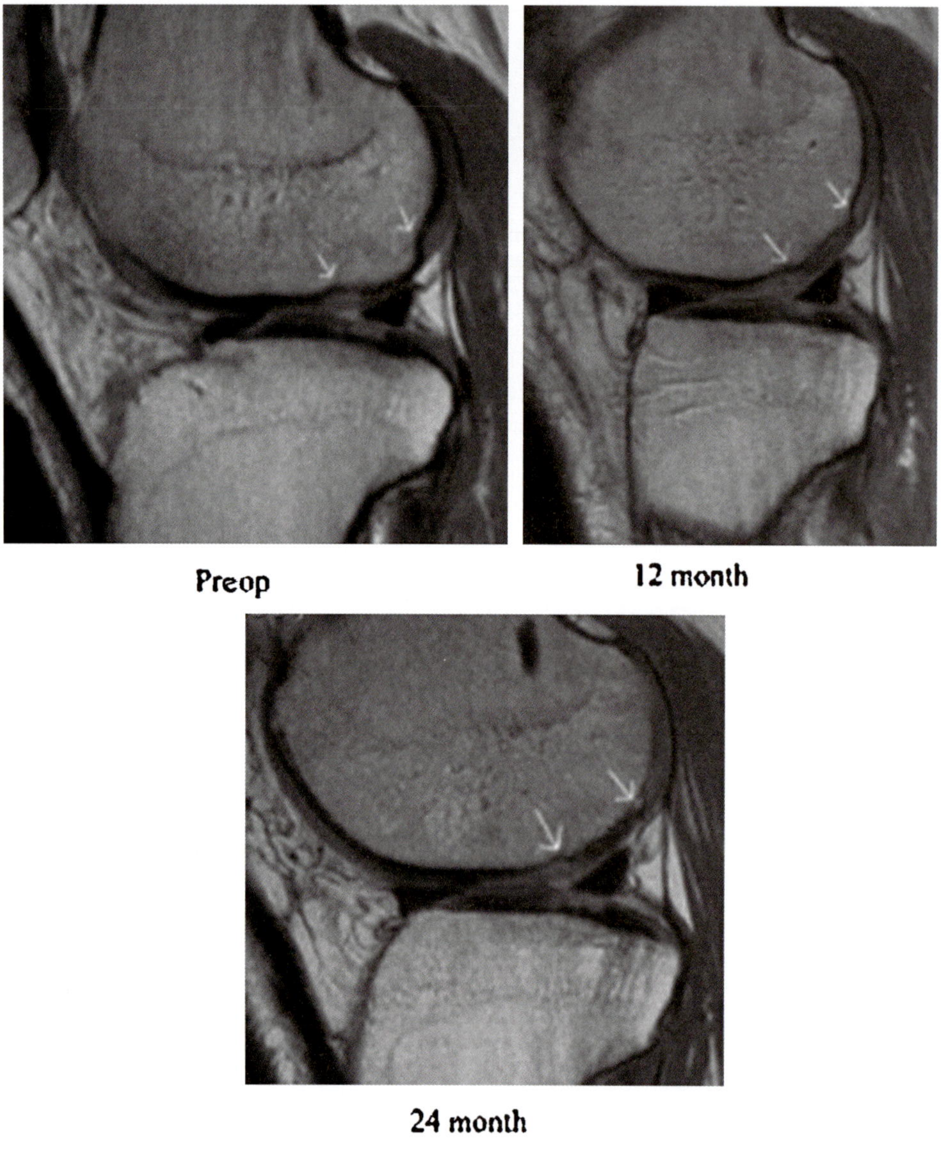

Fig. 9.4 Note that the sagittal plane is not identical from one image to the next secondary to slight positional differences of the knee in the MRI. The image plane that best demonstrated the lesion was chosen. (Reprinted with permission from Farr J, Yao JQ. Chondral defect repair with particulated juvenile cartilage allograft. Cartilage 2011;2(4):346–353.)

These truly represent a "burn no bridge" procedure, unlike reports that prior marrow stimulation may compromise subsequent revision surgeries.[27] Similarly, these procedures avoid the need to surgically create an osteochondral defect, as is necessary for osteochondral allograft transplantation. CAIS (DePuy/Mitek) and DeNovo NT (ISTO) use a strategy of cartilage–cartilage healing in the defect bed. This may help to avoid problems of bony healing as seen in failed osteochondral allograft procedures, including lack of bone incorporation, necrosis, and avascular necrosis–like collapse. Other potential advantages include (1) the use of fibrin fixation, which eliminates problems relating to flap hypertrophy, as seen with other techniques[28]; (2) CAIS (DePuy/Mitek) and DeNovo NT (ISTO) are single-stage procedures unlike techniques such as ACI; (3) DeNovo NT (ISTO) lacks any autogenous donor site morbidity; (4) the autograft tissue portion of CAIS (DePuy/Mitek) is obviously without charge (as compared with cultured cells or allograft).

A disadvantage specific to CAIS (DePuy/Mitek) is the potential for donor site morbidity at cell harvest. This risk is minimal and is, in theory, similar to the risk involved in ACI harvest. Potential disadvantages specific to DeNovo NT (ISTO) include the theoretical risk of disease transmission and/or immunological rejection, which is inherent to any allograft procedure. The risk of disease transmission is extremely low in allograft procedures due to stringent donor requirements by the FDA and standard allograft screening tests to ensure tissue safety.[7] Thirty years of cumulative knowledge has similarly shown that immune rejection is an extremely rare phenomenon with osteochondral allograft transplantation. No immune responses have been reported to the cartilage component of osteochondral allografts.[2] In addition, articular cartilage has been shown to be immune privileged, partly due to a lack of vascularity and the dense extracellular matrix of the tissue.[29]

Other clinicians have tried to treat chondral defects of the knee with particulated chondral or osteochondral tissue from mature donors. In particular, Stone et al implanted a paste of autologous osteochondral tissue into defects concomitantly treated with microfracture.[30] While good clinical results were reported, several animal studies have shown that a combination bone and cartilage paste forms both bone and cartilage, whereas cartilage pieces alone formed cartilage.[19,21,30] We believe that CAIS (DePuy/Mitek) and DeNovo NT (ISTO) potentially may improve upon the paste-grafting concept by using a homogeneous cartilage-only approach and by avoiding concomitant microfracture.

Despite the obvious limitations of short-term outcomes, the results of both CAIS (DePuy/Mitek) and DeNovo NT (ISTO) compare favorably to other procedures for similar cartilage lesions. Cole et al demonstrated that CAIS (DePuy/Mitek) is safe to use, with risks comparable to those of MFX. In that study, CAIS (DePuy/Mitek) patients had consistent and progressive improvement during the second year after surgery, when compared with the MFX group.[3] Similarly, the preliminary results of DeNovo NT (ISTO) compare favorably with 2-year postoperative KOOS pain scores for ACI and microfracture, and with IKDC subjective scores for ACI.[8,16,31] Comparison of MRI results from CAIS (DePuy/Mitek) and MFX patients suggests a difference in the biologic repair process. MRI from DeNovo NT (ISTO) patients also demonstrates good lesion fill at early follow-up.[2,8,10,26] Future study will require sophisticated imaging or second-look biopsies to determine whether the quality and quantity of hyaline-like fill correlates with subjective and objective clinical outcomes.

Conclusion

CAIS (DePuy/Mitek) and DeNovo NT (ISTO) appear to be promising new treatment options for the young patient with a symptomatic focal chondral defect in the knee. Further study is needed before there are evidence-based recommendations. Prospective randomized controlled studies will certainly help to refine the indications and contraindications for both CAIS (DePuy/Mitek) and DeNovo NT (ISTO).

References

1. Chondral Defect Repair with Particulated Juvenile Cartilage Allograft. Updated 2010; available at: http://www.zimmer.com/web/enUS/pdf/DeNovo_Chondral_Defect_Repair_White_Paper_05_2010.pdf
2. Farr J, Yao JQ. Chondral defect repair with particulated juvenile cartilage allograft. Cartilage 2011;2(4):346–353
3. Cole BJ, Farr J, Winalski CS, et al. Outcomes after a single-stage procedure for cell-based cartilage repair: a prospective clinical safety trial with 2-year follow-up. Am J Sports Med 2011;39(6):1170–1179
4. Hjelle K, Solheim E, Strand T, Muri R, Brittberg M. Articular cartilage defects in 1,000 knee arthroscopies. Arthroscopy 2002;18(7):730–734
5. Maletius W, Messner K. The effect of partial meniscectomy on the long-term prognosis of knees with localized, severe chondral damage. A twelve- to fifteen-year followup. Am J Sports Med 1996;24(3):258–262

6. Saris DBF, Vanlauwe J, Victor J, et al; TIG/ACT/01/2000&EXT Study Group. Treatment of symptomatic cartilage defects of the knee: characterized chondrocyte implantation results in better clinical outcome at 36 months in a randomized trial compared to microfracture. Am J Sports Med 2009;37(Suppl 1):10S–19S
7. Brittberg M, Lindahl A, Nilsson A, Ohlsson C, Isaksson O, Peterson L. Treatment of deep cartilage defects in the knee with autologous chondrocyte transplantation. N Engl J Med 1994;331(14):889–895
8. Zaslav K, Cole B, Brewster R, et al; STAR Study Principal Investigators. A prospective study of autologous chondrocyte implantation in patients with failed prior treatment for articular cartilage defect of the knee: results of the Study of the Treatment of Articular Repair (STAR) clinical trial. Am J Sports Med 2009;37(1):42–55
9. Gross AE, Shasha N, Aubin P. Long-term followup of the use of fresh osteochondral allografts for posttraumatic knee defects. Clin Orthop Relat Res 2005;(435):79–87
10. Hangody L, Kish G, Kárpáti Z, Szerb I, Udvarhelyi I. Arthroscopic autogenous osteochondral mosaicplasty for the treatment of femoral condylar articular defects. A preliminary report. Knee Surg Sports Traumatol Arthrosc 1997;5(4):262–267
11. Moseley JB, O'Malley K, Petersen NJ, et al. A controlled trial of arthroscopic surgery for osteoarthritis of the knee. N Engl J Med 2002;347(2):81–88
12. Steadman JR, Rodkey WG, Singleton SB, Briggs KK. Microfracture technique for full-thickness chondral defects: technique and clinical results. Oper Tech Orthop 1997;7:300–304
13. Farr J. DeNovo NT natural graft tissue. Poster presented at: 8th World Congress of the International Cartilage Repair Society 2009; Miami, FL
14. Asik M, Ciftci F, Sen C, Erdil M, Atalar A. The microfracture technique for the treatment of full-thickness articular cartilage lesions of the knee: midterm results. Arthroscopy 2008;24(11):1214–1220
15. Cole BJ, Pascual-Garrido C, Grumet RC. Surgical management of articular cartilage defects in the knee. J Bone Joint Surg Am 2009;91(7):1778–1790
16. Saris DBF, Vanlauwe J, Victor J, et al. Characterized chondrocyte implantation results in better structural repair when treating symptomatic cartilage defects of the knee in a randomized controlled trial versus microfracture. Am J Sports Med 2008;36(2):235–246
17. Knutsen G, Drogset JO, Engebretsen L, et al. A randomized trial comparing autologous chondrocyte implantation with microfracture. Findings at five years. J Bone Joint Surg Am 2007;89(10):2105–2112
18. Knutsen G, Engebretsen L, Ludvigsen TC, et al. Autologous chondrocyte implantation compared with microfracture in the knee. A randomized trial. J Bone Joint Surg Am 2004;86-A(3):455–464
19. Albrecht F, Roessner A, Zimmermann E. Closure of osteochondral lesions using chondral fragments and fibrin adhesive. Arch Orthop Trauma Surg 1983;101(3):213–217
20. Frisbie DD, Lu Y, Kawcak CE, DiCarlo EF, Binette F, McIlwraith CW. In vivo evaluation of autologous cartilage fragment-loaded scaffolds implanted into equine articular defects and compared with autologous chondrocyte implantation. Am J Sports Med 2009;37(Suppl 1):71S–80S
21. Lu Y, Dhanaraj S, Wang Z, et al. Minced cartilage without cell culture serves as an effective intraoperative cell source for cartilage repair. J Orthop Res 2006;24(6):1261–1270
22. Cheung HS, Cottrell WH, Stephenson K, Nimni ME. In vitro collagen biosynthesis in healing and normal rabbit articular cartilage. J Bone Joint Surg Am 1978;60(8):1076–1081
23. Adkisson HD IV, Martin JA, Amendola RL, et al. The potential of human allogeneic juvenile chondrocytes for restoration of articular cartilage. Am J Sports Med 2010;38(7):1324–1333
24. Stockwell RA. The interrelationship of cell density and cartilage thickness in mammalian articular cartilage. J Anat 1971;109(Pt 3):411–421
25. Namba RS, Meuli M, Sullivan KM, Le AX, Adzick NS. Spontaneous repair of superficial defects in articular cartilage in a fetal lamb model. J Bone Joint Surg Am 1998;80(1):4–10
26. Bonner KF, Daner W, Yao JQ. 2-year postoperative evaluation of a patient with a symptomatic full-thickness patellar cartilage defect repaired with particulated juvenile cartilage tissue. J Knee Surg 2010;23(2):109–114
27. Minas T, Gomoll AH, Rosenberger R, Royce RO, Bryant T. Increased failure rate of autologous chondrocyte implantation after previous treatment with marrow stimulation techniques. Am J Sports Med 2009;37(5):902–908
28. Gomoll AH, Probst C, Farr J, Cole BJ, Minas T. Use of a type I/III bilayer collagen membrane decreases reoperation rates for symptomatic hypertrophy after autologous chondrocyte implantation. Am J Sports Med 2009;37(Suppl 1):20S–23S
29. Gibson T, Brian Davis W, Curran R. The long-term survival of cartilage homografts in man. Br J Plast Surg 1959;11:177–187
30. Stone KR, Walgenbach AW, Freyer A, Turek TJ, Speer DP. Articular cartilage paste grafting to full-thickness articular cartilage knee joint lesions: a 2- to 12-year follow-up. Arthroscopy 2006;22(3):291–299
31. Crawford DC, Heveran CM, Cannon WD Jr, Foo LF, Potter HG. An autologous cartilage tissue implant NeoCart for treatment of grade III chondral injury to the distal femur: prospective clinical safety trial at 2 years. Am J Sports Med 2009;37(7):1334–1343

10

Nontraditional Modification to Articular Cartilage

Vasili Karas, Neil Ghodadra, Ellen Kroin, and Brian J. Cole

Biomechanical imbalance, trauma, and age-related degeneration often lead to chondral lesions, which may lead to overt osteoarthritis over time. Such cartilage pathology is frequently accompanied by persistent pain and loss of normal joint function. As a result, patients who suffer from biologically active articular cartilage lesions are often unable to function in high-level activities and may exhibit compromised activities of daily living. The limited potential for self-regeneration of hyaline cartilage has led to the emergence of new technologies to solve this difficult clinical problem. In the event that the chondral lesion remains superficial to the subchondral bone, repair relies on the proliferation of surrounding cells and cells within the synovium as lesions are not exposed to the cellular and protein components of circulating blood. Lesions that include the subchondral bone and expose the marrow cavity rely on components therein for regeneration and repair. Cartilage synthesized without exogenous intervention usually resembles type I fibrous cartilage, with inferior biomechanical properties when compared with native, hyaline cartilage replete with type II collagen.[1]

Treatment of arthritis and chondral lesions includes alleviation of pain and return of function through pharmacologic intervention and/or attempts at cartilage reparative, restorative, and reconstructive options.[2] Systemic pharmacologic treatments for degenerative arthritis aim to reduce inflammation and decrease associated pain. Topical treatments include nonsteroidal anti-inflammatories such as diclofenac gels that isolate the pathologic joint, localizing treatment and decreasing the possibility of systemic side effects. Traditional injectables such as cortisone injections and viscosupplementation have been found to decrease pain for short and medium time periods. Corticosteroids have been shown to provide a 30 to 50% decrease in pain that is most evident in the first 4 weeks after treatment.[3] Viscosupplementation with various formulations and molecular weights of hyaluronic acid has been shown to impart similar but longer-lasting results.[4]

It is our purpose to discuss the nontraditional and innovative nonsurgical treatments for articular cartilage pathology. Weight loss, physical therapy, oral anti-inflammatories, and corticosteroids are, at present, the standard of care for conservative treatment modalities for arthritis. The use of biologic injectables such as growth factors, platelet-rich plasma (PRP), autologous conditioned serum (ACS), and stem cell therapy is currently under investigation and will be the present focus. Although the clinical evidence supporting the use of these modalities is sparse, their potential is clear, as is the need for their continued development.

Growth Factors and Cytokines

Osteoarthritis is largely a cytokine-driven disease process. The synovial membrane, cartilage, and subchondral bone are all potential factors in cartilage degeneration as each is capable of producing large amounts of cytokines. A thorough understanding of the clinically relevant interactions between cytokines, mediators, growth factors, and mechanisms of action in this local environment is needed to ameliorate cartilage degeneration caused by the catabolic milieu present in osteoarthritis. Accompanying the increased interest in nontraditional treatment methods for articular cartilage disease is an increased interest in the use of cytokines as the basis for biological treatments such as PRP and ACS.

Growth factors are commonly defined as biologically active polypeptides that contribute to the regulation of growth and homeostasis of tissues throughout life.[5,6] The use of growth factors such as transforming growth factor (TGF), fibroblast growth factor (FGF), and bone morphogenic (BMP) to influence cell differentiation and anabolism is a possible solution in the context of osteoarthritis.[7–9] Recent basic science studies have shown an increasingly important role for growth factors in cartilage regeneration and have become the basis for the potential clinical benefits of modification of articular cartilage.[9,10]

TGF-β1 has been shown, in vitro, to stimulate the synthesis of extracellular matrix within cartilage, induce synovial proliferation, and increase mesenchymal stem cell (MSC) proliferation.[11–14] Positive effects of TGF-β1 have also been documented in cartilage defects within rabbit models.[15–19] Despite the positive effects of TGF-β1, safety concerns, specifically the presence of osteophytes and synovial fibrosis in murine and lapine studies, have limited extensive human testing.[14,20] Albeit on a smaller scale, compared with TGF-β1, TGF-β3 has been shown to stimulate extracellular matrix (ECM) formation in animal models[21–23] without these adverse effects.

BMP-2 is a close structural relative to both TGF-β1 and TGF-β3 and has been studied extensively in fracture care and spine surgery. The clinical success of BMP-2 in orthopedics has spurred basic science research investigating its potential effect on cartilage regeneration. In multiple studies, it has been shown in vitro to partially reverse dedifferentiated chondrocytes found in osteoarthritic models.[24] In addition, BMP-2 stimulates the synthesis and turnover of extracellular matrix, and specifically that of proteoglycans and type II collagen. Augmentation of a microfracture model with BMP-2 has also been reported in a rabbit model. Although surgical intervention is beyond the present scope, it is valuable to note that BMP-2 may guide differentiating cells to produce more hyaline-like cartilage.[25–27] Although the effects of BMP-2 on chondrocyte metabolism seem promising, synovial thickening, fibrosis, and, in some cases, osteophytes have been shown to develop after multiple injections.[28] In addition, a recent animal study suggests temporal limitations to the use of BMP-2.[29] Although the efficacy of BMP-2 seems promising, further studies are needed to develop the most efficacious dosing, timing, and route of administration.

BMP-7/OP-1 is the most investigated member of the TGF-β superfamily for its potential to regenerate articular cartilage. Not only does BMP-7 increase ECM synthesis, it decreases the activity of catabolic cytokines such as interleukin (IL-1), IL-6, IL-8, matrix matalloproteinase-1 (MMP-1), and MMP-7.[30] BMP-7 expression has been shown to decrease with age. Although decreased BMP-7 expression is a factor in cartilage breakdown, BMP-7 continues to have autocrine effects for both anabolism and catabolism.[31–34] Finally, although basic science studies suggest a beneficial effect from the administration of BMP-7, recent basic science and clinical literature has not shown a trend between endogenous levels of BMP-7 and higher symptomatic pain relief in patients with osteoarthritis.[35] The efficacy of BMP-7 seems to be clear; however, the need to develop the proper dosing, timing, and route of administration remains uncertain.

Insulin-like growth factor-I (IGF-1) has been investigated within the context of cartilage metabolism in both native and pathologic states.[30,36–39] IGF-1 has been shown to increase the anabolic response and decrease catabolism.[40] In contrast to evidence found in

BMP-7, IGF-1 shows a decreased responsiveness in aging and osteoarthritic cartilage.[41,42] Although IGF-1 may not be a viable option alone, it may offer a synergistic effect in conjunction with other growth factors.[36] Further studies are necessary to determine the optimal combination of growth factors.

Recent evidence suggests that platelet-derived growth factor (PDGF) has a possible place in cartilage repair based on its role in wound healing and stimulation of ECM proliferation in bone growth.[43–46] Multiple animal studies have shown that PDGF has an excellent safety profile when used in isolation. PDGF has had an increasingly prominent role in research and media as in vivo use of PDGF remains largely within the context of PRP. PRP has been used successfully in various clinical situations and has drawn national attention as it has shown promising results for tendon healing.

Blood-Derived Products

Although growth factors show promise, they must be carefully synthesized and stored and are thus very expensive to produce. As evidenced above, they may also have a synergistic effect and would thus require varied concentrations of multiple growth factors, a practice that is not sanctioned by the U.S. Food and Drug Administration. Thus, there has been a recent resurgence in interest in the use of the body's own combination of growth factors and cytokines using autologous blood as a medium from which to extract growth factor and cytokine-containing components such as platelets.

Autologous Conditioned Serum

Autologous conditioned serum (ACS) was developed in the mid-1990s and marketed under the name Orthokine (Arthrex, Inc., Naples, FL). It has been reported not only to be beneficial in the treatment of osteoarthritis, but also to be beneficial in rheumatoid arthritis, spinal disorders, and muscle injuries in humans.[47–51] To prepare an ACS injection, human whole blood from the patient is incubated with medical-grade glass beads or spheres, exposed to chromium sulfate, and placed into a centrifuge to separate into the plasma with platelets.[48] ACS is believed to be effective through its increased concentrations of cytokines and growth factors. Multiple studies have shown that the expression of IL-4, IL-10, IL-1Ra (receptor antagonist), fibroblastic growth factor-1, hepatocyte growth factor, and TGF-β1 are increased in human ACS. While there is an increase in these anti-inflammatory agents, there is no increase in proinflammatory cytokines-like IL-1β or TNF-α.[47]

In particular, IL-1Ra expression has been shown to increase as much as 140-fold in ACS. IL-1Ra is a competitive receptor antagonist of IL-1, a proinflammatory cytokine that triggers the destruction of hyaline cartilage and its matrix.[48] Thus IL-1Ra may play a role in the clinical improvement of osteoarthritis patients injected with ACS. IL-1 has also been identified as being the major mediator of cartilage loss in osteoarthritis. Currently, it is not clear if all biologically active IL-1 receptors need to be blocked to have a significant impact on treating conditions such as osteoarthritis; however, it is known that other anti-inflammatory cytokines that are expressed in ACS also affect IL-1 receptor signaling.[48] In gene therapy studies, it was found that IL-1Ra decreases synovial effusion, gross articular cartilage erosion, and synovial membrane vascularity as compared with placebo-treated joints.[47]

To induce the de novo production of IL-1Ra, aspirated venous blood is incubated with borosilicate glass spheres in a syringe. The anti-inflammatory cytokines, which are produced by peripheral blood leukocytes, accumulate and are recovered within the serum. The cytokine concentrations do vary between individual samples, and their synergistic action contributes to the effects.[48] After centrifugation, ACS can be injected into the osteoarthritic area in a series of six intraarticular injections twice a week for 3 weeks.[47,52]

Platelet Rich Plasma

The contemporary definition of PRP is a sample of plasma with a two-fold or more increase in platelet concentration or greater than 1.1 × 10^6 platelets/μL.[53] Presently, several different manufacturers have developed systems for PRP preparation for augmentation or as primary orthopedic treatments.[54] It is important to understand that preparations differ across manufacturers in final platelet count, presence of leukocytes and number of centrifugations for preparation.

The concept of PRP as a possible treatment for osteoarthritis derives from the platelet's role in wound healing[55] as platelets contain many of the cytokines and growth factors delineated above. In addition, platelets contain approximately another 1,500 proteins, some of which modulate the inflammatory response inherent in degenerative joint disease as well as the attraction of fibroblasts and stem cells to the site of injury.[56,57]

The use of PRP and its reported clinical success in treating various tendon pathologies throughout the body has led to increased interest in its potential role in cartilage repair. The use of PRP as treatment for articular cartilage repair is new, and thus there are sparse data on the clinical outcomes of its use. In the laboratory, injected PRP has been shown to increase production of chondrocytes and MSCs, leading to increased proliferation and synthesis of ECM, collagen II, and proteoglycans.[58,59] In animal models, damaged cartilage treated with PRP also demonstrated higher degree of degeneration when compared with control. In a recent trial of hyaluronic acid (HA) versus PRP for the treatment of osteoarthritis, Kon et al[60] compared the two treatment modalities over a 6-month time period to evaluate patient-reported outcomes. The study concluded that three weekly injections of autologous PRP, when compared with a series of three HA injections, showed more and longer efficacy in mitigating the symptoms of osteoarthritis. It also concluded that younger, more active patients with presumably a lesser degree of cartilage degeneration improve to a higher degree with PRP injections as compared with HA. These results are the most promising to date. However, a randomized controlled trial with more objective outcomes is needed to shed more definitive light on PRP as a treatment for cartilage degeneration.[60] A recent study reported a decrease in pain and an increase in function by the time patients reached 24 months.[61]

Stem Cell Therapy

Stem cell therapy serves as another possible method of treatment of articular cartilage defects. Not only do MSCs have the ability to self-renew but they possess the potential to differentiate into other specialized cells when placed in appropriate culture conditions.[62] For the purpose of treating cartilage defects, MSCs need to be differentiated toward chondrogenic lineage of cells and more specifically toward the formation of hyaline type II cartilage. Aside from MSC differentiation properties, they also have a trophic activity and secrete bioactive factors that have a protective immunoregulatory effect on the local tissue environment. Their anti-inflammatory and differentiation properties make MSCs good contenders for a possible tissue repair modality in osteoarthritis.[63]

Synovial membrane–derived, bone marrow–derived MSCs (BMSCs), and adipose-derived stem cells (ASCs) from adult tissues have the potential to form a hyaline-like cartilage matrix, with the latter being a more abundant and minimally morbid source (Hildner).[62] For example, recent studies suggest that the infrapatellar fat pad of adult knees is a good source of cells that can be induced to differentiate into chondrocytes that synthesize cartilage matrix molecules.[63] BMSCs and ASCs require a different growth factor treatment to differentiate into the sought-after material. Aggrecan upregulation in ASCs is seen when treated with BMP-6, while in BMSCs, TGF-β3 is needed instead. In addition, several studies have concluded that BMSCs are more easily differentiated toward the chondrogenic lineage than ASCs.[62,63]

Park et al showed that MSCs from both bone marrow and periosteum formed hyaline cartilaginous tissue when transplanted into cartilage defects in rats. This study also demonstrated that MSCs derived from bone marrow were superior to adipose-derived MSCs in forming hyaline cartilage in vivo.[64] Bone marrow, synovium, adipose tissue, and muscle of adult rabbits have also been studied to compare their in vivo chondrogenic potential. Results have shown that the potential of synovial and bone marrow MSCs to repair cartilage defects is higher than those from skeletal muscle and adipose tissue, and they produced more cartilage matrix than the other cells in the cartilage defects. More specifically, the MSCs taken from the synovial tissue had the greatest proliferation potential.[65]

Wakitani et al performed a clinical study using BMSCs resuspended in a collagen type I gel and transplanted with an autologous periosteal flap. This cell-containing scaffold was placed into osteoarthritic cartilage defects in the patients' medial femoral condyles. This was compared with patients who were transplanted with a cell-free scaffold in a similar defect.[66] Results showed that the cell-treated group's clinical scores were not significantly different 64 months after transplantation compared with the control group. In this situation, longer observation might be required and/or MSC transplantation may not be as effective in an osteoarthritic knee environment.[66] However, both the arthroscopic and histological scores were better in the group treated with the MSC transplant.[66] In addition, three case reports from the same group did report that the clinical symptoms in the patients with the MSC transplant had improved.[65]

In another clinical study, human bone marrow MSCs were used to treat a 20 × 30-mm full-thickness cartilage medial femoral condyle defect in a 31-year-old male athlete. Bone marrow was aspirated from the patient 4 weeks before surgery, and the cells were expanded in culture and then covered with an autologous periosteal flap once transferred to the defect. In a 7-month postsurgical evaluation, the defect was covered with a hyaline-like type of cartilage tissue. This smooth tissue also stained positively with Safranin-O. Clinically, the symptoms resolved significantly and the patient was able to return to full physical activity level with no pain or complications.[65,67]

◆ Conclusion

Since articular cartilage defects have limited intrinsic regenerative properties, there is an interest in providing nontraditional modifications to injured articular cartilage in patients. To explore methods to repair articular cartilage, transplantation of various progenitor cells other than chondrocytes is under investigation with renewed vigor to provide additional solutions to articular cartilage repair. Recent basic science and clinical research have initiated a paradigm shift in our understanding of the role of cytokines, growth factors, and stem cells in potential cartilage repair. Although results have been promising in animal studies, extensive human clinical studies are necessary to ascertain the benefit of the use of growth factors or blood-derived products to repair articular cartilage defects.

References

1. Nehrer S, Breinan HA, Ramappa A, et al. Matrix collagen type and pore size influence behaviour of seeded canine chondrocytes. Biomaterials 1997;18(11):769–776
2. McNickle AG, Provencher MT, Cole BJ. Overview of existing cartilage repair technology. Sports Med Arthrosc 2008;16(4):196–201
3. Bellamy N, Campbell J, Gee T, Robinson V, Bourne R, Wells G. Efficacy of intra-articular corticosteroid treatment in knee osteoarthritis. Cochrane Review 2006;64:494–494
4. Gigante A, Callegari L. The role of intra-articular hyaluronan (Sinovial) in the treatment of osteoarthritis. Rheumatol Int 2011;31(4):427–444
5. Quintana L, zur Nieden NI, Semino CE. Morphogenetic and regulatory mechanisms during developmental chondrogenesis: new paradigms for cartilage tissue engineering. Tissue Eng Part B Rev 2009;15(1):29–41
6. Gaissmaier C, Koh JL, Weise K. Growth and differentiation factors for cartilage healing and repair. Injury 2008;39(Suppl 1):S88–S96
7. Hill DJ, Logan A. Peptide growth factors and their interactions during chondrogenesis. Prog Growth Factor Res 1992;4(1):45–68
8. Hoffmann A, Gross G. BMP signaling pathways in cartilage and bone formation. Crit Rev Eukaryot Gene Expr 2001;11(1-3):23–45
9. Plaas A, Velasco J, Gorski DJ, et al. The relationship between fibrogenic TGFβ1 signaling in the joint and cartilage degradation in post-injury osteoarthritis. Osteoarthritis Cartilage 2011;19 (9):1081–1090
10. Puetzer JL, Petitte JN, Loboa EG. Comparative review of growth factors for induction of three-dimensional in vitro chondrogenesis in human mesenchymal stem cells isolated from bone marrow and adipose tissue. Tissue Eng Part B Rev 2010;16(4):435–444
11. Dickhut A, Dexheimer V, Martin K, Lauinger R, Heisel C, Richter W. Chondrogenesis of human mesenchymal stem cells by local transforming growth factor-beta delivery in a biphasic resorbable carrier. Tissue Eng Part A 2010;16(2):453–464
12. Fan H, Hu Y, Li X, et al. Ectopic cartilage formation induced by mesenchymal stem cells on porous gelatin-chondroitin-hyaluronate scaffold containing microspheres loaded with TGF-beta1. Int J Artif Organs 2006;29(6):602–611

13. Re'em T, Kaminer-Israeli Y, Ruvinov E, Cohen S. Chondrogenesis of hMSC in affinity-bound TGF-beta scaffolds. Biomaterials 2012;33(3):751–761
14. Blaney Davidson EN, van der Kraan PM, van den Berg WB. TGF-beta and osteoarthritis. Osteoarthritis Cartilage 2007;15(6):597–604
15. Wang W, Li B, Yang J, et al. The restoration of full-thickness cartilage defects with BMSCs and TGF-beta 1 loaded PLGA/fibrin gel constructs. Biomaterials 2010;31(34):8964–8973
16. Olivos-Meza A, Fitzsimmons JS, Casper ME, et al. Pretreatment of periosteum with TGF-beta1 in situ enhances the quality of osteochondral tissue regenerated from transplanted periosteal grafts in adult rabbits. Osteoarthritis Cartilage 2010;18(9):1183–1191
17. Guo X, Park H, Young S et al. Repair of osteochondral defects with biodegradable hydrogel composites encapsulating marrow mesenchymal stem cells in a rabbit model. Acta Biomater 2010;6(1):39–47
18. Fan H, Liu H, Zhu R, et al. Comparison of chondral defects repair with in vitro and in vivo differentiated mesenchymal stem cells. Cell Transplant 2007;16(8):823–832
19. Guo X, Zheng Q, Yang S, et al. Repair of full-thickness articular cartilage defects by cultured mesenchymal stem cells transfected with the transforming growth factor beta1 gene. Biomed Mater 2006;1(4):206–215
20. Bakker AC, van de Loo FA, van Beuningen HM, et al. Overexpression of active TGF-beta-1 in the murine knee joint: evidence for synovial-layer-dependent chondro-osteophyte formation. Osteoarthritis Cartilage 2001;9(2):128–136
21. Tang QO, Shakib K, Heliotis M, et al. TGF-beta3: A potential biological therapy for enhancing chondrogenesis. Expert Opin Biol Ther 2009;9(6):689–701
22. Fan H, Tao H, Wu Y, Hu Y, Yan Y, Luo Z. TGF-β3 immobilized PLGA-gelatin/chondroitin sulfate/hyaluronic acid hybrid scaffold for cartilage regeneration. J Biomed Mater Res A 2010;95(4):982–992
23. Lee CH, Cook JL, Mendelson A, Moioli EK, Yao H, Mao JJ. Regeneration of the articular surface of the rabbit synovial joint by cell homing: a proof of concept study. Lancet 2010;376(9739):440–448
24. Gouttenoire J, Valcourt U, Ronzière MC, Aubert-Foucher E, Mallein-Gerin F, Herbage D. Modulation of collagen synthesis in normal and osteoarthritic cartilage. Biorheology 2004;41(3–4):535–542
25. Claus S, Aubert-Foucher E, Demoor M, et al. Chronic exposure of bone morphogenetic protein-2 favors chondrogenic expression in human articular chondrocytes amplified in monolayer cultures. J Cell Biochem 2010;111(6):1642–1651
26. Lee CH, Cook JL, Mendelson A, Moioli EK, Yao H, Mao JJ. Regeneration of the articular surface of the rabbit synovial joint by cell homing: a proof of concept study. Lancet 2010;376(9739):440–448
27. Toh WS, Liu H, Heng BC, Rufaihah AJ, Ye CP, Cao T. Combined effects of TGFbeta1 and BMP2 in serum-free chondrogenic differentiation of mesenchymal stem cells induced hyaline-like cartilage formation. Growth Factors 2005;23(4):313–321
28. Postlethwaite AE, Raghow R, Stricklin G, Ballou L, Sampath TK. Osteogenic protein-1, a bone morphogenic protein member of the TGF-β superfamily, shares chemotactic but not fibrogenic properties with TGF-β. J Cell Physiol 1994;161(3):562–570
29. Krawczak DA, Westendorf JJ, Carlson CS, Lewis JL. Influence of bone morphogenetic protein-2 on the extracellular matrix, material properties, and gene expression of long-term articular chondrocyte cultures: loss of chondrocyte stability. Tissue Eng Part A 2009;15(6):1247–1255
30. Elshaier AM, Hakimiyan AA, Rappoport L, Rueger DC, Chubinskaya S. Effect of interleukin-1beta on osteogenic protein 1-induced signaling in adult human articular chondrocytes. Arthritis Rheum 2009;60(1):143–154
31. Chubinskaya S, Hurtig M, Rueger DC. OP-1/BMP-7 in cartilage repair. Int Orthop 2007;31(6):773–781
32. Gavenis K, Schneider U, Wallich R, Mueller-Rath R, Schmidt-Rohlfing B, Andereya S. Effects of low concentrated BMP-7 administered by co-cultivation or plasmid transfection on human osteoarthritic chondrocytes. Int J Artif Organs 2010;33(6):339–347
33. Hunter DJ, Pike MC, Jonas BL, Kissin E, Krop J, McAlindon T. Phase 1 safety and tolerability study of BMP-7 in symptomatic knee osteoarthritis. BMC Musculoskelet Disord 2010;11:232
34. Hunter DJ, Pike MC, Jonas BL, Kissin E, Krop J, McAlindon T. Phase 1 safety and tolerability study of BMP-7 in symptomatic knee osteoarthritis. BMC Musculoskelet Disord 2010;11:232
35. Schmal H, Niemeyer P, Zwingmann J, Stoffel F, Südkamp NP, Mehlhorn AT. Association between expression of the bone morphogenetic proteins 2 and 7 in the repair of circumscribed cartilage lesions with clinical outcome. BMC Musculoskelet Disord 2010;11:170
36. Chubinskaya S, Hakimiyan A, Pacione C, et al. Synergistic effect of IGF-1 and OP-1 on matrix formation by normal and OA chondrocytes cultured in alginate beads. Osteoarthritis Cartilage 2007;15(4):421–430
37. Busschers E, Holt JP, Richardson DW. Effects of glucocorticoids and interleukin-1 beta on expression and activity of aggrecanases in equine chondrocytes. Am J Vet Res 2010;71(2):176–185
38. Allen KD, Adams SB Jr, Mata BA, et al. Gait and behavior in an IL1β-mediated model of rat knee arthritis and effects of an IL1 antagonist. J Orthop Res 2011;29(5):694–703
39. Ley C, Svala E, Nilton A, et al. Effects of high mobility group box protein-1, interleukin-1β, and interleukin-6 on cartilage matrix metabolism in three-dimensional equine chondrocyte cultures. Connect Tissue Res 2011;52(4):290–300
40. Davies LC, Blain EJ, Gilbert SJ, Caterson B, Duance VC. The potential of IGF-1 and TGFbeta1 for promoting "adult" articular cartilage repair: an in vitro study. Tissue Eng Part A 2008;14(7):1251–1261
41. Kapoor M, Martel-Pelletier J, Lajeunesse D, Pelletier JP, Fahmi H. Role of proinflammatory cytokines in the pathophysiology of osteoarthritis. Nat Rev Rheumatol 2011;7(1):33–42
42. Fortier LA, Miller BJ. Signaling through the small G-protein Cdc42 is involved in insulin-like growth factor-I resistance in aging articular chondrocytes. J Orthop Res 2006;24(8):1765–1772
43. Eli N, Oragui E, Khan W. Advances in meniscal tissue engineering. Ortop Traumatol Rehabil 2011;13(4):319–326
44. Brandl A, Angele P, Roll C, Prantl L, Kujat R, Kinner B. Influence of the growth factors PDGF-BB,

TGF-beta1 and bFGF on the replicative aging of human articular chondrocytes during in vitro expansion. J Orthop Res 2010;28(3):354–360

45. Gomez-Camarillo MA, Almonte-Becerril M, Vasquez Tort M, Tapia-Ramirez J, Kouri Flores JB. Chondrocyte proliferation in a new culture system. Cell Prolif 2009;42(2):207–218

46. Mishima Y, Lotz M. Chemotaxis of human articular chondrocytes and mesenchymal stem cells. J Orthop Res 2008;26(10):1407–1412

47. Frisbie DD, Kawcak CE, Werpy NM, Park RD, McIlwraith CW. Clinical, biochemical, and histologic effects of intra-articular administration of autologous conditioned serum in horses with experimentally induced osteoarthritis. Am J Vet Res 2007;68(3):290–296

48. Wehling P, Moser C, Frisbie D, et al. Autologous conditioned serum in the treatment of orthopedic diseases: the orthokine therapy. BioDrugs 2007;21(5):323–332

49. Milano G, Deriu L, Sanna Passino E, et al. The effect of autologous conditioned plasma on the treatment of focal chondral defects of the knee. An experimental study. Int J Immunopathol Pharmacol 2011;24(1, Suppl 2):117–124

50. Fortier LA, Barker JU, Strauss EJ, McCarrel TM, Cole BJ. The role of growth factors in cartilage repair. Clin Orthop Relat Res 2011;469(10):2706–2715

51. Georg R, Maria C, Gisela A, Bianca C. Autologous conditioned plasma as therapy of tendon and ligament lesions in seven horses. J Vet Sci 2010;11(2):173–175

52. Baltzer AW, Moser C, Jansen SA, Krauspe R. Autologous conditioned serum (Orthokine) is an effective treatment for knee osteoarthritis. Osteoarthritis Cartilage 2009;17(2):152–160

53. Miller Y, Bachowski G, Benjamin R, Eklund D, Hibbard A, Lightfoot T. Practice Guidelines for Blood Transfusion: A Compilation from Recent Peer-Reviewed Literature. Washington, DC: American Red Cross; 2007

54. Hall MP, Band PA, Meislin RJ, Jazrawi LM, Cardone DA. Platelet-rich plasma: current concepts and application in sports medicine. J Am Acad Orthop Surg 2009;17(10):602–608

55. Nurden AT, Nurden P, Sanchez M, Andia I, Anitua E. Platelets and wound healing. Front Biosci 2008;13:3532–3548

56. Qureshi AH, Chaoji V, Maiguel D, et al. Proteomic and phospho-proteomic profile of human platelets in basal, resting state: insights into integrin signaling. PLoS ONE 2009;4(10):e7627

57. Senzel L, Gnatenko DV, Bahou WF. The platelet proteome. Curr Opin Hematol 2009;16(5):329–333

58. Mishra A, Tummala P, King A, et al. Buffered platelet-rich plasma enhances mesenchymal stem cell proliferation and chondrogenic differentiation. Tissue Eng Part C Methods 2009;15(3): 431–435

59. Akeda K, An HS, Okuma M, et al. Platelet-rich plasma stimulates porcine articular chondrocyte proliferation and matrix biosynthesis. Osteoarthritis Cartilage 2006;14(12):1272–1280

60. Kon E, Mandelbaum B, Buda R, et al. Platelet-rich plasma intra-articular injection versus hyaluronic acid viscosupplementation as treatments for cartilage pathology: from early degeneration to osteoarthritis. Arthroscopy 2011;27(11):1490–1501

61. Filardo G, Kon E, Buda R, et al. Platelet-rich plasma intra-articular knee injections for the treatment of degenerative cartilage lesions and osteoarthritis. Knee Surg Sports Traumatol Arthrosc 2011;19(4): 528–535

62. Hildner F, Albrecht C, Gabriel C, Redl H, van Griensven M. State of the art and future perspectives of articular cartilage regeneration: a focus on adipose-derived stem cells and platelet-derived products. J Tissue Eng Regen Med 2011; 5(4):e36–e51

63. Toghraie FS, Chenari N, Gholipour MA, et al. Treatment of osteoarthritis with infrapatellar fat pad derived mesenchymal stem cells in Rabbit. Knee 2011;18(2):71–75

64. Park J, Gelse K, Frank S, von der Mark K, Aigner T, Schneider H. Transgene-activated mesenchymal cells for articular cartilage repair: a comparison of primary bone marrow-, perichondrium/periosteum- and fat-derived cells. J Gene Med 2006;8(1):112–125

65. Koga H, Muneta T, Nagase T, et al. Comparison of mesenchymal tissues-derived stem cells for in vivo chondrogenesis: suitable conditions for cell therapy of cartilage defects in rabbit. Cell Tissue Res 2008;333(2):207–215

66. Wakitani S, Goto T, Young RG, Mansour JM, Goldberg VM, Caplan AI. Repair of large full-thickness articular cartilage defects with allograft articular chondrocytes embedded in a collagen gel. Tissue Eng 1998;4(4):429–444

67. Kuroda R, Ishida K, Matsumoto T, et al. Treatment of a full-thickness articular cartilage defect in the femoral condyle of an athlete with autologous bone-marrow stromal cells. Osteoarthritis Cartilage 2007;15(2):226–231

11

Osteochondral Autograft Transplantation/Mosaicplasty

Brett McCoy and Anthony Miniaci

Healthy articular cartilage transmits load to subchondral bone while minimizing friction between articulating surfaces.[1] Articular cartilage has minimal inherent healing potential, and the natural history of untreated lesions is progressive degenerative changes and deterioration in functional outcomes scores.[2–6] The treatment of patients with full-thickness chondral lesions remains a difficult task for physicians.

Multiple treatment options exist to address full-thickness lesions. Early procedures such as abrasion chondroplasty and microfracture targeted bone marrow stimulation to elicit a fibrocartilaginous "healing" response. These procedures had promising short-term results, but long-term outcomes have been less predictable.[7,8] Osteochondral autograft transplantation (OAT/mosaicplasty) is a technique that addresses these lesions with the goal of preserving hyaline cartilage. Initial treatments predominantly addressed post-traumatic tibiofemoral and talar pathology but have been subsequently described for multiple etiologies in varying anatomic areas.[9–21]

Reports of osteochondral grafts date back to the early 20th century.[22] In 1985, Yamashita et al described the transplantation of autologous osteochondral grafts for the treatment of large lesions.[23] This technique had notable limitations, including donor site morbidity and difficulty matching the native contour of the condyle. Allograft transplantation has also been performed, but availability can be an issue and it carries a theoretical risk of immunologic rejection and infection.[24] The use of multiple smaller autologous osteochondral grafts emerged as an option to minimize donor morbidity and more accurately match the native contour without the inherent risks of an allograft.

The mosaicplasty technique was initially tested in canine and equine models with promising results.[25] Further studies in the goat model demonstrated successful incorporation of the graft with 86% chondrocyte viability at a 6-month follow-up.[26] These histological results were reaffirmed in a clinical study with longer follow-up.[27] Clinical use began in 1992, and long-term studies have demonstrated successful results.[28–31] When compared with autologous chondrocyte implantation (ACI), mosaicplasty offers the benefit of a single-stage procedure with lower cost and a shorter duration for graft adaptation and remodeling.[32,33]

◆ Diagnosis

The etiology of chondral lesions includes both traumatic injury and repetitive microtrauma. Patients will frequently present with

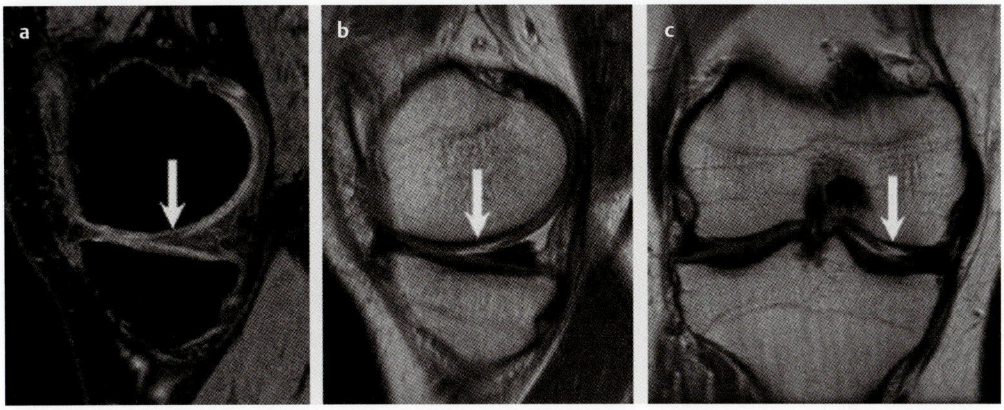

Fig. 11.1 Magnetic resonance imaging demonstrating chondral injury. Sagittal fat-saturated spoiled gradient-recalled-echo image (**a**), sagittal fast-spin echo intermediate-weighted image (**b**), and coronal image (**c**).

pain, swelling, and mechanical symptoms. Concomitant pathology such as meniscal or ligamentous injury may be the predominant factor in the initial symptomatology. The chondral defect may evoke a more insidious clinical picture. Thus, a high suspicion for chondral lesions should be maintained during clinical and imaging evaluation. It is also important to note that full-thickness chondral defects are common in athletes, and many of these are asymptomatic.[34]

The physical examination should include observation of the patient's gait and overall limb alignment. The assessment should include evaluation for an effusion, patellar maltracking, crepitance, and tenderness over the affected area. Plain radiographs should include anteroposterior, Rosenberg, lateral, and patellar views.[35] These films should be scrutinized for evidence of degenerative changes, osteochondritis dissecans (OCD), or loose bodies. If concern for malalignment exists, long-standing views can be obtained.

Bone scan (technetium-99 isotope) and computed tomography (CT) (with or without arthrography) have limited utility in diagnosing chondral defects. Magnetic resonance imaging (MRI) remains the preferred advanced imaging modality.[36] The most sensitive sequence is the T1-weighted fat-suppressed three-dimensional spoiled echo gradient images.[37] This technique utilizes the high spatial resolution of T1-weighted images and optimizes the signal-to-noise ratio via gradient echo techniques (**Fig. 11.1**).

Further advances in MRI, such as isotropic resolution reconstruction, may allow for improved preoperative assessment of chondral lesions, but, despite the sophistication of current MRI techniques, articular lesions can be accurately defined only at the time of initial arthroscopy. It is important to counsel the patient about the possibility of mosaicplasty (via open or arthroscopic means) before the surgery.

Indications

Mosaicplasty is indicated for symptomatic focal, unipolar, full-thickness lesions (chondral and osteochondral) of the knee, including patients with OCD lesions in situ or with the fragment missing (**Fig. 11.2**). The knee

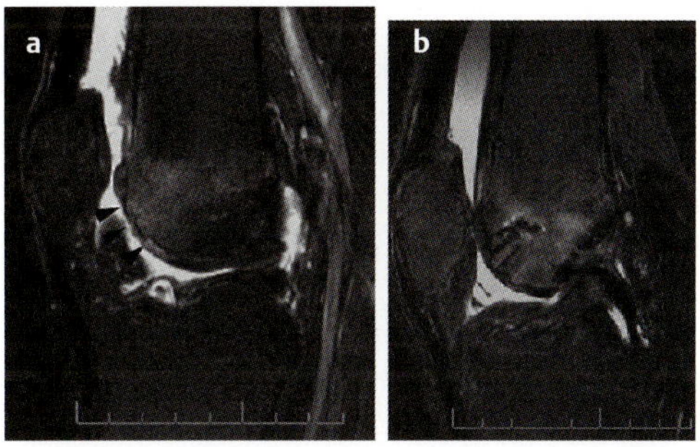

Fig. 11.2 Magnetic resonance imaging pre- (**a**) and postoperatively (**b**) demonstrating treatment of a full-thickness lesion with mosaicplasty.

Table 11.1 Indications and contraindications for osteochondral autologous transplantation

Indications	Contraindications
Full-thickness lesion between 1 and 4 cm^2	Previous total meniscectomy
Symptomatic patient	Noncompliant patient
Contact-bearing surface	Advanced age
Acceptable alignment	Malalignment
Stable joint	Unstable joint
Osteochondritis dissecans Fragment in situ Fragment missing	

should be stable and normally aligned. The lesions should be greater than 1 cm^2 and less than 4 to 5 cm^2 due to limitations of donor availability.[38] The defect should extend < 10 mm into the subchondral bone. Larger lesions may be amenable to treatment with mosaicplasty in conjunction with an alternative technique such as microfracture or ACI, although limited clinical data exist at this time (**Table 11.1**).[39,40]

◆ Technical Considerations

Positioning/Preparation

Patient positioning depends on surgeon preference and the location of the lesion. In general, the patient should be supine and the limb positioned to accommodate 120 degrees of flexion to ensure perpendicular access to more posterior lesions. The decision for an open versus arthroscopic procedure should be dictated by the location of the lesion and the surgeon experience. Several cadaveric studies demonstrate similar graft suitability in open and arthroscopic procedures.[41,42]

Open procedures can be accomplished via a vertical mini-arthrotomy (anterolateral or medial parapatellar) for femoral lesions. For tibial or patellar lesions, a standard medial parapatellar arthrotomy enhances visualization. Patellar lesions can also be addressed with a lateral parapatellar arthrotomy in conjunction with a tibial tubercle osteotomy, which protects the graft and functions as a concomitant realignment procedure (if clinically indicated).

If performed arthoscopically, a post or padded leg holder can be utilized per surgeon preference. The perpendicularity of portal placement should be assessed with an 18-gauge spinal needle before formal establishment. The contralateral leg can be positioned as desired, but it should undergo sterile prep for larger lesions as it may be needed as a site to obtain additional grafts. Arthroscopic portals should be established in a vertical direction to allow incorporation into an arthrotomy, if necessary. For arthroscopic procedures, the anteromedial and anterolateral portals should be established ~ 1 cm off the patellar tendon and will yield three to four 4.5-mm grafts. Accessory portals can be established proximally to obtain a total of 9 to 12 plugs depending on the size of the femur. If more graft is necessary the contralateral knee is an appropriate donor site.

After identification of an appropriate-size defect, the recipient site should be prepared. Any loose tissue should be excised and the rim should be debrided to a clean, stable margin using various tools (arthroscopic resector, curet, or scalpel blade). The edges should be oriented at 90 degrees. After the stable edges are obtained, a rasp or burr can be applied to the base of the lesion to expose subchondral bone. This will allow fibrocartilage ingrowth between the plugs placed. The graft chisel can then be placed over the lesion to accurately determine the location of the plugs and the number required. The chisel can gently score the recipient sites as a reference for plug placement.

Donor Harvest

The ideal donor site is easily accessible and provides appropriate functional tissue quality with minimal morbidity. Traditionally, the sites include the medial and lateral margins of the femoral trochlea and the intercondylar notch (**Fig. 11.3**). One study noted lower contact pressures in the medial trochlea (when compared with lateral) and recommended this as the ideal site for harvest.[43] The intercondylar notch has several notable shortcomings, including thinner cartilage and

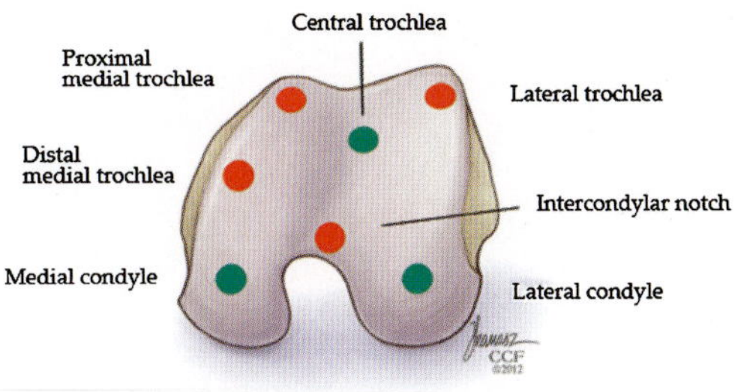

Fig. 11.3 Locations for graft harvest (*red circles*) and recipient sites (*green circles*).

a concave contour that will not match recipient sites on the femoral condyles but may adequately address central trochlear defects. Cadaveric CT studies utilizing topographic mapping noted that the medial and lateral patellar groove were a better topographic match than the intercondylar notch for lesions of the weight-bearing aspect of the medial and lateral femoral condyle.[44,45] Grafts harvested from the intercondylar notch were also less perpendicular.[41] The posterior condyle has also been suggested as a potential donor site, but cadaveric data found unsuitable grafts based on the angle of harvest and should not be considered as a routine harvest site.[46] In our experience, the lateral condyle is the most accessible area for graft harvest.

After preparation of the donor site, multiple systems exist for graft harvest and include both reusable and disposable types. The diameter of the harvested plugs varies. Donor site morbidity is a concern with larger plugs (> 6 mm). Animal studies with larger plugs have demonstrated the formation of cavitary lesions and sclerotic walled cysts that can result in collapse adjacent to the donor site, which can result in osteoarthritic changes.[47] Smaller plugs minimize donor site morbidity and result in fibrocartilaginous fill of the defects.[48] The difficulty with small plugs (< 3 mm) pertains to fragility and difficulty handling the graft. Manipulation can also be problematic and an increased risk of fragmentation during insertion has been reported. The authors suggest that a 4.5-mm-diameter plug is an "ideal" graft with minimal donor morbidity, reasonable ease of handling, and less concern for fragmentation.

Grafts should be harvested manually as power trephination has been shown to negatively impact chondrocyte viability.[49] The grafts are harvested with double-edged tubular cutting chisels that will allow for accuracy in both length and diameter. If the base of the graft is asymmetric, it can be modified to create a flat surface and thus has a more consistent length measurement. After harvest, the grafts should be placed in saline-soaked gauze and the donor sites can be filled to potentially minimize hemarthroses. In a canine study, compressed collagen demonstrated the best fibrocartilaginous fill during histologic evaluation of the donor sites.[50]

Graft Insertion

The different systems for mosaicplasty require a varying amount of insertional force and some degree of "toggling" during graft removal.[51] Clinicians should remain aware of the principles of an ideal system, which preserves the maximal amount of viable tissue with minimal tissue trauma. The grafts should be placed gently as excess forces have been demonstrated to negatively impact the chondrocyte viability.[52,53] If the recipient hole is shorter than the graft, excess force will be required to achieve congruency; thus, the recommendation is equal length.[54,55] The stability of the press fit plug is dependent on several factors. In a porcine model, grafts were found to be more stable with larger diameters and shorter dilation length, and single grafts were superior to multiple grafts. No difference was noted between grafts aligned in a row versus a circular pattern.[56]

The grafts are anticipated to expand 0.1 to 0.2 mm after harvest. Thus, a conical dilator is used to help prepare the tunnel to minimize the stresses required to insert the graft. When the dilator is placed in the next recipient hole it will compress the bone adjacent to the previously placed graft.

Congruency of the transplanted graft with the adjacent native articular cartilage is a crucial technical aspect of the procedure (**Fig. 11.4**). Huang et al[57] demonstrated a limited tolerance for incongruity in a sheep model, noting that all grafts countersunk > 2 mm had cartilage necrosis or overgrowth. In a cadaveric study, grafts that were 1 mm proud experienced a 21% increase in peak contact pressure.[58] In the setting of tissue loss, graft congruency can be more difficult. For example, if a lesion has 5-mm depth of tissue loss and the donor plug has a length of 20 mm, then drilling to 15 mm will achieve the ideal congruency. In other words, the graft may remain proud of the recipient drill hole in the setting of tissue loss to obtain congruency with adjacent cartilage.

The reproduction of joint congruency requires accurately positioning the plugs to match the native contour of the articular surface. These grafts will be predominantly placed in convex locations. Starting at the periphery of the lesion and working toward the center helps to avoid a "flat" graft (**Figs. 11.4** and **11.5**). A "flat" graft increases the risk of fibrocartilaginous overgrowth,

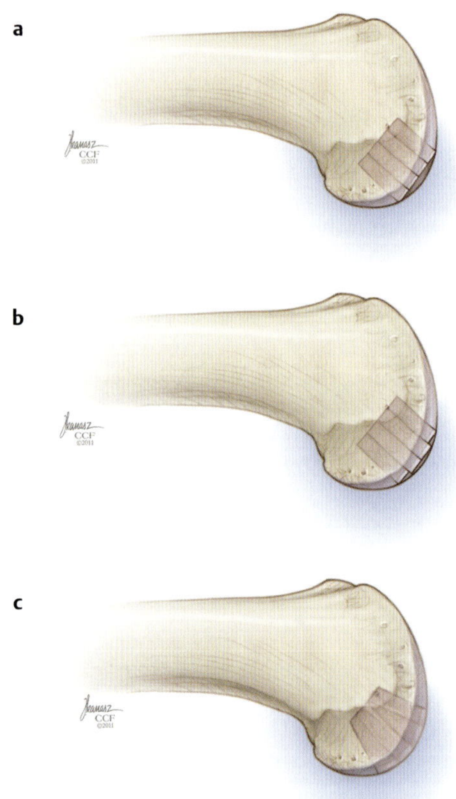

Fig. 11.4 Improper graft placement falls to restore the contour (**a**) or the curvature (**b**). Proper graft placement (**c**) with restoration of both the contour and curvature due to slight obliquity.

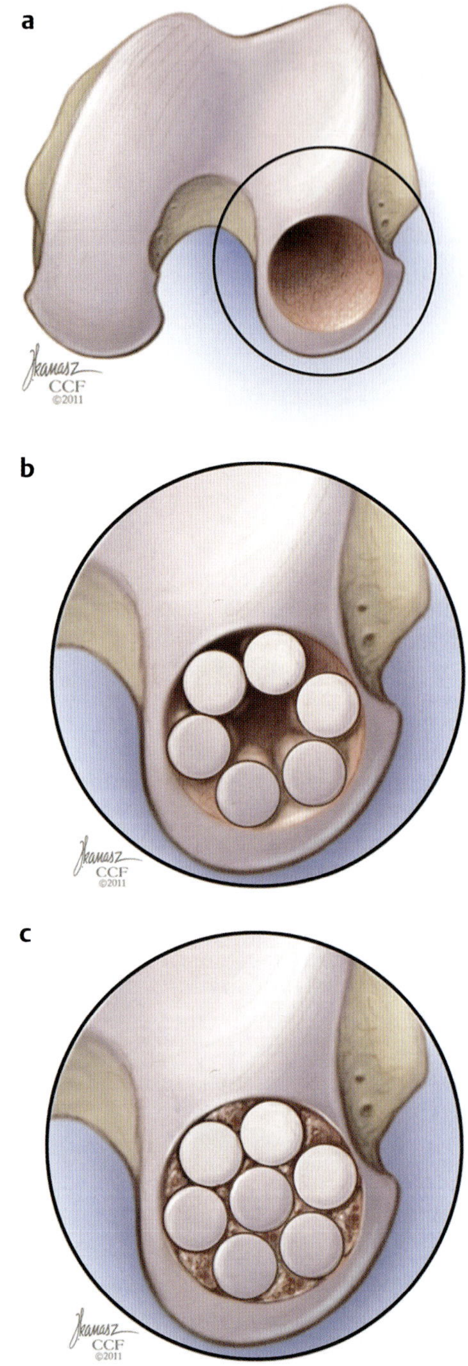

Fig. 11.5 A depiction of a common pattern for order of insertion for an osteochondral lesion (**a**) of the medial femoral condyle. Placement is peripheral (**b**) followed by central (**c**).

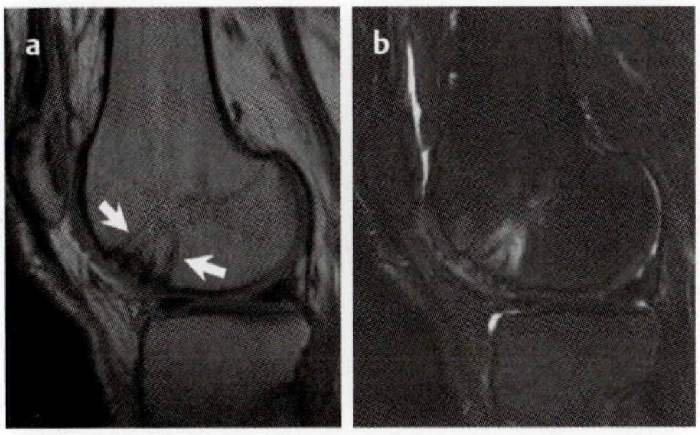

Fig. 11.6 Magnetic resonance imaging demonstrating graft convergence (*white arrows*) because perpendicular placement was not obtained. T1 (**a**), T2 (**b**).

which supplants the beneficial component of hyaline cartilage preservation. Grafts are generally 15 to 20 mm in depth, but the central grafts may be longer than peripherally placed grafts. The grafts should be placed in a perpendicular or slightly oblique fashion with an attempt to avoid graft convergence (**Fig. 11.6**).

In an OCD lesion where the fragment is missing, the procedure will be similar to posttraumatic defects (**Fig. 11.7**). If the fragment is intact, mosaicplasty can be utilized to confer stability to the lesion and allow for vascular inflow and the theoretical benefit of improved healing. A Kirschner wire can be used to stabilize the graft during the procedure. Alternatively, a screw can be placed to lag an unstable fragment while plugs are placed peripherally. The screw can then be removed and replaced with a plug (**Fig. 11.8**). The central plug should be of adequate length to reach the cancellous bone deep to

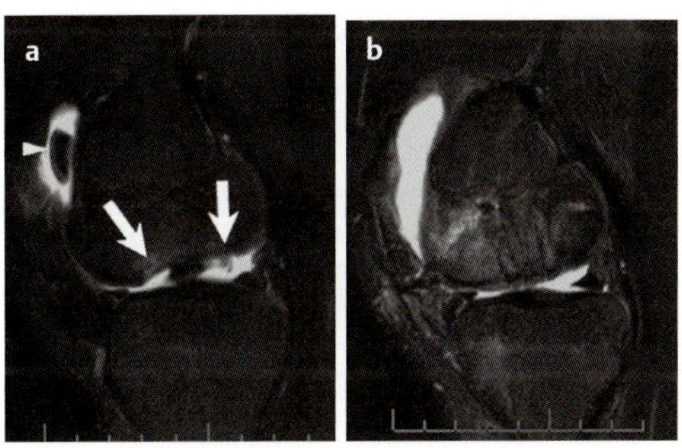

Fig. 11.7 Magnetic resonance imaging pre- (**a**) and postoperative (**b**) after mosaicplasty.

the lesion. The lesion should be probed to assess stability and debrided if the plugs do not adequately stabilize it. The plug from the recipient site can be placed in the donor site if dilation is not performed.[59]

Postoperative cyst formation deep to the grafts has several theoretical causes. They include trapped or communicating synovial fluid, graft necrosis, and increased graft motion. Adequate planning can help eliminate some of these potential risks; for example, avoiding power during graft harvesting and placement reduces the risk of thermal necrosis. A press fit graft will eliminate motion and synovial communication. When sized properly the graft will abut the bottom of the recipient hole and have good contact along the peripheral margin (**Fig. 11.9**).

Postoperative Course

Weight bearing before graft incorporation can be detrimental to the patient's outcome. Toe-touch weight bearing is advocated for the first 6 weeks. Range of motion of the knee can be beneficial during this time period. We recommend a brace that allows motion from 0 to 90 degrees. The patient should be encouraged to perform isometric quadriceps exercises, calf pumps, and straight-leg raises. The use of continuous passive motion after surgery has ample basic science support, but a systematic review notes a lack of well-conducted clinical trials and thus is not currently recommended.[60] Gradual weight bearing is instituted at 6 weeks if radiographs are acceptable. Return to athletics requires minimal range-of-motion deficit and quadriceps strength comparable to the contralateral extremity.

MRI with newer matrix assessment techniques can be a useful tool for evaluating healing but requires an experienced interpreter. The appearance will vary on the basis of the technique utilized and the time interval from intervention.[36,61] Some studies suggest that persistent edema on MRI is common after osteochondral grafting with minimal relationship to clinical outcome.[62,63] In a recent study, correlation was noted between long-term clinical outcome scores and MRI findings.[31]

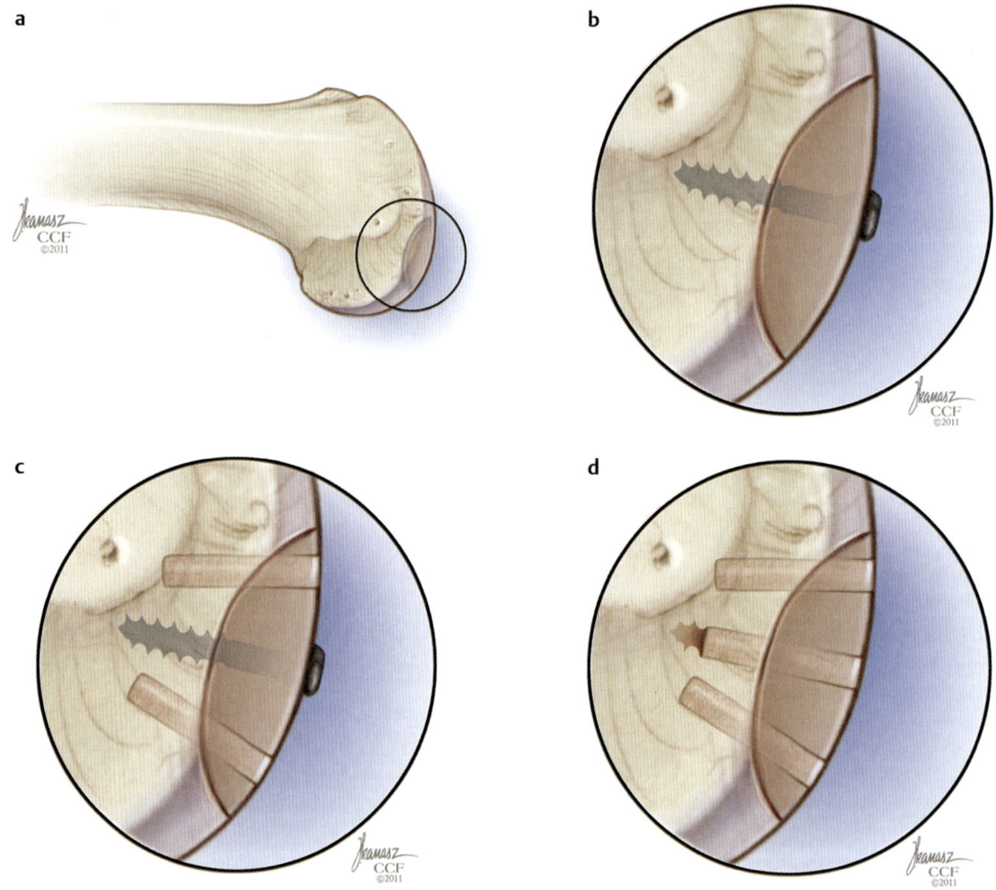

Fig. 11.8 Osteochondritis dissecans in situ (unstable) (**a**) fixed first with lag screw (**b**) followed by peripheral grafts (**c**) and then replacement of the lag screw with a graft (**d**).

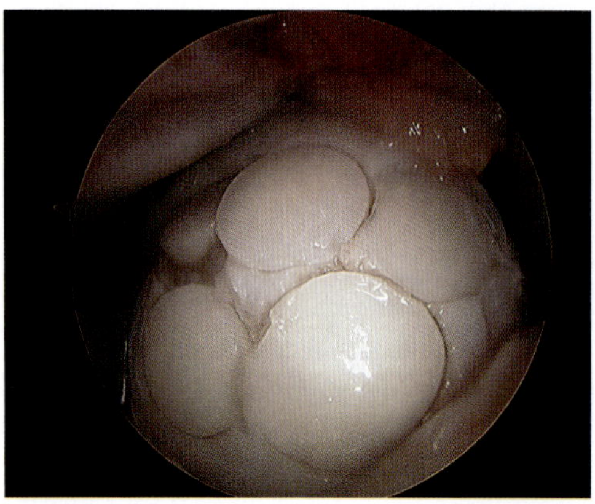

Fig. 11.9 Intraoperative image of mosaicplasty.

Outcomes

The clinical outcome data regarding mosaicplasty are predominantly drawn from prospective cohorts, case-control studies, and case series. Multiple studies demonstrate good results after mosaicplasty during short- and midterm follow-up.[64–67] Solheim et al noted a decrease in objective outcome scores between the 1-year and 5- to 9-year follow-up.[28] However, 88% of the patients stated that they would undergo the procedure again. Marcacci et al noted a decrease in sports activities during a 2- to 7-year follow-up after mosaicplasty (**Table 11.2**).[30]

Table 11.2 A summary of clinical studies evaluating mosaicplasty

Study	Type	Number	Age (y)	Mean follow-up	Complications	Return to sport	Outcomes
Solheim et al[28]	Prospective	69	33	7 y	Two hemarthroses One DVT Three superficial wound issues One septic arthritis	N/A	Lysholm 48 preop, 81 at 12 mo, 68 after 12 mo VAS 62 preop, 24 at 12 mo, 32 after 12 mo
Marcacci et al[30]	Prospective	30	29.3	7+ y	No serious complications	2 y: Same level 73% Decreased level 13% 7 y: Same level 23% Decreased level 47%	ICRS 76% good or excellent IKDC 72 (35 preop) Tegner 5.6 (2.9 preop)
Gudas et al[38]	Prospective randomized	60	24.3	37.1 mo	Two superficial infections	Same level 93% mosaicplasty 52% microfracture	HSS and ICRS good or excellent 96% mosaicplasty 52% microfracture
Muller et al[64]	Retrospective	15	27–52	42 mo	Three reoperations (two persistent pain) (one septic arthritis)	92%	IKDC 58.3 Lysholm 80.9 Subjective 7.2 (4.7 preop)
Lahav et al[65]	Retrospective	16	N/A	10 mo	N/A	N/A	KOOS 80.6 IKDC 68.2 Subjective 8 (3.1 preop)
Oztürk et al[66]	Retrospective	19	33.1	32.4 mo	None	N/A	Lysholm 87.5 (45.8 preop) 85% good or excellent results
Marcacci et al[67]	Prospective	37	29.5	24–48 mo	N/A	73% same level 14% lower level	ICRS 78% good or excellent results
Hangody et al[68]	Case series	354	24.3	9.6 y	5% donor site morbidity 8% hemorrhage Two cases of septic arthritis	63% same level 27% lower level	Good to excellent results 92% talar 91% femoral 86% tibial 74% patellofemoral
Gudas et al[75]	Prospective randomized	50	14.3	4.2 y	One superficial infection	Same level: ~1 y 84% mosaicplasty 33% microfracture ~4 y 81% mosaicplasty 14% microfracture	ICRS good or excellent 91% mosaicplasty 56% microfracture
Bentley et al[77]	Prospective randomized	100	31.3	19 mo	One DVT One arthrofibrosis One superficial infection		Good to excellent functional results 88% ACI 69% mosaicplasty

Abbreviations: y, year; DVT, deep venous thrombosis; preop, preoperative; mo, month; ICRS, International Cartilage Repair Society; IKDC, International Knee Documentation Society; HSS, Hospital for Special Surgery; KOOS, Knee and Osteoarthritis Outcome Score; ACI, autologous chondrocyte implantation.

In one of the larger series with long-term follow-up, Hangody and Füles reported on the clinical outcome of 831 patients who underwent mosaicplasty for small to medium-sized lesions.[48] Good to excellent results were noted in 94% of talar, 92% of femoral condylar, 87% of tibial, and 79% of patellofemoral/trochlear lesions, respectively. Subsequent follow-up at a mean of 9.6 years demonstrated a slight deterioration in clinical results but good to excellent results in 92% of talar, 91% of femoral condylar, 86% of tibial, and 74% of patellofemoral lesions.[68] Donor site morbidity assessed with the Bandi score was noted in 3% of the patients. Paul et al noted that when harvesting grafts from asymptomatic knees for talar lesions increased, body mass index was a risk factor for higher morbidity at the donor site.[69] This morbidity was identified by changes in the Lysholm and WOMAC scores. Age, number of grafts, and diameter of the grafts were not statistically significant. Morbidity from graft harvest can be difficult to discern in the setting of the intervention; however, it should be suspected if there is persistent pain or mechanical symptoms that do not correlate with the treated lesion.

In addition to clinical outcome measures, direct visualization and histology data also demonstrate good results after mosaicplasty. Hangody et al noted congruent gliding surfaces and survival of hyaline cartilage during second-look arthroscopy in 81 of 98 patients.[25] Barber and Chow found viable chondrocytes and graft survival in all patients at 1-year follow-up during arthroscopic biopsy in a small series.[27] A similar study noted excellent congruency and survival of hyaline cartilage with fibrocartilage fill between the plugs.[70] During total knee arthroplasty, Huntley et al harvested 4.5-mm plugs and noted with laser scanning microscopy that, despite good survival centrally, one-third of the lateral margin underwent cell death.[71] The clinical implication of this study has not been realized.

In a recent systematic review, Benthien et al note that level I and II evidence is needed to determine the appropriate manner to treat cartilage defects in the knee.[72] Multiple treatment options exist with promising results; thus, comparative studies are useful. In a systematic review of chondral defects in the athlete's knee, ACI and OAT had better clinical outcomes than microfracture.[73] Microfracture was noted to have a deterioration of results over time and was less effective for larger lesions.[74] Gudas et al demonstrated in a randomized control trial that OAT returned 92% of the young athletes to an equivalent preinjury level versus only 52% with microfracture.[75] In the treatment of OCD lesions, a prospective randomized trial also demonstrated that the results of microfracture deteriorate over time when compared with mosaicplasty.[75]

The results comparing ACI with mosaicplasty are varied. A recent Cochrane review identified three studies that directly compared ACI with mosaicplasty.[76] In a prospective randomized trial, Bentley et al noted that ACI was superior to mosaicplasty in clinical outcome scoring and second-look arthroscopy, but this was limited to post hoc subgroup analysis of medial condylar lesions.[77] They noted that mosaicplasty failed in all five cases for patellar lesions. In a separate multicenter randomized control trial, ACI and mosaicplasty were found to be clinically equivalent.[33] In one study, the timing of return to athletics was quickest with OAT and slowest with ACI.[73]

The conflicting data regarding patient outcomes reaffirm the belief that more high-quality studies are needed to compare the various treatment options for chondral defects. Long-term data will also further delineate the role each treatment has on preventing degenerative changes.

◆ Future Directions

Technical advances have helped mosaicplasty evolve as a procedure. In a recent study, computer-assisted surgery helped improve the accuracy and precision of harvest and insertion angles.[78] Mosaicplasty may also have a role in conjunction with unicompartmental arthroplasty or resurfacing procedures in an effort to preserve as much native anatomy as possible.[79]

Biologic interventions continue to pose great potential benefit to cartilaginous lesions. These interventions may augment mosaicplasty procedures. In animal studies evaluating substances such as hepatocyte growth factor, bone morphogenic protein (BMP-2), and hyaluronate sodium have demonstrated promise as adjuncts for graft healing and chondrocyte survival.[80-82] There remains a great deal of uncertainty regarding the long-term prognosis of joints affected by articular cartilage injuries. Analysis of cartilage adjacent to the site of intervention has been performed in a laboratory setting and may help in discerning the overall milieu of the joint, and thus the role interventions may play in the long-term health of the knee.[83]

Summary

Osteochondral autograft transplantation has a place in the algorithm for the treatment of chondral injuries in the knee. There is increasing clinical evidence that it may be superior to microfracture in a young patient with a focal chondral lesion.

References

1. Hasler EM, Herzog W, Wu JZ, Müller W, Wyss U. Articular cartilage biomechanics: theoretical models, material properties, and biosynthetic response. Crit Rev Biomed Eng 1999;27(6):415–488
2. Piasecki DP, Spindler KP, Warren TA, Andrish JT, Parker RD. Intraarticular injuries associated with anterior cruciate ligament tear: findings at ligament reconstruction in high school and recreational athletes. An analysis of sex-based differences. Am J Sports Med 2003;31(4):601–605
3. Mandelbaum BR, Browne JE, Fu F, et al. Articular cartilage lesions of the knee. Am J Sports Med 1998;26(6):853–861
4. Buckwalter JA, Mankin HJ. Articular cartilage: degeneration and osteoarthritis, repair, regeneration, and transplantation. Instr Course Lect 1998; 47:487–504
5. Kujala UM, Kettunen J, Paananen H, et al. Knee osteoarthritis in former runners, soccer players, weight lifters, and shooters. Arthritis Rheum 1995;38(4):539–546
6. Drawer S, Fuller CW. Propensity for osteoarthritis and lower limb joint pain in retired professional soccer players. Br J Sports Med 2001;35(6):402–408
7. Mithoefer K, Williams RJ III, Warren RF, et al. The microfracture technique for the treatment of articular cartilage lesions in the knee. A prospective cohort study. J Bone Joint Surg Am 2005;87(9): 1911–1920
8. Steadman JR, Briggs KK, Rodrigo JJ, Kocher MS, Gill TJ, Rodkey WG. Outcomes of microfracture for traumatic chondral defects of the knee: average 11-year follow-up. Arthroscopy 2003;19(5):477–484
9. Nagura I, Fujioka H, Kokubu T, Kurosaka M. Autologous osteochondral plug transplantation for osteochondrosis of the second metatarsal head: a case report. J Med Case Reports 2011;5:308
10. Krych AJ, Lorich DG, Kelly BT. Treatment of focal osteochondral defects of the acetabulum with osteochondral allograft transplantation. Orthopedics 2011;34(7):e307–e311
11. Nam D, Shindle MK, Buly RL, Kelly BT, Lorich DG. Traumatic osteochondral injury of the femoral head treated by mosaicplasty: a report of two cases. HSS J 2010;6(2):228–234
12. Park TS, Kim TS, Cho JH. Arthroscopic osteochondral autograft transfer in the treatment of an osteochondral defect of the humeral head: report of one case. J Shoulder Elbow Surg 2006;15(6):e31–e36
13. Ozyurekoglu T. Multiple osteochondral autograft transfer to the proximal interphalangeal joint: case report. J Hand Surg Am 2010;35(6):931–935
14. Kircher J, Patzer T, Magosch P, Lichtenberg S, Habermeyer P. Osteochondral autologous transplantation for the treatment of full-thickness cartilage defects of the shoulder: results at nine years. J Bone Joint Surg Br 2009;91(4):499–503
15. Fansa AM, Murawski CD, Imhauser CW, Nguyen JT, Kennedy JG. Autologous osteochondral transplantation of the talus partially restores contact mechanics of the ankle joint. Am J Sports Med 2011;39(11):2457–2465
16. Liu W, Liu F, Zhao W, Kim JM, Wang Z, Vrahas MS. Osteochondral autograft transplantation for acute osteochondral fractures associated with an ankle fracture. Foot Ankle Int 2011;32(4):437–442
17. Iwasaki N, Kato H, Ishikawa J, Saitoh S, Minami A. Autologous osteochondral mosaicplasty for capitellar osteochondritis dissecans in teenaged patients. Am J Sports Med 2006;34(8):1233–1239
18. Tsuda E, Ishibashi Y, Yamamoto Y, Maeda S, Kimura Y, Sato H. Osteochondral autograft transplantation for advanced stage Freiberg disease in adolescent athletes: a report of 3 cases and surgical procedures. Am J Sports Med 2011;39(11): 2470–2475
19. Taketomi S, Hiraoka H, Nakagawa T, et al. Osteochondral autograft for medial femoral condyle chondral lesions in a patient with multiple epiphyseal dysplasia: long-term result. J Orthop Sci 2011; May 11 (Epub ahead of print)
20. Lee YK, Lee M, Lee JM. Osteochondral autograft transplantation for osteochondritis dissecans of the scaphoid: case report. J Hand Surg Am 2011; 36(5):820–823
21. Gobbi A, Francisco RA, Lubowitz JH, Allegra F, Canata G. Osteochondral lesions of the talus: randomized controlled trial comparing chondroplasty, microfracture, and osteochondral autograft transplantation. Arthroscopy 2006;22(10): 1085–1092
22. Lexer E. Substitution of whole or half joints from freshly amputated extremities by free plastic operation. Surg Gynecol Obstet 1908;6:601–607
23. Yamashita F, Sakakida K, Suzu F, Takai S. The transplantation of an autogeneic osteochondral fragment for osteochondritis dissecans of the knee. Clin Orthop Relat Res 1985;(201):43–50
24. Bugbee WD, Convery FR. Osteochondral allograft transplantation. Clin Sports Med 1999;18(1):67–75
25. Hangody L, Feczkó P, Bartha L, Bodó G, Kish G. Mosaicplasty for the treatment of articular defects of the knee and ankle. Clin Orthop Relat Res 2001;(391, Suppl):S328–S336

26. Lane JG, Massie JB, Ball ST, et al. Follow-up of osteochondral plug transfers in a goat model: a 6-month study. Am J Sports Med 2004;32(6):1440–1450
27. Barber FA, Chow JC. Arthroscopic osteochondral transplantation: Histologic results. Arthroscopy 2001;17(8):832–835
28. Solheim E, Hegna J, Oyen J, Austgulen OK, Harlem T, Strand T. Osteochondral autografting (mosaicplasty) in articular cartilage defects in the knee: results at 5 to 9 years. Knee 2010;17(1): 84–87
29. Hangody L, Vásárhelyi G, Hangody LR, et al. Autologous osteochondral grafting—technique and long-term results. Injury 2008;39(Suppl 1):S32–S39
30. Marcacci M, Kon E, Delcogliano M, Filardo G, Busacca M, Zaffagnini S. Arthroscopic autologous osteochondral grafting for cartilage defects of the knee: prospective study results at a minimum 7-year follow-up. Am J Sports Med 2007; 35(12):2014–2021
31. Tetta C, Busacca M, Moio A, et al. Knee osteochondral autologous transplantation: long-term MR findings and clinical correlations. Eur J Radiol 2010;76(1):117–123
32. Derrett S, Stokes EA, James M, Bartlett W, Bentley G. Cost and health status analysis after autologous chondrocyte implantation and mosaicplasty: a retrospective comparison. Int J Technol Assess Health Care 2005;21(3):359–367
33. Dozin B, Malpeli M, Cancedda R, et al. Comparative evaluation of autologous chondrocyte implantation and mosaicplasty: a multicentered randomized clinical trial. Clin J Sport Med 2005; 15(4):220–226
34. Flanigan DC, Harris JD, Trinh TQ, Siston RA, Brophy RH. Prevalence of chondral defects in athletes' knees: a systematic review. Med Sci Sports Exerc 2010;42(10):1795–1801
35. Rosenberg TD, Paulos LE, Parker RD, Coward DB, Scott SM. The forty-five-degree posteroanterior flexion weight-bearing radiograph of the knee. J Bone Joint Surg Am 1988;70(10):1479–1483
36. Hayter C, Potter H. Magnetic resonance imaging of cartilage repair techniques. J Knee Surg 2011; 24(4):225–240
37. Chung CB, Frank LR, Resnick D. Cartilage imaging techniques: current clinical applications and state of the art imaging. Clin Orthop Relat Res 2001;(391, Suppl):S370–S378
38. Gudas R, Kalesinskas RJ, Kimtys V, et al. A prospective randomized clinical study of mosaic osteochondral autologous transplantation versus microfracture for the treatment of osteochondral defects in the knee joint in young athletes. Arthroscopy 2005;21(9):1066–1075
39. Lane JG, Healey RM, Chen AC, Sah RL, Amiel D. Can osteochondral grafting be augmented with microfracture in an extended-size lesion of articular cartilage? Am J Sports Med 2010;38(7): 1316–1323
40. Gudas R, Simonaitytė R, Cekanauskas E, Mickevičius T. Concomitant autologous chondrocyte implantation with osteochondral grafting for treatment of a massive osteochondral defect in the bilateral knees of a child. Medicina (Kaunas) 2011;47(3):170–173
41. Diduch DR, Chhabra A, Blessey P, Miller MD. Osteochondral autograft plug transfer: achieving perpendicularity. J Knee Surg 2003;16(1):17–20
42. Keeling JJ, Gwinn DE, McGuigan FX. A comparison of open versus arthroscopic harvesting of osteochondral autografts. Knee 2009;16(6):458–462
43. Garretson RB III, Katolik LI, Verma N, Beck PR, Bach BR, Cole BJ. Contact pressure at osteochondral donor sites in the patellofemoral joint. Am J Sports Med 2004;32(4):967–974
44. Bartz RL, Kamaric E, Noble PC, Lintner D, Bocell J. Topographic matching of selected donor and recipient sites for osteochondral autografting of the articular surface of the femoral condyles. Am J Sports Med 2001;29(2):207–212
45. Ahmad CS, Cohen ZA, Levine WN, Ateshian GA, Mow VC. Biomechanical and topographic considerations for autologous osteochondral grafting in the knee. Am J Sports Med 2001;29(2):201–206
46. Thaunat M, Beaufils P. The posterior femoral condyles: a potential donor site for mosaic-like osteochondral autograft? Arthroscopy 2010;26(11): 1503–1509
47. Hurtig MB, et al. THe effect of graft size and number on the outcome of mosaic arthroplasty resurfacing: an experimental model in sheep. Proceedings of the Arthroscopy Association of North America. 1999;18:16–17
48. Hangody L, Füles P. Autologous osteochondral mosaicplasty for the treatment of full-thickness defects of weight-bearing joints: ten years of experimental and clinical experience. J Bone Joint Surg Am 2003;85-A(Suppl 2):25–32
49. Evans PJ, Miniaci A, Hurtig MB. Manual punch versus power harvesting of osteochondral grafts. Arthroscopy 2004;20(3):306–310
50. Feczkó P, Hangody L, Varga J, et al. Experimental results of donor site filling for autologous osteochondral mosaicplasty. Arthroscopy 2003;19(7): 755–761
51. Barber FA, Herbert MA, McGarry JE, Barber CA. Insertion force of articular cartilage transplantation systems. J Knee Surg 2008;21(3):200–204
52. Pylawka TK, Wimmer M, Cole BJ, Virdi AS, Williams JM. Impaction affects cell viability in osteochondral tissues during transplantation. J Knee Surg 2007;20(2):105–110
53. Patil S, Butcher W, D'Lima DD, Steklov N, Bugbee WD, Hoenecke HR. Effect of osteochondral graft insertion forces on chondrocyte viability. Am J Sports Med 2008;36(9):1726–1732
54. Kordás G, Szabó JS, Hangody L. The effect of drill-hole length on the primary stability of osteochondral grafts in mosaicplasty. Orthopedics 2005;28(4):401–404
55. Kock NB, Van Susante JL, Buma P, Van Kampen A, Verdonschot N. Press-fit stability of an osteochondral autograft: Influence of different plug length and perfect depth alignment. Acta Orthop 2006;77(3):422–428
56. Kordás G, Szabó JS, Hangody L. Primary stability of osteochondral grafts used in mosaicplasty. Arthroscopy 2006;22(4):414–421
57. Huang FS, Simonian PT, Norman AG, Clark JM. Effects of small incongruities in a sheep model of osteochondral autografting. Am J Sports Med 2004;32:1842–1848
58. Harris JD, Solak KK, Siston RA, Litsky A, Richards J, Flanigan DC. Contact pressure comparison of proud osteochondral autograft plugs versus proud synthetic plugs. Orthopedics 2011;34(2):97

59. Nakagawa Y, Matsusue Y, Nakamura T. A novel surgical procedure for osteochondritis dissecans of the lateral femoral condyle: Exchanging osteochondral plugs taken from donor and recipient sites. Arthroscopy 2002;18(1):E5

60. Fazalare JA, Griesser MJ, Siston RA, Flanigan DC. The use of continuous passive motion following knee cartilage defect surgery: a systematic review. Orthopedics 2010;33(12):878

61. Choi YS, Potter HG, Chun TJ. MR imaging of cartilage repair in the knee and ankle. Radiographics 2008;28(4):1043–1059

62. Kokkinakis M, Kafchitsas K, Rajeev A, Mortier J. Is MRI useful in the early follow-up after autologous osteochondral transplantation? Acta Orthop Belg 2008;74(5):636–642

63. Nemec SF, Marlovits S, Trattnig S. Persistent bone marrow edema after osteochondral autograft transplantation in the knee joint. Eur J Radiol 2009;71(1):159–163

64. Muller S, Breederveld RS, Tuinebreijer WE. Results of osteochondral autologous transplantation in the knee. Open Orthop J 2010;4:111–114

65. Lahav A, Burks RT, Greis PE, Chapman AW, Ford GM, Fink BP. Clinical outcomes following osteochondral autologous transplantation (OATS). J Knee Surg 2006;19(3):169–173

66. Oztürk A, Ozdemir MR, Ozkan Y. Osteochondral autografting (mosaicplasty) in grade IV cartilage defects in the knee joint: 2- to 7-year results. Int Orthop 2006;30(3):200–204

67. Marcacci M, Kon E, Zaffagnini S, et al. Multiple osteochondral arthroscopic grafting (mosaicplasty) for cartilage defects of the knee: prospective study results at 2-year follow-up. Arthroscopy 2005;21(4):462–470

68. Hangody L, Dobos J, Baló E, Pánics G, Hangody LR, Berkes I. Clinical experiences with autologous osteochondral mosaicplasty in an athletic population: a 17-year prospective multicenter study. Am J Sports Med 2010;38(6):1125–1133

69. Paul J, Sagstetter A, Kriner M, Imhoff AB, Spang J, Hinterwimmer S. Donor-site morbidity after osteochondral autologous transplantation for lesions of the talus. J Bone Joint Surg Am 2009;91(7):1683–1688

70. Rădulescu RA, Cirstoiu CF, Bădilă AE. Arthroscopical and histological study of cartilaginous lesions treated by mosaicplasty. J Med Life 2010;3(4):407–411

71. Huntley JS, Bush PG, McBirnie JM, Simpson AH, Hall AC. Chondrocyte death associated with human femoral osteochondral harvest as performed for mosaicplasty. J Bone Joint Surg Am 2005;87(2):351–360

72. Benthien JP, Schwaninger M, Behrens P. We do not have evidence based methods for the treatment of cartilage defects in the knee. Knee Surg Sports Traumatol Arthrosc 2011;19(4):543–552

73. Harris JD, Brophy RH, Siston RA, Flanigan DC. Treatment of chondral defects in the athlete's knee. Arthroscopy 2010;26 (6):841–852

74. Magnussen RA, Dunn WR, Carey JL, Spindler KP. Treatment of focal articular cartilage defects in the knee: a systematic review. Clin Orthop Relat Res 2008;466(4):952–962

75. Gudas R, Simonaityte R, Cekanauskas E, Tamosiūnas R. A prospective, randomized clinical study of osteochondral autologous transplantation versus microfracture for the treatment of osteochondritis dissecans in the knee joint in children. J Pediatr Orthop 2009;29(7):741–748

76. Wasiak J, Clar C, Villanueva E. Autologous cartilage implantation for full thickness articular cartilage defects of the knee. Cochrane Database Syst Rev 2006;3(3):CD003323

77. Bentley G, Biant LC, Carrington RW, et al. A prospective, randomised comparison of autologous chondrocyte implantation versus mosaicplasty for osteochondral defects in the knee. J Bone Joint Surg Br 2003;85(2):223–230

78. Koulalis D, Di Benedetto P, Citak M, O'Loughlin P, Pearle AD, Kendoff DO. Comparative study of navigated versus freehand osteochondral graft transplantation of the knee. Am J Sports Med 2009;37(4):803–807

79. Unnithan A, Jimulia T, Mohammed R, Learmonth DJ. Unique combination of patellofemoral joint arthroplasty with Osteochondral Autograft Transfer System (OATS)—a case series of six knees in five patients. Knee 2008;15(3):187–190

80. Siebert CH, Schneider U, Sopka S, Wahner T, Miltner O, Niedhart C. Ingrowth of osteochondral grafts under the influence of growth factors: 6-month results of an animal study. Arch Orthop Trauma Surg 2006;126(4):247–252

81. Tibesku CO, Daniilidis K, Szuwart T, Jahn UR, Schlegel PM, Fuchs-Winkelmann S. Influence of hepatocyte growth factor on autologous osteochondral transplants in an animal model. Arch Orthop Trauma Surg 2011;131(8):1145–1151

82. Tytherleigh-Strong G, Hurtig M, Miniaci A. Intra-articular hyaluronan following autogenous osteochondral grafting of the knee. Arthroscopy 2005;21(8):999–1005

83. Strauss EJ, Goodrich LR, Chen CT, Hidaka C, Nixon AJ. Biochemical and biomechanical properties of lesion and adjacent articular cartilage after chondral defect repair in an equine model. Am J Sports Med 2005;33(11):1647–1653

12

Improved Preservation of Fresh Osteochondral Allografts for Clinical Use

Aaron M. Stoker, Joeseph T. Garrity, Clark T. Hung, James P. Stannard, and James L. Cook

Fresh osteochondral allografts (OCAs) have been used clinically to treat cartilage defects in the knee for over 30 years.[1] The major advantages of OCAs over other currently available "biologic" treatment options for cartilage defects of the knee include implantation of hyaline cartilage and bone in a graft that is site- and size-matched with tissue architecture and material properties that can withstand the loads normally transmitted to the joint. Fresh OCA grafting has a reported 5- to 10-year functional survival rate of 75 to 85% for treatment of focal defects of the femoral condyle (FC).[2–4] Additionally, the longevity of the fresh OCA tissue after implantation has been documented to be as long as 25 years,[3] indicating that this procedure can provide a long-term solution for treatment of osteochondral defects. The factor most consistently reported to influence the long-term success of OCAs is the viability of chondrocytes in the transplanted tissue.[3,5,6] However, in the United States, fresh OCA tissue has only been available commercially since 1998, so much of the long-term data are from centers that could harvest tissue from cadaveric donors, process and store the tissue, and perform the transplant procedure at a single site. In these centers, tissues were harvested from the donor and used for OCA transplant within 24 to 72 hours,[1,3] which allowed for optimal maintenance of chondrocyte viability and tissue biochemical and biomechanical properties. However, concerns regarding disease transmission stimulated the implementation of a mandatory disease and contamination testing period of 14 days once tissue banks in the United States made OCAs commercially available.

With the 14-day testing period requirement, tissue banks had to develop protocols for storing OCAs in a way that would maintain sterility and chondrocyte viability until they could be delivered and implanted.

Initially, OCAs were stored in lactated Ringer solution (LRS) at 4°C based on standard protocol.[1] However, chondrocyte viability in OCAs stored in LRS at 4°C rapidly declines to ~ 60% of at-harvest levels by day 7,[7,8] and ~ 20% by day 14[7] after harvest. This essentially renders the grafts unusable for clinical patients. To maintain chondrocyte viability in OCAs for clinically relevant time periods, researchers have applied tissue preservation, rather than storage, methods to this problem by using cell culture protocols for OCAs.[7–18]

Use of various culture media preparations to preserve OCAs at 4°C has been reported to significantly improve chondrocyte viability compared with LRS.[7,14,17,19] These media preparations can be separated into those that include fetal bovine serum (FBS)[12,20,21] and those that do not.[22] When OCA tissues were preserved in culture media without FBS, chondrocyte viability was maintained at 54 to 97% of at-harvest levels through 14 days of storage.[7,8] However, chondrocyte viability dropped significantly to only 15 to 70% of at-harvest levels by day 28 of storage.[7,12,14,23] When FBS was added to media, chondrocyte viability was maintained at 80 to 86% and 45 to 80% of at-harvest levels at days 14 and 28 postharvest, respectively. Chondrocyte viability levels were significantly higher with FBS when directly compared with FBS-free media at all time points.[7,12,13,20] As such, the addition of FBS to media used to preserve OCAs at 4°C is the current standard of care for most tissue banks. However, even with the use of FBS, there is still a significant loss of cell viability by day 28 postharvest, especially in the superficial zone of the cartilage.[13,19,20,22] With the mandatory disease testing period, this leaves a "clinical window for use" of only 14 days, which can be problematic when considering sizing, shipping, and scheduling requirements for tissue banks, surgeons, and patients. In addition, concerns regarding batch-to-batch variability and potential for zoonotic disease transmission and contamination associated with FBS make its use less desirable. Therefore, recent research efforts have focused on methods for improving chondrocyte viability of OCAs to extend their clinical window for use.

A significant increase in apoptotic gene expression has been observed in OCA tissues stored at 4°C,[24] indicating that loss of chondrocyte viability observed during storage is at least partially due to apoptosis. There is evidence that indirectly blocking apoptosis using etanercept to block TNF-α signaling, a proapoptotic pathway, significantly improves chondrocyte viability in the superficial zone.[13] However, total chondrocyte viability in the tissue was not significantly different from untreated controls at 28 days. Similarly, another study found that storing OCA tissue with ZVAD-fmk, a potent apoptosis inhibitor, did not improve chondrocyte viability through 28 days of storage.[7] These data indicate that blocking the apoptosis pathway alone does not effectively increase the clinical window for use of OCAs.

As with the transition from LRS to cell culture medium and the addition of FBS, researchers continue to look to cell culturing procedures to improve maintenance of cell viability of OCA tissue during storage. Several groups have looked at cryopreservation of the tissue at subzero temperatures in various cryoprotectant formulations,[8,25–34] commonly used for long-term storage of cells for culture. However, no study has shown consistent maintenance of chondrocyte viability in OCA tissue, especially in OCAs of sizes that are typically used clinically. Further, cryopreserved tissues have performed poorly when transplanted into in vivo animal models.[32,34] Therefore, cryopreservation of OCA tissue is not currently considered a viable option for clinical use.

Conversely, numerous studies have reported promising results with respect to chondrocyte viability when OCAs are preserved at 37°C, the standard temperature for in vitro cell and tissue culture, when compared with 4°C.[11,16,19,35–38] OCAs preserved at 37°C have consistently had significantly higher chondrocyte viability at day 28 compared with those preserved at 4°C.[19,35,36]

Importantly, these studies have also indicated that supplementation with FBS may not be required to maintain high levels of chondrocyte viability of the OCA preserved at 37°C tissue.[35–37] In one study, there was no significant difference in chondrocyte viability between OCAs preserved with or without FBS at 37°C through 28 days after harvest.[36] Another group used insulin, transferrin, and selenious acid to supplement the media in place of FBS, and reported no reduction in chondrocyte viability through 28 days in storage.[37] Our group has reported maintenance of viable chondrocyte density at harvest levels through 56 days of preservation of OCAs at 37°C using a proprietary medium that does not include FBS (U.S. Patent pending).[35] Therefore, it is possible to preserve osteochondral tissues with clinically applicable chondrocyte viability levels at 37°C without FBS supplementation, removing the possibility for zoonotic transfer of disease and the potential variability in storage media composition that can arise from differences in the variable composition of FBS[39] while effectively tripling the clinical window of use for OCAs.

While the data for 37°C preservation of OCAs have been promising, it has not been adopted as the standard of care protocol for tissue banks. This is at least in part due to the concern that preservation of tissues at 37°C will increase the risk of bacterial and fungal contamination. While higher contamination rates have not been reported to date, there has not been a study specifically designed to determine if OCA preservation at 37°C results in higher rates of contamination, to the authors' knowledge. The other significant hurdle for 37°C storage of OCAs is the associated costs of equipment, supplies, personnel, and training. The transition from preservation in refrigerators to CO_2 incubators would be substantial. Therefore, the associated gain in OCA quality would have to be significant for tissue banks to make this investment and transition. If a safe and effective methodology for OCA preservation could be developed that did not require these large investments, its adoption might be more appealing to tissue banks.

Despite advances in OCA preservation methodology, a remaining problem affecting clinical efficacy of this procedure is related to the significant interspecimen variability in chondrocyte viability of OCAs at the time of implantation.[22] For example, chondrocyte viability levels at 28 days postharvest in OCAs preserved using the same protocol in the same laboratory have ranged from 27 ± 13% to 70 ± 11%.[12–14] Similarly, the highest levels of chondrocyte viability at 28 days postharvest, 60 to 70% have been reported using methods with and without FBS and at both 4 and 37°C.[12–14,36] While this variability may be related to differences in harvest timing or technique, animal model used, or viability assessment technique, the inter- and intra-study variability in chondrocyte viability of OCAs is indicative of true variability in OCA tissues retrieved from a spectrum of organ donors by different harvest teams in different locations, which can have significant impact on graft quality and patient outcomes. Currently, OCA chondrocyte viability at the time of implantation is based on published reference ranges based on postharvest time point. This means that for the typical OCA sent from the tissue bank at 21 to 28 days postharvest, chondrocyte viability could be less than 25% to greater than 90% depending on all associated variables. Unfortunately, the tissue bank, surgeon, and patient currently have no idea what the viability of each graft is at the time of implantation. Because chondrocyte viability is known to influence outcomes for OCA transplantation, it is critical that we develop methods for determining viability in each graft. If the viability of OCAs can be accurately determined nondestructively before transplantation into the patient, samples with unacceptable viability can be discarded, improving success rates and decreasing costs associated with OCA surgery.

In an attempt to address the current limitations in consistently providing tissues of appropriate quality and quantity for use in osteochondral allografting procedures, our goal was to develop a preservation methodology that would maximize chondrocyte viability, minimize disease transmission and contamination potential, allow for nondestructive viability testing, and avoid large financial costs associated with preservation technique. Our hypothesis was that OCA tissues could be preserved at room temperature (~ 25°C) without CO_2 supplementation for at least 56 days with chondrocyte viability maintained at > 70% of at-harvest levels. Further, we hypothesized that we could nondestructively determine chondrocyte viability in OCAs using a novel assay technique that would strongly correlate ($r > 0.7$) to the "gold standard" technique of fluorescent microscopy.

◆ Materials and Methods

Tissue Harvest and Culture

All procedures were performed under Animal Care and Use Committee approval. During the course of two studies, medial and lateral FCs from both knees of 14 adult canine cadavers were aseptically harvested within 4 hours of euthanasia performed for reasons unrelated to this study. The FCs were either used as time 0 (at harvest) controls ($n = 7$) or separated into one of the five test groups based on proprietary media composition (M-1, M-2, and M-3) and container condition (C-1, C-2, C-3) (U.S. Patent pending) such that each FC from a single animal was placed into a distinct group. The following media and container condition groupings were assessed for study one: M-1/C-1 ($n = 7$), M-1/C-2 ($n = 4$), M-1/C-3 ($n = 7$), and M-2/C-1 ($n = 5$). For study two the media and container condition groupings were M-1/C-1 ($n = 5$), M-1/C-3 ($n = 5$), and M-3/C-3 ($n = 8$), resulting in the five different OCA storage groups. The M-1 medium was designed to provide basic tissue nutrition, the M-2 medium was designed to be antidegradative, and the M-3 medium was designed to be anti-inflammatory. Tissues were stored at 25°C without CO_2 supplementation in 60 mL of media for 63 days. The media were changed every 7 days and saved for biomarker analyses. At the end of the storage, osteochondral plugs were evaluated for tissue viability.

Chondrocyte Viability Analysis

Chondrocyte viability was determined using a proprietary metabolic assay (U.S. Patent pending) and the fluorescent live tissue stain Calcein AM (Invitrogen, Grand Island, NY). For the metabolic assay, OCAs were incubated in a proprietary solution for 24 hours. After 24 hours, a media sample was analyzed for level of fluorescence (550/590 nm ex/em) at a standard sensitivity using a Synergy HT plate reader (BioTek, Winoaski, VT). Increased fluorescence in the media is indicative of cell metabolism and viability. On the last day of storage, OCA tissues were assessed for viable chondrocyte density using the fluorescent live cell stain Calcein AM and fluorescent microscopy. Osteochondral tissues were incubated in stain for 25 minutes at 25°C. Images were taken at either 4× (study 1) or 10× (study 2) magnification. Green-staining live cells were manually counted through the depth of the tissue, and the area of the tissue analyzed was determined. Chondrocyte viability of the tissue was expressed as the ratio of live cells (LC)/area (μm^2). Because the focal depth of 4× images was significantly different from the focal depth of 10× images, the viability could not be compared between the 4× and 10× images, and analysis was only performed between samples that were taken at the same magnification. Percentage of day 0 viable chondrocyte density was determined by comparing the mean day 0 viable chondrocyte density to the viable chondrocyte density for each sample in all groups.

Media Analysis

Media were assessed for concentrations using a novel panel of biomarkers (U.S. Patent pending) including vascular endothelial growth factor (VEGF), matrix metalloproteinases (MMP)-2, -3, -9, and -13, interleukin (IL)-6, IL-8, keratinocyte chemoattractant (KC), and monocyte chemotactic protein (MCP)-1 using Luminex multiplex assays (R&D System, Minneapolis, MN, and Millipore, Billerica, MA) according to the manufacturers' protocols.

Data Analysis

All data were analyzed using SigmaPlot 11 (San Jose, CA). Direct comparisons between groups for chondrocyte viability were analyzed using ANOVA and the Tukey post hoc test. To determine correlation between media biomarker concentrations, fluorescence values, OCA chondrocyte viability, and data from the last day of storage were compared by Pearson product-moment correlation. Significance was set at $p < 0.05$ and correlations were considered strong at $r > 0.7$.

◆ Results

Day 0 and Day 63 Tissue Viability

Mean chondrocyte viability levels and ranges are listed for each group at day 63 for each magnification in **Table 12.1**. For the samples analyzed at 4× magnification, time 0 (at-harvest) and the M-3/C-3 group had significantly higher viable chondrocyte density (LC/mm^2) compared with the M-1/C-1 and M-2/C-1 ($p \leq 0.039$ and $p \leq 0.004$, respectively) groups at day 63. The M-1/C-3 group had significantly ($p = 0.018$) higher viable chondrocyte density compared with the M-2/C-1 group at day 63. The M-3/C-3 group had the highest mean viable chondrocyte density and the lowest variability of all groups. The sample size was smaller for the 10× magnification groups, and there was not a significant

Table 12.1 Viable chondrocyte density (LC/mm²) for each preservation protocol (viable cell density was determined using images taken at 4× or 10× magnification)

Media	Container condition	4× Tissue viability (LC/mm²)		10× Tissue viability (LC/mm²)	
		Mean	Range	Mean	Range
M-1	C-1	0.916	0.038–3.24	0.623	0.0–1.38
M-1	C-2	2.115	1.29–2.62		
M-1	C-3	2.217	0.1–3.49[a]	0.804	0.213–1.36
M-2	C-1	0.0423	0.0–0.117		
M-3	C-3	3.195	2.99–3.31[b]	1.137	0.97–1.29
Day 0		2.901	0.5–5.35[b]	1.13	1.02–1.25

[a]Significantly higher than M-2/C-1.
[b]Significantly higher than M-1/C-1 and M-2/C-1.

difference among groups. However, in agreement with study 1 data, the M-3/C-3 group had the highest mean viable chondrocyte density with the lowest variability, with values similar to time 0 controls. When the viable chondrocyte density data were considered as a percentage of day 0 viable chondrocyte density (**Fig. 12.1**), the time 0 and M-3/C-3 were significantly ($p < 0.05$) higher than the M-1/C-1 and M-2/C-1 groups at day 63. These data indicate that media composition and preservation "environment" significantly affect OCA chondrocyte viability during storage at 25°C. Container condition C-3 in combination with media M-3 maintained chondrocyte viable cell density at 104% of time 0 (at-harvest) levels for 63 days postharvest (**Table 12.1** and **Fig. 12.1**).

Day 0 and Day 63 Cell Distribution

The time 0 and M-3/C-3 groups had viable chondrocytes distributed throughout the cartilage tissue. The M-1/C-3 group had lower

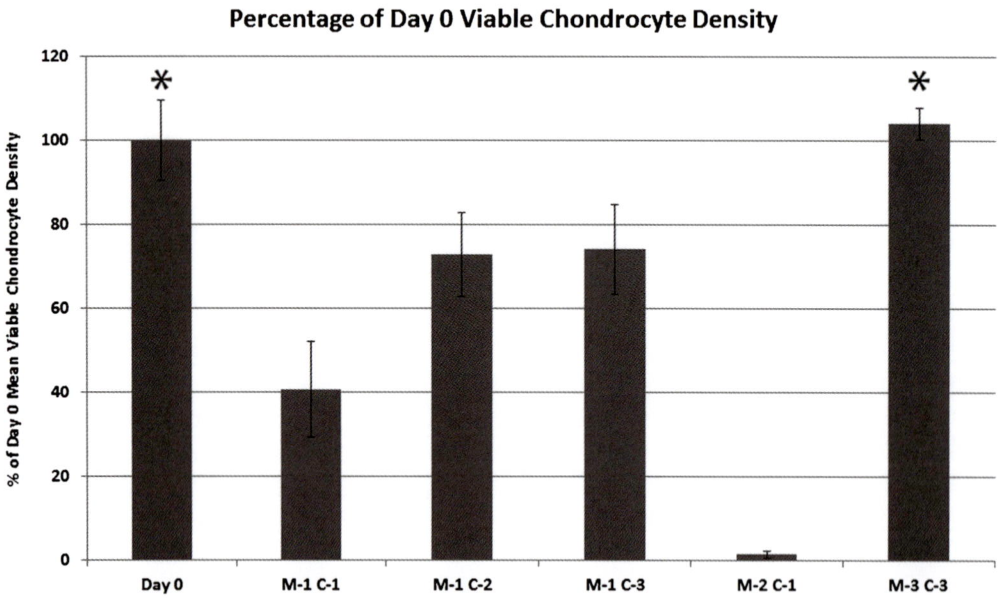

Fig. 12.1 Mean (±se) percentage of day 0 viable chondrocyte density for each group on day 63 (*significantly higher than M-1 C-1 and M-2 C-1 groups).

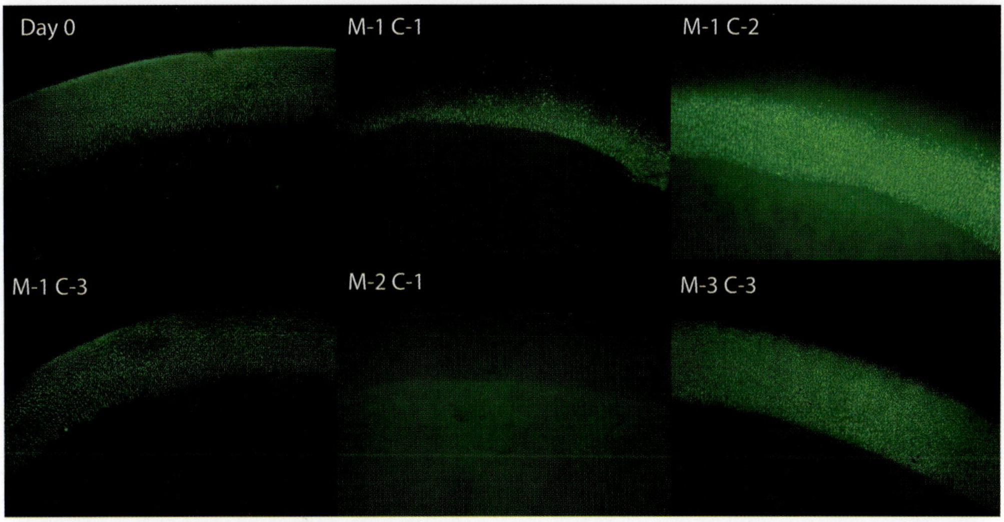

Fig. 12.2 Representative 4× images for viable chondrocyte density throughout OCAs at day 63 for each group compared with time 0 (at-harvest) controls.

numbers of viable chondrocytes in the superficial zone of the cartilage compared with the M-3/C-3 group but higher than the other M-1 groups. The M-1/C-1 and M-1/C-2 groups had very low numbers of viable chondrocytes in the superficial and middle zones of the cartilage. The M-2/C-1 group had very few viable chondrocytes in any region of the cartilage. These data indicate that the M-3/C-3 group more consistently maintained chondrocyte viability in the superficial zone compared with the other storage media and container conditions (**Fig. 12.2**).

Metabolic Assay Analysis

M-1/C-3 and M-3/C-3 groups had significantly ($p < 0.001$) higher tissue metabolic activity compared with all other groups based on the level of fluorescence in the media (**Fig. 12.3**).

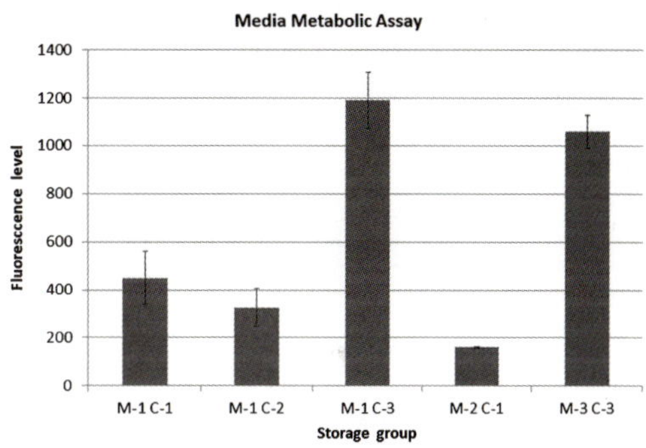

Fig. 12.3 Mean (±se) fluorescence level using our novel metabolic assay for chondrocyte viability in osteochondral allografts for groups on day 63 of preservation (*significantly higher than M-1 C-1, M-1 C-2, and M-2 C-1 groups).

Media Protein Analysis

The OCA tissues released detectable levels of MMP-2, MMP-3, MMP-13, KC, IL-6, IL-8, MCP-1, and VEGF throughout the 63-day study period, indicating that the tissues remained metabolically active throughout preservation. The M-3 medium was designed to reduce inflammation during storage, and the concentration of KC, IL-6, IL-8, MCP-1, and VEGF in the storage media was significantly ($p \leq 0.05$) lower in the M-3/C-3 group compared with the M-1/C-3 group throughout the study (**Fig. 12.4**).

Correlation Analyses

In study 1, significant ($p < 0.001$) moderate to strong positive correlations to viable chondrocyte density were found for the novel metabolic assay data ($r = 0.724$),

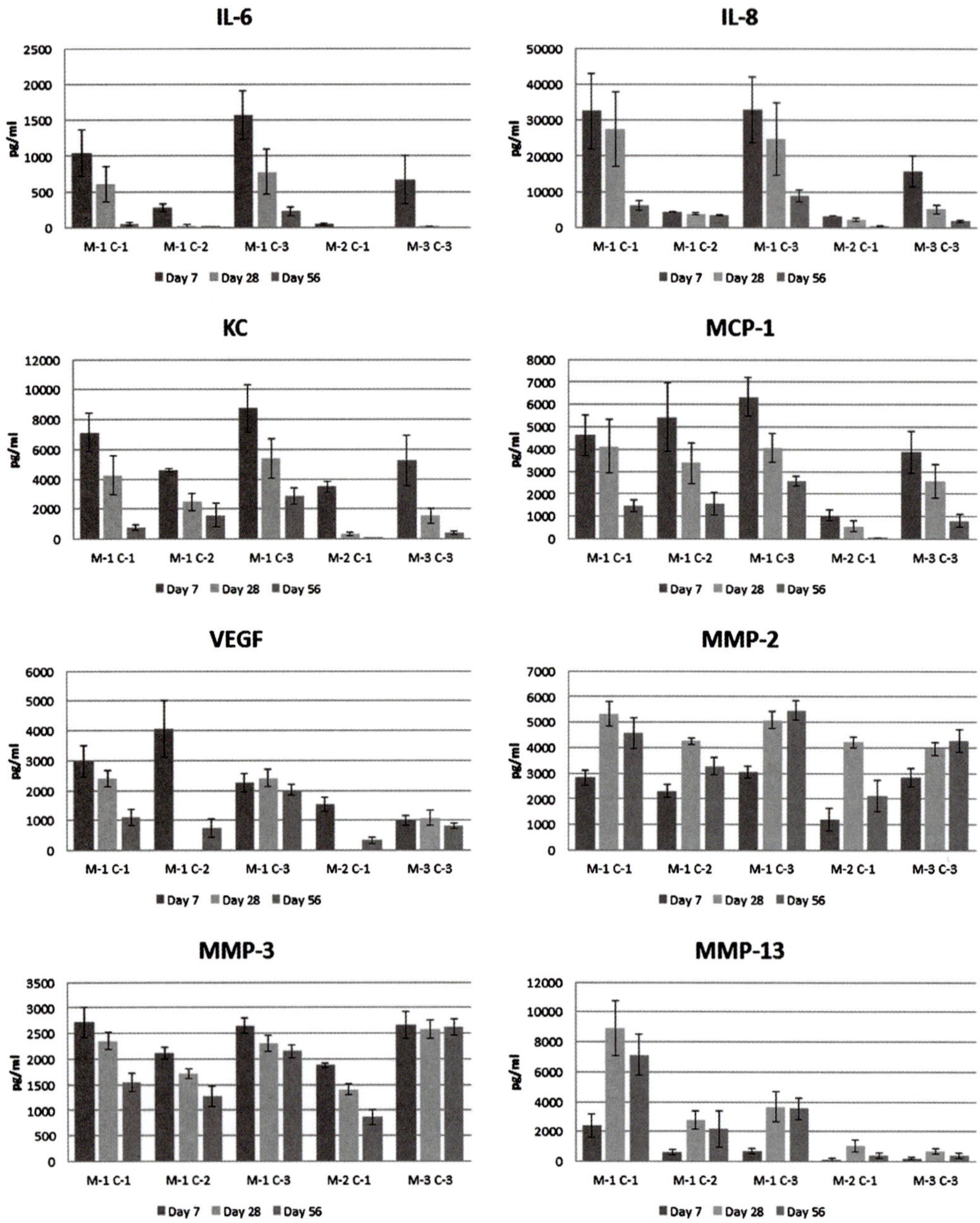

Fig. 12.4 Mean (±se) media protein concentration (pg/mL) from study 1 and study 2 on days 7, 28, and 63 of storage for IL-6, IL-8, KC, MCP-1, VEGF, MMP-2, MMP-3, and MMP-13. *Abbreviations:* IL, interleukin; KC, keratinocyte chemoattractant; MCP, monocyte chemotactic protein; VEGF, vascular endothelial growth factor; MMP, matrix metalloproteinase.

IL-8 ($r = 0.598$), VEGF ($r = 0.655$), KC ($r = 0.738$), MCP-1 ($r = 0.822$), MMP-2 ($r = 0.699$), and MMP-3 ($r = 0.682$). There were not significant or strong correlations to viable chondrocyte density for IL-6 ($r = 0.385$, $p = 0.0694$) and MMP-13 ($r = 0.203$, $p = 0.319$). In study 2, a significant ($p < 0.001$) and strong ($r = 0.761$) positive correlation to viable chondrocyte density was found for the novel metabolic assay data; however, there were not significant or strong correlations between viable chondrocyte density and any of the protein biomarkers analyzed. These data indicate that our novel metabolic assay for OCA tissues correlates strongly with OCA viable chondrocyte density, such that it could be used to determine the quality of individual OCAs before shipping for clinical use (**Fig. 12.5**).

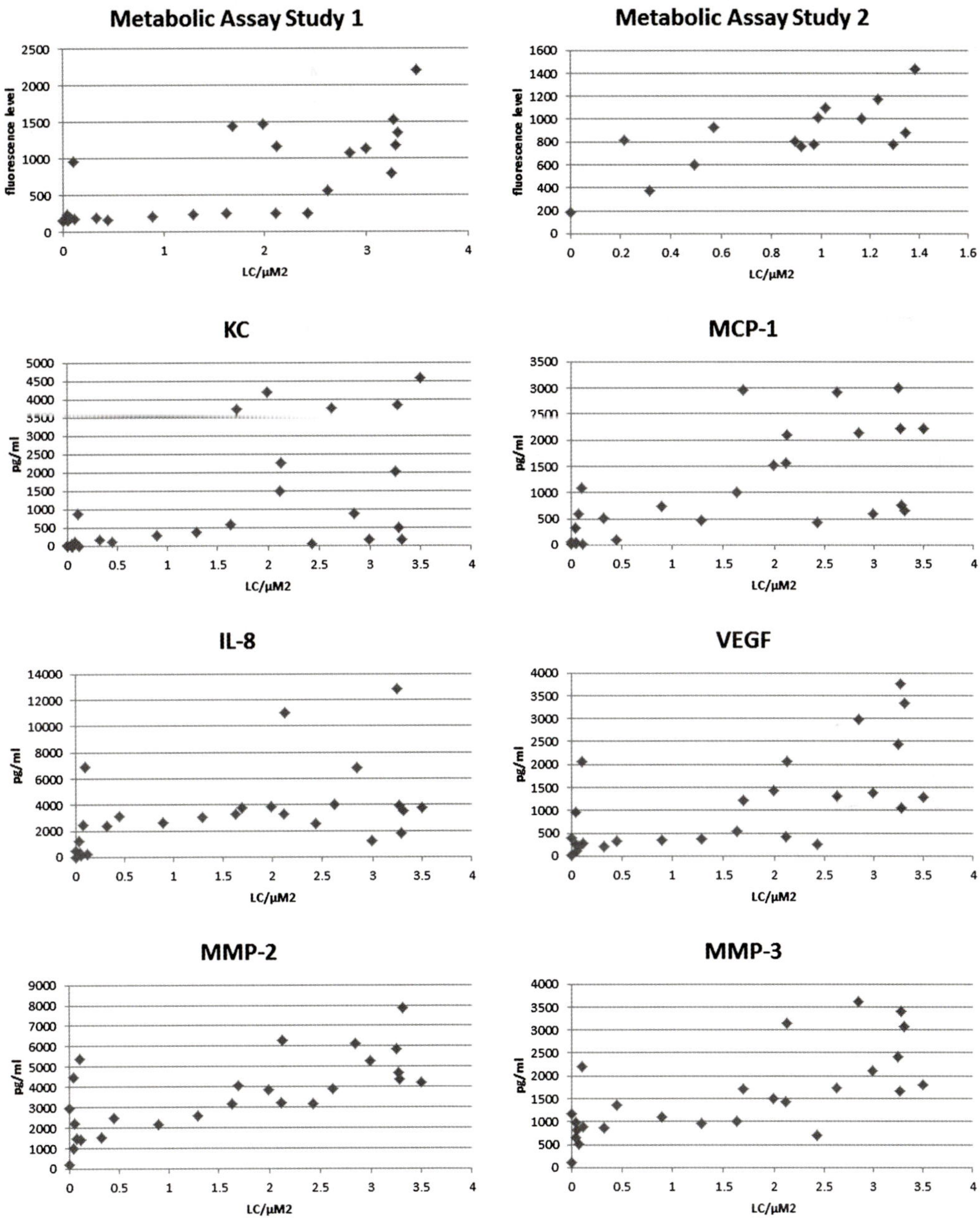

Fig. 12.5 Correlation of tissue viability (LC/mm^2) to metabolic assay fluorescence level in study 1 ($r = 0.724$) and study 2 ($r = 0.761$), and study 1 media protein concentrations (pg/mL) for KC ($r = 0.738$), MCP-1 ($r = 0.822$), IL-8 ($r = 0.598$), VEGF ($r = 0.655$), MMP-2 ($r = 0.699$), and MMP-3 ($r = 0.682$). *Abbreviations:* KC, keratinocyte chemoattractant; MCP, monocyte chemotactic protein; IL, interleukin; VEGF, vascular endothelial growth factor; MMP, matrix metalloproteinase.

Discussion

These data show that osteochondral tissues can be preserved at room temperature for at least 63 days with viable chondrocyte density maintained at > 90% of at-harvest levels, and that chondrocyte viability can be accurately assessed without sacrificing donor tissue using a novel metabolic assay on the preservation media. The practical implications of this work are that OCAs can be effectively preserved without the use of refrigerators or incubators such that the clinical window of use is three times longer than the current standard. Importantly, this preservation system allows for determination of viability of each graft so that tissue bank, surgeon, and patient can be confident about the quality of the graft to be implanted.

Using tissue banks' current protocols for preservation of osteochondral tissues, there is consistently a significant and progressive decrease in chondrocyte viability over time in storage. Further, the rate of cell death appears to increase after day 14 postharvest. This is of particular significance as this is the time point when grafts first become available for clinical use in the United States based on mandatory disease testing. Chondrocyte viability has been reported to be a critical factor for long-term success for OCA transplantation procedures, with most surgeons using a 70% level as a relative cut-off point for a "high quality" OCA. Therefore, current recommendations for clinical use of OCA tissues are to transplant the graft as soon after it has been cleared based on disease testing, and not longer than 28 days after harvest based on 28-day viability levels. This short 2-week window in conjunction with sizing and scheduling issues, as well as loss of tissues that fail disease testing, results in a significant shortage of OCAs available for clinical use. Therefore, the ability to better maintain chondrocyte viability in stored OCA tissues over longer periods of time postharvest is a critical need and the subject of numerous studies.

The data presented in this study and reported in the literature clearly indicate that preservation of OCAs at physiologic (37°C) or room (25°C) temperatures significantly improves chondrocyte viability of osteochondral tissues compared with the currently used 4°C protocols. Further, the data presented here show that OCA chondrocyte viability can be maintained at time 0 levels for 63 days at room temperature with optimized media and container conditions. This improvement in chondrocyte viability could significantly expand the window for clinical use of OCA grafts and therefore increase availability, ease scheduling issues, and improve success rates for the OCA procedures in the United States.

Another factor reported to affect success of OCA procedures is the viability of the bone portion of the graft.[2,16] However, unlike the cartilage portion of the OCA, it is suggested that devitalization of the bone component results in improved graft incorporation and decreased failure rates. It is theorized that because bone is not an immunologically privileged tissue, host immune responses to viable bone can result in poor incorporation and eventual collapse of the graft.[40] In support of this theory, OCA grafts stored for longer periods of time before transplantation had better graft incorporation than those stored for shorter time periods based on MRI assessment.[2] Bone devitalizes more rapidly in OCA tissues stored at 37°C compared with 4°C,[16] providing more impetus for use of temperatures higher than 4°C for preservation of OCA tissues. It is important to note that much of the long-term data regarding OCAs is based on patients who had grafts implanted within 7 days of harvest.[1,3] Therefore, conclusions regarding the impact of viability of the bone portion of OCAs on outcome using current protocols cannot be clearly made without further research.

No matter what preservation protocol is used, there can be significant variability in chondrocyte viability among stored osteochondral tissues at any time point.[8] Currently, differentiation between low-viability and high-viability OCAs is not performed before implantation. Therefore, chondrocyte viability in the OCA tissues sent to surgeons for clinical use is a "best guess" based on ranges determined by historical data, and the surgeon is not informed with respect to the quality of the graft being transplanted into the patient. We have developed, and now validated, a nondestructive method for accurately assessing chondrocyte viability of stored osteochondral tissues. By testing a sample of the preservation media immediately before shipping, chondrocyte viability can be determined and used to prevent poor-quality grafts from being used in patients. However, the level of chondrocyte viability required for consistent long-term success of OCA grafts has not been clearly determined. Therefore, further study is required to determine clear criteria of success for OCA transplantation and then correlate our viability assay to those criteria for success.

While it is clear that preservation of osteochondral tissues at room temperature can maintain high levels of chondrocyte viability for extended periods of time, the safety of this preservation protocol needs to be fully evaluated before this method can be recommended for use by tissue banks. In this study, we did not perform microbial testing to determine risk of bacterial and fungal contamination compared with tissues stored at 4°C using current media and container conditions. However, in vitro and in vivo studies designed to address this question are currently ongoing in our laboratory. In addition, these ongoing in vivo studies will evaluate long-term functional success of OCAs stored for 30 and 60 days using our optimized preservation protocol in comparison to the current standard of care.

Conclusion

Over the last decade, significant research has been performed to develop and improve protocols for preservation of osteochondral tissue before transplantation into patients for treatment of cartilage defects. This work has resulted in preservation protocols that allow for maintenance of OCA tissues for time periods sufficient for clinical use based on disease-testing requirements in the United States. However, graft quality and the window for clinical use of these tissues could be greatly enhanced from current levels. If the preservation protocol reported here can be validated for safety and functional outcome, it could then be employed in tissue banks throughout the world, decreasing the number of grafts discarded and improving the quality of life for thousands of patients affected by cartilage defects.

References

1. Gross AE, Silverstein EA, Falk J, Falk R, Langer F. The allotransplantation of partial joints in the treatment of osteoarthritis of the knee. Clin Orthop Relat Res 1975;(108):7–14
2. Williams RJ III, Ranawat AS, Potter HG, Carter T, Warren RF. Fresh stored allografts for the treatment of osteochondral defects of the knee. J Bone Joint Surg Am 2007;89(4):718–726
3. Gross AE, Kim W, Las Heras F, Backstein D, Safir O, Pritzker KP. Fresh osteochondral allografts for posttraumatic knee defects: long-term followup. Clin Orthop Relat Res 2008;466(8):1863–1870
4. Farr J, Cole B, Dhawan A, Kercher J, Sherman S. Clinical cartilage restoration: evolution and overview. Clin Orthop Relat Res 2011;469(10):2696–2705
5. Bakay A, Csönge L, Papp G, Fekete L. Osteochondral resurfacing of the knee joint with allograft. Clinical analysis of 33 cases. Int Orthop 1998;22(5): 277–281
6. Beaver RJ, Mahomed M, Backstein D, Davis A, Zukor DJ, Gross AE. Fresh osteochondral allografts for post-traumatic defects in the knee. A survivorship analysis. J Bone Joint Surg Br 1992;74(1): 105–110
7. Teng MS, Yuen AS, Kim HT. Enhancing osteochondral allograft viability: effects of storage media composition. Clin Orthop Relat Res 2008;466(8):1804–1809
8. Ball ST, Amiel D, Williams SK, et al. The effects of storage on fresh human osteochondral allografts. Clin Orthop Relat Res 2004;(418):246–252
9. Black J, Shadle CA, Parsons JR, Brighton CT. Articular cartilage preservation and storage. II. Mechanical indentation testing of viable, stored articular cartilage. Arthritis Rheum 1979;22(10):1102–1108
10. Brighton CT, Shadle CA, Jimenez SA, Irwin JT, Lane JM, Lipton M. Articular cartilage preservation and storage. I. Application of tissue culture techniques to the storage of viable articular cartilage. Arthritis Rheum 1979;22(10):1093–1101
11. Thomas VJ, Jimenez SA, Brighton CT, Brown N. Sequential changes in the mechanical properties of viable articular cartilage stored in vitro. J Orthop Res 1984;2(1):55–60
12. Pennock AT, Wagner F, Robertson CM, Harwood FL, Bugbee WD, Amiel D. Prolonged storage of osteochondral allografts: does the addition of fetal bovine serum improve chondrocyte viability? J Knee Surg 2006;19(4):265–272
13. Linn MS, Chase DC, Healey RM, Harwood FL, Bugbee WD, Amiel D. Etanercept enhances preservation of osteochondral allograft viability. Am J Sports Med 2011;39(7):1494–1499
14. Williams SK, Amiel D, Ball ST, et al. Prolonged storage effects on the articular cartilage of fresh human osteochondral allografts. J Bone Joint Surg Am 2003;85-A(11):2111–2120
15. Amiel D, Harwood FL, Hoover JA, Meyers M. A histological and biochemical assessment of the cartilage matrix obtained from in vitro storage of osteochondral allografts. Connect Tissue Res 1989;23(1):89–99

16. Bastian JD, Egli RJ, Ganz R, Hofstetter W, Leunig M. Chondrocytes within osteochondral grafts are more resistant than osteoblasts to tissue culture at 37°C. J Invest Surg 2011;24(1):28–34
17. Williams JM, Virdi AS, Pylawka TK, Edwards RB III, Markel MD, Cole BJ. Prolonged-fresh preservation of intact whole canine femoral condyles for the potential use as osteochondral allografts. J Orthop Res 2005;23(4):831–837
18. Pennock AT, Robertson CM, Wagner F, Harwood FL, Bugbee WD, Amiel D. Does subchondral bone affect the fate of osteochondral allografts during storage? Am J Sports Med 2006;34(4):586–591
19. McCarty WJ, Pallante AL, Rone RJ, Bugbee WD, Sah RL. The proteoglycan metabolism of articular cartilage in joint-scale culture. Tissue Eng Part A 2010;16(5):1717–1727
20. Williams RJ III, Dreese JC, Chen CT. Chondrocyte survival and material properties of hypothermically stored cartilage: an evaluation of tissue used for osteochondral allograft transplantation. Am J Sports Med 2004;32(1):132–139
21. Malinin T, Temple HT, Buck BE. Transplantation of osteochondral allografts after cold storage. J Bone Joint Surg Am 2006;88(4):762–770
22. Allen RT, Robertson CM, Pennock AT, et al. Analysis of stored osteochondral allografts at the time of surgical implantation. Am J Sports Med 2005;33(10):1479–1484
23. Csönge L, Bravo D, Newman-Gage H, et al. Banking of osteochondral allografts, Part II. Preservation of Chondrocyte Viability During Long-Term Storage. Cell Tissue Bank 2002;3(3):161–168
24. Robertson CM, Allen RT, Pennock AT, Bugbee WD, Amiel D. Upregulation of apoptotic and matrix-related gene expression during fresh osteochondral allograft storage. Clin Orthop Relat Res 2006;442:260–266
25. Ohlendorf C, Tomford WW, Mankin HJ. Chondrocyte survival in cryopreserved osteochondral articular cartilage. J Orthop Res 1996;14(3):413–416
26. Gole MD, Poulsen D, Marzo JM, Ko SH, Ziv I. Chondrocyte viability in press-fit cryopreserved osteochondral allografts. J Orthop Res 2004;22(4):781–787
27. Jomha NM, Law GK, McGann LE. Storage of porcine articular cartilage at high subzero temperatures. Cell Tissue Bank 2006;7(1):55–60
28. Acosta CA, Izal I, Ripalda P, Forriol F. Cell viability and protein composition in cryopreserved cartilage. Clin Orthop Relat Res 2007;460:234–239
29. Judas F, Rosa S, Teixeira L, Lopes C, Ferreira Mendes A. Chondrocyte viability in fresh and frozen large human osteochondral allografts: effect of cryoprotective agents. Transplant Proc 2007;39(8):2531–2534
30. Brockbank KG, Chen ZZ, Song YC. Vitrification of porcine articular cartilage. Cryobiology 2010; 60(2):217–221
31. Jomha NM, Weiss AD, Fraser Forbes J, Law GK, Elliott JA, McGann LE. Cryoprotectant agent toxicity in porcine articular chondrocytes. Cryobiology 2010;61(3):297–302
32. Forriol F, Longo UG, Alvarez E, et al. Scanty integration of osteochondral allografts cryopreserved at low temperatures with dimethyl sulfoxide. Knee Surg Sports Traumatol Arthrosc 2011;19(7): 1184–1191
33. Chase D, Gortz S, Pallante A, et al. Histological and Biochemical Characterization of Fresh vs. Frozen Osteochondral Allografts after Six Months In Vivo in a Goat Model Paper presented at: Annual Meeting of the Orthopaedic Research Society 2011; Long Beach, CA
34. Chase D, Görtz S, Pallante AL, et al. Histological and Biochemical Characterization of Fresh vs. Frozen Osteochondral Allografts after Six Months In Vivo in a Goat Model. Orthop Res Soc (ORS) Meeting 2011; Long Beach, CA (Poster 2098)
35. Stoker A, Garrity J, Sims H, Cook J. Optimization of Osteochondral Allograft Preservation to Extend the Usable Life Span of Harvested Tissue. Paper presented at: Annual Meeting of the Orthopaedic Research Society 2011; Long Beach, CA
36. Pallante AL, Bae WC, Chen AC, Görtz S, Bugbee WD, Sah RL. Chondrocyte viability is higher after prolonged storage at 37 degrees C than at 4 degrees C for osteochondral grafts. Am J Sports Med 2009; 37(Suppl 1):24S–32S
37. Bian L, Stoker AM, Marberry KM, Ateshian GA, Cook JL, Hung CT. Effects of dexamethasone on the functional properties of cartilage explants during long-term culture. Am J Sports Med 2010;38(1): 78–85
38. López C, Ajenjo N, Muñoz-Alonso MJ, Farde P, León J, Gómez-Cimiano J. Determination of viability of human cartilage allografts by a rapid and quantitative method not requiring cartilage digestion. Cell Transplant 2008;17(7):859–864
39. van der Valk J, Brunner D, De Smet K, et al. Optimization of chemically defined cell culture media—replacing fetal bovine serum in mammalian in vitro methods. Toxicol In Vitro 2010;24(4): 1053–1063
40. Friedlaender GE, Horowitz MC. Immune responses to osteochondral allografts: nature and significance. Orthopedics 1992;15(10):1171–1175

13

Osteochondral Allograft Transplantation in the Knee

William Bugbee, Marco Cavallo, and Sandro Giannini

The concept of treating articular cartilage diseases with bone and cartilage substitution in the knee has now a history of more than a century, since the first joint transplantation described by Lexer in 1908.[1,2] Animal and clinical studies concerning transplantation and immunology were performed in the 1960s, demonstrating that transplanted fresh cadaver cartilage is viable.[3–5] In the 1970s, Gross and colleagues began reporting on their experience with osteochondral allograft (OCA) for post-traumatic and periarticular tumor reconstruction.[6,7] In the 1980s, Meyers et al first applied this technique to specific chondral and osteochondral diseases such as chondromalacia, osteoarthritis, and osteonecrosis,[8] developing the shell-shaped graft. Later in the 1990s, Garrett first reported on the use of allograft plugs for the treatment of osteochondritis dissecans (OCD) of the knee.[9] In the past 20 years a large number of basic scientific and clinical studies have been performed by several investigators. These studies and the increasing availability of fresh allografts have led to an increasing popularity of fresh allografts and the inclusion of this procedure as part of the "cartilage repair paradigm" for the treatment of chondral or osteochondral lesions in the knee.[10–12]

Allograft Recovery, Processing, and Storage

Historically, in North America, fresh OCA procedures were performed at university-based centers that had associated tissue banks, which independently established recovery, processing, and release protocols. Fresh OCAs were typically stored in lactated Ringer solution and transplanted fresh within 1 week after the donor's death. Under this model ~100 allografts per year were implanted in North America in the 1980s and 1990s. Beginning around 1998, commercially supplied allografts became available in the United States through several tissue banks that established new protocols under the oversight of the Food and Drug Administration. Commercial distribution of grafts required a prolonged storage interval (10 to 45 days) to allow for completion of recovery and testing protocols. This resulted in an increase in the number of allografting procedures performed in the United States to ~2,000 per year.

Allograft tissue recovery is performed within 12 to 24 hours of the donor's death.[13] Suitable donors are generally between 15 and 35 years of age with macroscopically healthy articular cartilage. As the transplantation procedure is based on cartilage substitution, a process that maintains allograft cartilage tissue health during storage is mandatory. Many studies have been performed to identify the ideal storage media and to evaluate the effects of hypothermic storage on chondrocytes and extracellular matrix (ECM).[14–19]

OCA can be stored frozen, cryopreserved, or fresh. Each of these options affects chondrocyte viability, immunogenicity, and length of time to transplantation. Frozen grafts showed a chondrocyte survivorship of less than 5% because of the freezing process at −80°C.[20] As chondrocytes are responsible for maintenance of the ECM, studies have shown that the matrix in these frozen allografts tends to deteriorate over time.[21,22] Along with the decreased chondrocyte viability, frozen allografts showed decreased immunogenicity.[23]

With cryopreservation it is possible to maintain chondrocyte viability during this freezing process by adding glycerol and dimethyl sulfoxide to the tissue. Theoretically, the addition of these chemicals prevents ice formation within cells. Multiple studies have reported variable results, with chondrocyte survival ranging from 20 to 70%.[24–27] Unfortunately, viable cells were found only at the surface of the articular cartilage layer.[28] Fresh allografts proved to have the highest rates of chondrocyte viability of the three different methods of storage.[19,25,29,30] Fresh grafts are usually placed in tissue culture medium at 4°C (or potentially 37°C). Chondrocyte viability is significantly affected by length of storage, with little effect from storage times less than 1 week.[31,32] The time of storage before implantation is a key point. Studies have shown a time-dependent decreased chondrocyte viability and degradation of biomechanical properties of fresh grafts stored for more than 14 days.[33–35] Currently the trend of the tissue banks is to hold transplants for a minimum of 14 days, to allow completion of microbiologic and serologic testing before release.[36]

More recently a new off-the-shelf alternative to classic OCA has been developed and released in the U.S. market: The Chondrofix Osteochondral Allograft (Zimmer, Inc., Warsaw, IN). This product is an OCA consisting of decellularized hyaline cartilage and cancellous bone, recovered by an accredited tissue bank, processed to be sterile and viral-inactivated, hydrated, precut, and ready for implantation. The relative advantages include an off-the-shelf availability, sterility, and ease of use, whereas potential limits are availability in sizes only up to 15 mm and the absence of viable cells within the graft. Currently no published peer-reviewed data are available; however, preliminary experience suggests a place for this product in the pool of newer alternatives for chondral and osteochondral repair or replacement.

Biologic Response to Implanted OCA

Intact hyaline cartilage is a relatively immunoprivileged tissue as it is not vascularized and its cellular portion is embedded in the ECM, inaccessible to the host immune system. Conversely, the osseous component of the graft is laden with potentially immunogenic cells and proteins, which can be partially mechanically removed by graft lavage before implantation. Several studies have demonstrated that the osseous portion of the graft is replaced with time by host bone, the process of creeping substitution, which may or may not lead to complete replacement of allograft bone by host bone.[13,37,38] In another study, larger grafts (> 10 cm²) were noted to be far more likely to elicit a systemic immune response.[39] These studies have led to our practice of transplanting the minimal bone volume necessary for osseous restoration or fixation, to facilitate this integration process.

Indication for Allografts

The structural features and multishaping possibilities make OCA suitable for the treatment of a wide spectrum of diseases, which can be grouped into two main paradigms (**Table 13.1**). The first treatment paradigm includes complex reconstruction procedures to address such conditions as posttraumatic deformity, degenerative lesions associated with intra-articular fracture malunion, most commonly of the tibial plateau,[40,41] and unicompartmental arthrosis or multifocal

Table 13.1 Indications for OCA in the knee

Complex reconstructions	Cartilage repair
Posttraumatic and degenerative lesions associated with articular fracture malunion	Chondral or osteochondral defects larger than 2 cm^2
Unicompartmental or multifocal osteoarthritis	OCD
Massive OCD	Revision procedure in case of previous cartilage repair failures
Osteonecrosis	

Abbreviations: OCA, osteochondral allografts; OCD, osteochondritis dissecans.

chondrosis, including patellofemoral degeneration.[42–44] This group also includes massive type 3 or 4 OCD,[9,10] osteonecrosis,[45] as well as other diseases primarily affecting the subchondral bone. The second treatment paradigm addresses conditions primarily affecting articular cartilage. These include large chondral defects treated primarily with allografts or defects that have been previously treated by another cartilage repair technique such as microfracture, osteochondral autograft transfer, or autologous chondrocyte implantation that have failed or that have developed compromise of the subchondral bone.

Although the knee is the most common joint for osteochondral allografting, experience in other joints has been reported. Several case series have been reported in the ankle joint. Good results have been shown with the use of allografts in the treatment of large osteochondral lesions of the talus.[46–50] Mixed results have been demonstrated for bipolar shell grafting for ankle osteoarthritis.[51,52] Experience with allografts has also been described in the hip or in the shoulder, as treatment of femoral or humeral head osteonecrosis, or for osteochondral lesions associated with shoulder instability.[53–55]

Surgical Techniques

As allografting procedure in the knee is used to treat a wide spectrum of diseases, the surgical technique is strictly related to the characteristics of the lesion and the surface to be grafted. Common to all the techniques is the use of a tourniquet and a leg positioner able to hold the knee in varying degrees of flexion (70 to 130 degrees). Before the surgical incision on the patient it is mandatory to inspect the graft, to verify its integrity and appropriate sizing. Generally a midline incision is performed, and the joint is entered medially or laterally depending on the lesion location. Retractors are positioned with care in protecting menisci, healthy cartilage, and cruciate ligaments. The lesion is then visualized, mapped, and treated with the most appropriate technique. Most femoral condyle lesions can be treated with plug or dowel grafts but occasionally a shell (posterior femoral condyle) or small fragment graft (tibial plateau or patella) is necessary.

Femoral Plug Grafts

The ideal lesion candidate for this treatment is a chondral or osteochondral defect in an easily accessible surface of the knee (**Fig. 13.1**). Once plugs have been chosen as the ideal technique the lesion is mapped with multiple diameter-sizing dowels to plan the reconstruction. In case of wide lesion multiple plugs can be used to resurface the entire affected area, and in this case it is mandatory to proceed sequentially with the plugs in anteroposterior or posteroanterior direction. The position identified with the sizing dowel is fixed with a guide wire and the lesion is drilled with a reamer of the same diameter to 5 to 7 mm in depth to minimize the bone component of the graft. In selected cases of massive subchondral bone disruption the reaming depth can exceed 7 mm to position the plug onto a healthy bony tissue. After the guide wire removal, measurements are taken at the four poles of the reamed lesion. The plug is then harvested from the graft with a graft-harvesting reamer in the corresponding area of the lesion site to better match the condyle curvature. The depth measurements are transferred to the plug and the excess of bone is resected. After high-pressure lavage to remove marrow elements, the plug is positioned onto the recipient site and carefully tamped into place or compressed into the

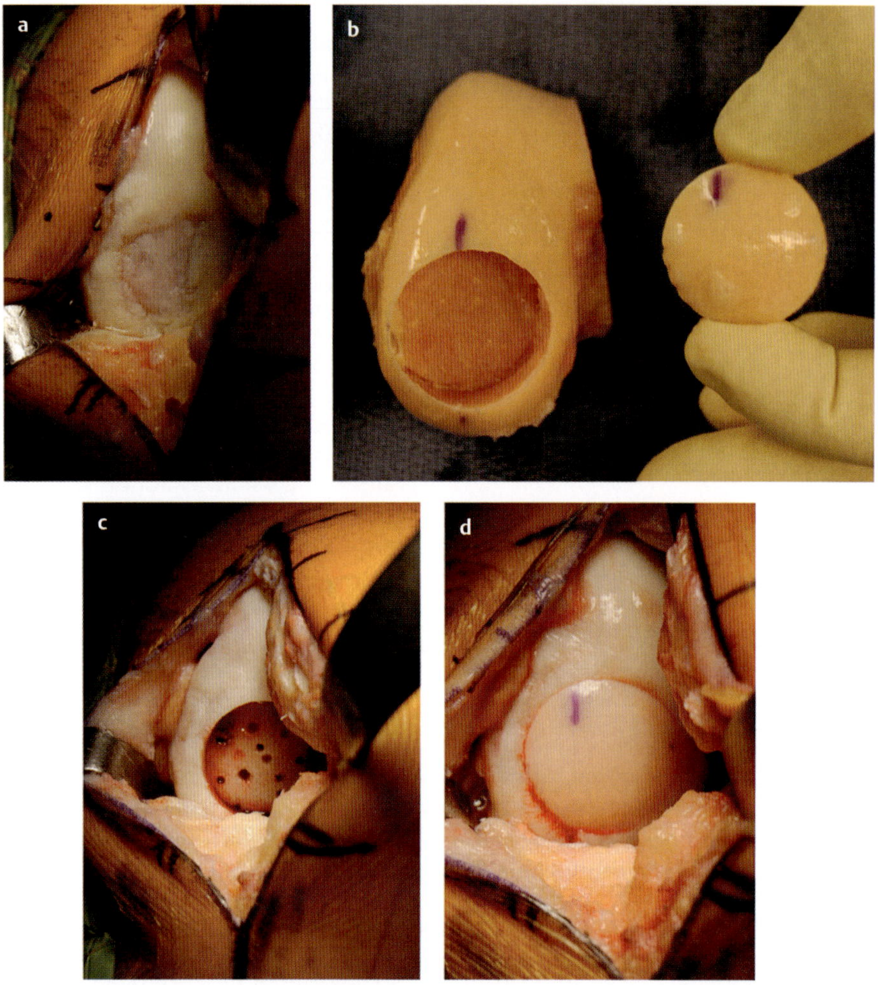

Fig. 13.1 Surgical technique for osteochondral allograft plug. (**a**) Intraoperative photo of a type 4 osteochondritis dissecans lesion of the lateral femoral condyle. (**b**) Allograft plug after harvest from femoral condyle. (**c**) Lesion after preparation with a cylindrical reamer and drilling of sclerotic bone at the base of the lesion. (**d**) Allograft secured in place with press-fit technique.

site with the help of passive joint range-of-motion forces. It is important to press-fit the graft with low-energy impacts to minimize chondrocyte injury.[56,57] If necessary, fixation can be augmented with bioabsorbable screws or chondral darts. When additional plugs are required that are either juxtaposed or partially overlapped, care should be taken not to dislodge the first graft when reaming for the second. A standard closure concludes the procedure. Postoperatively, patients are maintained in a partial weight-bearing status for a period of 4 to 12 weeks and followed radiographically at regular intervals until radiographic evidence of graft incorporation.

Femoral Shell Grafts

In cases of complex lesions or lesions in inaccessible areas, such as the posterior portion of the condyle, a shell allograft is indicated. In this technique, a flat surface is created at the site of the lesion area with saw or burr. The surface is measured and, after marking the graft with the same measures, the portion of the condyle suitable for reconstruction is resected with freehand technique. Once preliminarily positioned onto the recipient area, the graft can be further trimmed and sized until appropriate geometry is restored. The aim of the reconstruction is to anatomically restore proper anatomical dimensions of the lesion site, and in this setting fluoroscopy can be useful. As the shell grafts are uncontained, stable fixation is mandatory and is usually achieved by bioabsorbable or 3-mm cannulated screws.

Tibial Plateau Grafts

In cases of posttraumatic deformity, especially after tibial plateau fracture when bone loss may occur, small fragment allografts are particularly suitable (**Fig. 13.2**). The surgical technique is similar to tibial resurfacing in unicompartmental knee arthroplasty (UKA). The meniscus is often damaged, so it should be transplanted with the tibial graft. The damaged meniscus is removed, and using a UKA jig or a freehand technique the tibial hemiplateau is prepared. Bone resection should be limited at the minimum required. With the knee in extension position the gap between the femoral condyle and the prepared tibial plateau is measured to provide a preliminary idea of the graft thickness necessary to restore the anatomic plateau height. The allograft is shaped following the three-dimensional measures of the prepared area. The graft is positioned into the recipient area with particular care for the meniscus, and both dynamically and with the help of fluoroscopy, the restoration of the general knee balance and kinematics,

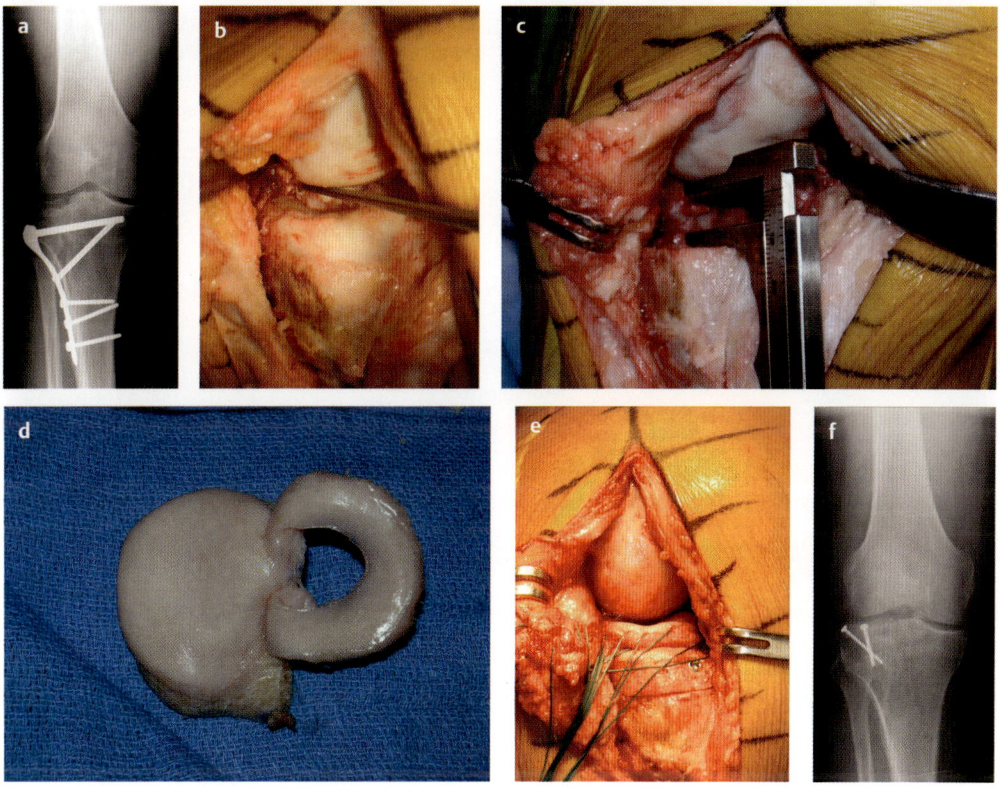

Fig. 13.2 Lateral tibial plateau allograft for posttraumatic malunion. (**a**) Preoperative radiograph. Note disruption of the subchondral bone. (**b**) Intraoperative photograph of the lateral plateau lesion. Note central defect. (**c**) After resection of tibial plateau a gap measurement is made to estimate required graft thickness. (**d**) Allograft tibial plateau with attached meniscus. (**e**) Plateau and meniscus graft in place with sutures for meniscal repair. (**f**) Postoperative radiograph at 6 months demonstrating graft healing.

tibial plateau height, and varus-valgus angulation are checked. Frequently trimming of graft height is required. Once an optimal fit and kinematics have been obtained after multiple small revisions of the graft, screw fixation is performed, and the meniscus is repaired in standard fashion.

◆ Results

Clinical results of fresh OCA in the knee joint have shown encouraging long-term results, with overall success rates from 50 to 95%.[10,30,32,41,45,58] The most commonly treated lesion location is the femoral condyle. **Table 13.2** outlines the major studies reporting outcomes of osteochondral allografting of isolated lesions of the femoral condyle.

In posttraumatic reconstruction of the tibial plateau, Shasha et al in 2003[41] reported the long-term outcome of 65 patients treated with fresh tibial OCA. At a mean of 12 years, 44 patients had an intact graft, while 21 had conversion to a total knee arthroplasty (TKA) at an average of 10 years. The reported survival rate was 95% at 5 years, 80% at 10 years, and 65% at 15 years.

The "San Diego experience" with osteochondral allografting in the knee extends almost 30 years. In 1983, an institutional review board (IRB)-approved osteochondral allografting program was established for the evaluation and treatment of complex or advanced articular cartilage disease. This comprehensive program comprised scientific and clinical components, including retrieval studies. Over the past 30 years, we have collected outcomes data on all patients undergoing fresh osteochondral allografting with the purpose of better defining the indications and understanding clinical outcome. Since 1983, 515 patients have undergone 576 knee allografting procedures. Of those, 328 patients (354 knees) currently have minimum of 2-year follow-up and 187 patients (222 knees) do not have 2-year follow-up data (59 patients are less than 2 years from surgery). The following results include only the 354 knees with minimum of 2-year follow-up. The mean follow-up period was 86 months (range, 24 to 309 months). Patient characteristics and details regarding the allograft are presented in **Table 13.3**. Objective clinical outcome showed improvements in both pain and function on all measures used (**Table 13.4**). Subjectively, 96% of patients reported satisfaction (73% extremely satisfied), 93% reported less pain, and 94% reported better function as a result of the

Table 13.2 Clinical studies of osteochondral allografting of the femoral condyle

Study	Mean follow-up time (years)	Number of knees	Diagnosis	Failure rate	Graft survival rate
McCulloch et al[59] 2007	2.9	25	Various	1/25 (4%)	96%
LaPrade et al[60] 2009	3	23	Various	None	100%
Davidson et al[61] 2007	3.4	10	Various		
Williams et al[62] 2007	4	19	Various	4/19 (21%)	79%
Görtz et al[58] 2010	5.6	28	Osteonecrosis	3/28 (11%)	89%
Emmerson et al[10] 2007	7.7	65	OCD	10/65 (15%)	5 yr—91% 10 yr—76% 15 yr—76%
Gross et al[32] 2005	10	60	Posttraumatic, OCD	12/60 (20%)	5 yr—95% 10 yr—85% 15 yr—74%
Levy et al[63] 2012	14.6	122	Various	31 (24%)	10 yr—82% 15 yr—74% 20 yr—66%

Abbreviation: OCD, osteochondritis dissecans.

Table 13.3 Patient characteristics and allograft details for 354 knees in the authors' series

Variable	Mean (SD) or %	Range
Age (y)	34 (11.8)	14–68
Male	53.1%	
Diagnosis		
OCD	26.8%	
Degenerative chondral lesion	22.6%	
Traumatic chondral injury	14.7%	
Osteoarthritis	12.4%	
Osteonecrosis	9.3%	
Fracture	7.1%	
Failed OCA	7.1%	
Previous surgery on affected joint	90.7%	
Number of previous surgeries	2.6 (1.8)	1–13
Graft location		
Femoral condyle (medial)	35.3%	
Femoral condyle (lateral)	18.4%	
Tibial plateau (medial)	1.1%	
Tibial plateau (lateral)	2.5%	
Patella	7.6%	
Trochlea	5.4%	
Combo (two locations)	26.3%	
Combo (three locations)	3.4%	
Number of grafts	1.5 (0.7)	1–4
Total graft area (cm^2)	10.1 (7)	1.2–57.5

Abbreviations: OCD, osteochondritis dissecans; OCA, osteochondral allograft; SD, standard deviation.

Table 13.4 Results of objective outcome measures

Measure	Preoperative mean (SD) or %	Postoperative mean (SD) or %	*p* Value[a]
Modified D'Aubigne and Postel	12.1 (2.1)	16.0 (2.3)	< 0.001
Excellent	–	32.0%	
Good	10.9%	42.3%	
Fair	48.1%	20.0%	
Poor	41.0%	5.7%	
IKDC pain	6.2 (2.3)	3.2 (2.7)	< 0.001
IKDC function	3.4 (1.9)	7.3 (2.3)	< 0.001
KS-F	65.8 (21.4)	82.4 (19.3)	< 0.001

Abbreviations: IKDC, International Knee Documentation Committee; KS-F, Knee Society Function; SD, standard deviation.
[a]Paired *t*-test.

allograft procedure. Almost all patients (92%) stated they would have the surgery again under similar circumstances.

There were 72 knees (20%) that underwent a reoperation that included removal or revision of the allograft and were defined as clinical failures. The 72 failures included 41 TKAs, 23 revision allografts, 4 partial knee arthroplasties, 2 patellectomies, and 2 knee fusions. The mean time to failure was 40 months (range, 3 to 165 months). Survivorship was 82% at 5 years, 72% at 10 years, and 70% at 25 years (**Fig. 13.3**). The best outcome, by diagnosis, was seen in the patients with OCD (12% failures). Of the OCD nonfailures, the mean modified D'Aubigne and Postel score[64] improved from 12.9 to 16.7, the mean International Knee Documentation Committee (IKDC)[65] pain score improved from 5.2 to 2.2, the mean IKDC function score improved from 3.9 to 8.1, and the mean Knee Society Functional (KS-F) score[66] improved from 76 to 92 (all $p < 0.001$). The worst outcome was seen in patients with osteoarthritis (48% failures). Of the osteoarthritis nonfailures, the mean modified D'Aubigne and Postel score improved from 11.7 to 15.2 ($p < 0.001$), the mean IKDC pain score improved from 5.9 to 3.9 ($p = 0.047$), the mean IKDC function score improved from 3.7 to 6.1 ($p = 0.002$), and the mean KS-F score improved from 67 to 79 ($p = 0.113$). Overall, of the 282 nonfailing knees, 29% underwent further surgery such as arthroscopy that did not include removal or revision of the allograft.

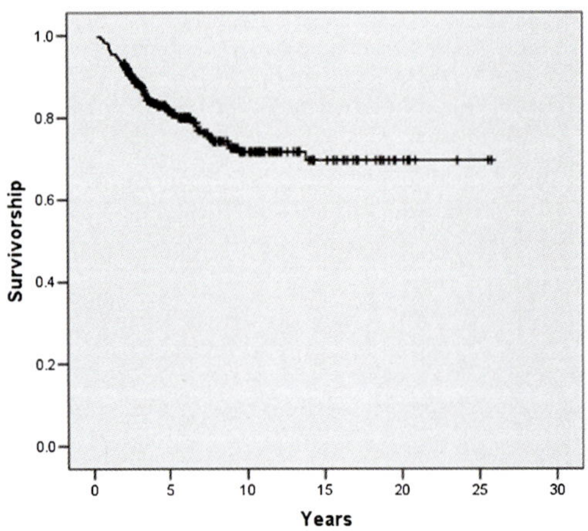

Fig. 13.3 Survivorship of 354 osteochondral allografts.

Conclusion

The treatment of osteochondral defects continues to be a difficult problem for both patients and clinicians. Fresh osteochondral allografting has a long history, and the encouraging clinical data have resulted in this surgical option being increasingly used in cartilage repair and knee reconstruction procedures. Questions still remain regarding details of patient selection and nuances of surgical technique. The expansion in the use of OCA for the management of joint diseases is, however, limited by various factors, including donor availability, safety issues, and immunology. Further work is necessary to optimize tissue banking recovery and storage protocols, as well as to understand mechanisms of graft failure and the immune response elicited by osteochondral grafts. Further goals of ongoing basic research include the use of intact allografts as scaffolds for cell-based technologies, and allografts as a chondrocyte cell source for tissue-engineered repair.

References

1. Lexer E. Substitution of whole or half joints from freshly amputated extremities by free plastic operation. Surg Gynecol Obstet 1908;6:601–607
2. Lexer E. Joint transplantations and arthroplasty. Surg Gynecol Obstet 1925;40:782–809
3. Campbell CJ, Ishida H, Takahashi H, Kelly F. The transplantation of articular cartilage: an experimental study in dogs. J Bone Joint Surg Am 1963;45:1579–1592
4. Craigmyle MBL. An autoradiographic and histochemical study of long-term cartilage grafts in the rabbit. J Anat 1958;92(3):467–472
5. Depalma AF, Tsaltas TT, Mauler GG. Viability of osteochondral grafts as determined by uptake of S35. J Bone Joint Surg Am 1963;45:1565–1578
6. Locht RC, Gross AE, Langer F. Late osteochondral allograft resurfacing for tibial plateau fractures. J Bone Joint Surg Am 1984;66(3):328–335
7. Bell RS, Davis A, Allan DG, Langer F, Czitrom AA, Gross AE. Fresh osteochondral allografts for advanced giant cell tumors at the knee. J Arthroplasty 1994;9(6):603–609
8. Meyers MH, Akeson W, Convery FR. Resurfacing of the knee with fresh osteochondral allograft. J Bone Joint Surg Am 1989;71(5):704–713
9. Garrett JC. Fresh osteochondral allografts for treatment of articular defects in osteochondritis dissecans of the lateral femoral condyle in adults. Clin Orthop Relat Res 1994(303):33–37
10. Emmerson BC, Görtz S, Jamali AA, Chung C, Amiel D, Bugbee WD. Fresh osteochondral allografting in the treatment of osteochondritis

dissecans of the femoral condyle. Am J Sports Med 2007;35 (6):907–914

11. Görtz S, Bugbee WD. Allografts in articular cartilage repair. J Bone Joint Surg Am 2006;88(6):1374–1384
12. Sgaglione NA, Chen E, Bert JM, Amendola A, Bugbee WD. Current strategies for nonsurgical, arthroscopic, and minimally invasive surgical treatment of knee cartilage pathology. Instr Course Lect 2010;59:157–180
13. Burchardt H. The biology of bone graft repair. Clin Orthop Relat Res 1983;174(174):28–42
14. Williams SK, Amiel D, Ball ST, et al. Prolonged storage effects on the articular cartilage of fresh human osteochondral allografts. J Bone Joint Surg Am 2003;85-A(11):2111–2120
15. Allen RT, Robertson CM, Pennock AT, et al. Analysis of stored osteochondral allografts at the time of surgical implantation. Am J Sports Med 2005; 33(10):1479–1484
16. Pennock AT, Robertson CM, Wagner F, Harwood FL, Bugbee WD, Amiel D. Does subchondral bone affect the fate of osteochondral allografts during storage? Am J Sports Med 2006;34(4):586–591
17. Robertson CM, Allen RT, Pennock AT, Bugbee WD, Amiel D. Upregulation of apoptotic and matrix-related gene expression during fresh osteochondral allograft storage. Clin Orthop Relat Res 2006;442:260–266
18. Pennock AT, Wagner F, Robertson CM, Harwood FL, Bugbee WD, Amiel D. Prolonged storage of osteochondral allografts: does the addition of fetal bovine serum improve chondrocyte viability? J Knee Surg 2006;19(4):265–272
19. Pallante AL, Bae WC, Chen AC, Görtz S, Bugbee WD, Sah RL. Chondrocyte viability is higher after prolonged storage at 37°C than at 4°C for osteochondral grafts. Am J Sports Med 2009;37(Suppl 1): 24S–32S
20. Enneking WF, Mindell ER. Observations on massive retrieved human allografts. J Bone Joint Surg Am 1991;73(8):1123–1142
21. Enneking WF, Campanacci DA. Retrieved human allografts: a clinicopathological study. J Bone Joint Surg Am 2001;83-A(7):971–986
22. Acosta CA, Izal I, Ripalda P, Forriol F. Cell viability and protein composition in cryopreserved cartilage. Clin Orthop Relat Res 2007;460:234–239
23. Görtz S, Bugbee WD. Allografts in articular cartilage repair. Instr Course Lect 2007;56:469–480
24. Rodrigo JJ, Thompson E, Travis C. Deep-freezing versus 4 degrees preservation of avascular osteocartilaginous shell allografts in rats. Clin Orthop Relat Res 1987;218(218):268–275
25. Schachar N, McAllister D, Stevenson M, Novak K, McGann L. Metabolic and biochemical status of articular cartilage following cryopreservation and transplantation: a rabbit model. J Orthop Res 1992;10(5):603–609
26. Jamali AA, Hatcher SL, You Z. Donor cell survival in a fresh osteochondral allograft at twenty-nine years. A case report. J Bone Joint Surg Am 2007;89(1):166–169
27. Bobić V. Arthroscopic osteochondral autograft transplantation in anterior cruciate ligament reconstruction: a preliminary clinical study. Knee Surg Sports Traumatol Arthrosc 1996;3(4):262–264
28. Judas F, Rosa S, Teixeira L, Lopes C, Ferreira Mendes A. Chondrocyte viability in fresh and frozen large human osteochondral allografts: effect of cryoprotective agents. Transplant Proc 2007;39(8):2531–2534
29. Malinin TI, Wagner JL, Pita JC, Lo H. Hypothermic storage and cryopreservation of cartilage. An experimental study. Clin Orthop Relat Res 1985;187(197):15–26
30. Bugbee WD, Convery FR. Osteochondral allograft transplantation. Clin Sports Med 1999;18(1):67–75
31. Sammarco VJ, Gorab R, Miller R, Brooks PJ. Human articular cartilage storage in cell culture medium: guidelines for storage of fresh osteochondral allografts. Orthopedics 1997;20(6):497–500
32. Gross AE, Shasha N, Aubin P. Long-term followup of the use of fresh osteochondral allografts for posttraumatic knee defects. Clin Orthop Relat Res 2005;435(435):79–87
33. Ball ST, Amiel D, Williams SK, et al. The effects of storage on fresh human osteochondral allografts. Clin Orthop Relat Res 2004;418(418):246–252
34. Pearsall AW IV, Tucker JA, Hester RB, Heitman RJ. Chondrocyte viability in refrigerated osteochondral allografts used for transplantation within the knee. Am J Sports Med 2004;32(1):125–131
35. Malinin T, Temple HT, Buck BE. Transplantation of osteochondral allografts after cold storage. J Bone Joint Surg Am 2006;88(4):762–770
36. Vangsness CT Jr, Garcia IA, Mills CR, Kainer MA, Roberts MR, Moore TM. Allograft transplantation in the knee: tissue regulation, procurement, processing, and sterilization. Am J Sports Med 2003;31(3):474–481
37. Gross AE. Use of fresh osteochondral allografts to replace traumatic joint defects. In: Czitrom AA, Gross AE, eds. Allografts in Orthopaedic Practice. Baltimore: Williams & Wilkins; 1992;67–82
38. McDermott AG, Langer F, Pritzker KP, Gross AE. Fresh small-fragment osteochondral allografts. Long-term follow-up study on first 100 cases. Clin Orthop Relat Res 1985(197):96–102
39. Stevenson S, Shaffer JW, Goldberg VM. The humoral response to vascular and nonvascular allografts of bone. Clin Orthop Relat Res 1996(326):86–95
40. Ghazavi MT, Pritzker KP, Davis AM, Gross AE. Fresh osteochondral allografts for post-traumatic osteochondral defects of the knee. J Bone Joint Surg Br 1997;79(6):1008–1013
41. Shasha N, Krywulak S, Backstein D, Pressman A, Gross AE. Long-term follow-up of fresh tibial osteochondral allografts for failed tibial plateau fractures. J Bone Joint Surg Am 2003;85-A(Suppl 2): 33–39
42. Jamali AA, Emmerson BC, Chung C, Convery FR, Bugbee WD. Fresh osteochondral allografts: results in the patellofemoral joint. Clin Orthop Relat Res 2005;437(437):176–185
43. Gross AE, Silverstein EA, Falk J, Falk R, Langer F. The allotransplantation of partial joints in the treatment of osteoarthritis of the knee. Clin Orthop Relat Res 1975;108(108):7–14
44. Beaver RJ, Mahomed M, Backstein D, Davis A, Zukor DJ, Gross AE. Fresh osteochondral allografts for post-traumatic defects in the knee. A survivorship analysis. J Bone Joint Surg Br 1992;74(1):105–110
45. Flynn JM, Springfield DS, Mankin HJ. Osteoarticular allografts to treat distal femoral osteonecrosis. Clin Orthop Relat Res 1994;303(303):38–43

46. Gross AE, Agnidis Z, Hutchison CR. Osteochondral defects of the talus treated with fresh osteochondral allograft transplantation. Foot Ankle Int 2001;22(5):385–391

47. Raikin SM. Fresh osteochondral allografts for large-volume cystic osteochondral defects of the talus. J Bone Joint Surg Am 2009;91(12):2818–2826

48. Görtz S, De Young AJ, Bugbee WD. Fresh osteochondral allografting for osteochondral lesions of the talus. Foot Ankle Int 2010;31(4):283–290

49. Hahn DB, Aanstoos ME, Wilkins RM. Osteochondral lesions of the talus treated with fresh talar allografts. Foot Ankle Int 2010;31(4):277–282

50. El-Rashidy H, Villacis D, Omar I, Kelikian AS. Fresh osteochondral allograft for the treatment of cartilage defects of the talus: a retrospective review. J Bone Joint Surg Am 2011;93(17):1634–1640

51. Meehan R, McFarlin S, Bugbee W, Brage M. Fresh ankle osteochondral allograft transplantation for tibiotalar joint arthritis. Foot Ankle Int 2005; 26(10):793–802

52. Giannini S, Buda R, Grigolo B, et al. Bipolar fresh osteochondral allograft of the ankle. Foot Ankle Int 2010;31(1):38–46

53. Meyers MH. Resurfacing of the femoral head with fresh osteochondral allografts. Long-term results. Clin Orthop Relat Res 1985;197(197):111–114

54. Johnson DL, Warner JJ. Osteochondritis dissecans of the humeral head: treatment with a matched osteochondral allograft. J Shoulder Elbow Surg 1997;6(2):160–163

55. Chapovsky F, Kelly JD IV. Osteochondral allograft transplantation for treatment of glenohumeral instability. Arthroscopy 2005;21(8):1007

56. Pylawka TK, Wimmer M, Cole BJ, Virdi AS, Williams JM. Impaction affects cell viability in osteochondral tissues during transplantation. J Knee Surg 2007;20(2):105–110

57. Borazjani BH, Chen AC, Bae WC, et al. Effect of impact on chondrocyte viability during insertion of human osteochondral grafts. J Bone Joint Surg Am 2006;88(9):1934–1943

58. Görtz S, De Young AJ, Bugbee WD. Fresh osteochondral allografting for steroid-associated osteonecrosis of the femoral condyles. Clin Orthop Relat Res 2010;468(5):1269–1278

59. McCulloch PC, Kang RW, Sobhy MH, Hayden JK, Cole BJ. Prospective evaluation of prolonged fresh osteochondral allograft transplantation of the femoral condyle: minimum 2-year follow-up. Am J Sports Med 2007;35(3):411–420

60. LaPrade RF, Botker J, Herzog M, Agel J. Refrigerated osteoarticular allografts to treat articular cartilage defects of the femoral condyles. A prospective outcomes study. J Bone Joint Surg Am 2009;91 (4):805–811

61. Davidson PA, Rivenburgh DW, Dawson PE, Rozin R. Clinical, histologic, and radiographic outcomes of distal femoral resurfacing with hypothermically stored osteoarticular allografts. Am J Sports Med 2007;35(7):1082–1090

62. Williams RJ III, Ranawat AS, Potter HG, Carter T, Warren RF. Fresh stored allografts for the treatment of osteochondral defects of the knee. J Bone Joint Surg Am 2007;89(4):718–726

63. Levy YD, Görtz S, Pulido PA, McCauley JC, Bugbee WD. Long-tem follow-up of fresh osteochondral allografting of the femoral condyle. Paper No. 709, presented at the annual meeting of the American Academy of Orthopaedic Surgeons; San Francisco, CA; February 7–11, 2012

64. D'Aubigne RM, Postel M. Functional results of hip arthroplasty with acrylic prosthesis. J Bone Joint Surg Am 1954;36-A(3):451–475

65. Irrgang JJ, Anderson AF, Boland AL, et al. Development and validation of the international knee documentation committee subjective knee form. Am J Sports Med 2001;29(5):600–613

66. Insall JN, Dorr LD, Scott RD, Scott WN. Rationale of the Knee Society clinical rating system. Clin Orthop Relat Res 1989(248):13–14

14

The Use of Scaffolds in the Treatment of Osteochondral Lesions in the Knee: Current Concepts and Future Trends

Aad A. M. Dhollander, Victor R. Guevara Sánchez, Karl F. Almqvist, René Verdonk, Gust Verbruggen, and Peter C. M. Verdonk

The treatment of chondral and osteochondral lesions has become a major interest to orthopedic surgeons because most lesions do not heal spontaneously and may predispose the joint to the subsequent development of secondary osteoarthritis.[1] This poor repair capacity of articular cartilage has led to the development of various surgical techniques.[2] Several bone marrow-stimulating procedures directed at the recruitment of bone marrow cells have been widely used to treat local cartilage defects. In this type of procedure, mesenchymal stem cells (MSCs) migrate in the fibrin network of the blood clot.[3] However, this fibrin clot is not mechanically stable to withstand the tangential forces.[4] The most popular and frequently used is the microfracture technique including abrading the tidemark and creating small holes perpendicular to the subchondral bone plate to allow bleeding into the defect.[5] Microfracture usually results in a fibrous–fibrohyaline unstructured repair tissue. This tissue lacks the biomechanical and viscoelastic features of hyaline cartilage. The potential short-term improvement in symptoms is usually followed by repair tissue failure and potentially by gradual deterioration to osteoarthritis and return of symptoms.[6]

Since 1987, autologous chondrocytes have been implanted in chondral lesions of the human knee.[7] To this end, cartilage is harvested from a non–weightbearing area of the knee joint—for example, the intercondylar notch or the edge of the trochlea—then digested, and the isolated chondrocytes are propagated in vitro in the monolayer culture condition. This procedure, proposed by Brittberg et al in 1994 and called autologous chondrocyte implantation (ACI), is gaining wide scientific and clinical support for use in the repair of focal articular cartilage lesions.[7,8] However, during in vitro propagation of the chondrocytes, dedifferentiation of the cells can occur, and afterwards these fibroblast-like chondrocytes show different biosynthetic properties than the original cartilage cells in the knee joint.[9]

Recently, there has been an increasing interest and awareness of the importance of subchondral bone for its role in cartilage repair. One should carefully consider the subchondral bone in the treatment of articular surface damage, in the evaluation of the results over time and in the determination of the patient's prognosis. In fact, the conditions of articular cartilage and its supporting bone are tightly coupled and should be viewed as a connected osteochondral unit.[10] The ultimate aim of the treatment is the restoration of normal knee function by regenerating hyaline cartilage in the defect and complete integration of the regenerated cartilage with the surrounding cartilage and underlying bone. The treatment should restore the physiological properties of the entire osteochondral unit.

However, none of the currently available treatment options achieve this goal. Ideally, future cartilage repair strategies should (1) be easy and quick to implant, (2) reduce surgical morbidity, (3) not require harvesting of other tissues (e.g., periosteum), (4) exhibit enhanced cell proliferation and maturation, (5) have easier phenotype maintenance, and (6) allow for efficient and complete integration with surrounding articular cartilage.[2]

Tissue engineering could be an alternative and promising option for the treatment of cartilage defects. Tissue engineering is based on three basic ingredients: scaffolds, growth factors, and cells.[11] Scaffolds should be biodegradable to anchor, deliver, and orient the cells. Growth factors are the instructional cues to guide cartilage growth and differentiation. Finally, cells must be present that are capable of proliferating, producing cartilage matrix, and ultimately reacting as normal cartilage does.[2]

◆ Scaffolds Design Criteria

Ideally, scaffolds should be versatile in terms of applications and be suitable for resurfacing full-thickness lesions as well as repairing partial-thickness lesions. The latest recommendations on scaffolds' biochemical and structural requirements and their corresponding function for tissue engineering are as follows[2,12]:

1. Biologic compatibility: Inciting ideally no or minimal inflammatory response.
2. Noncytotoxic: The scaffold should not be cytotoxic to the cells or surrounding tissues.
3. Three-dimensional (3D) matrix architecture: Allowing a physiologically relevant environment to hold cells and support cell function.
4. Void space: Highly porous and interconnected pores that allow cell infiltration, diffusion of nutrients and humoral factors as well as waste products to promote cellular proliferation and the production of extracellular matrix.
5. Surface chemistry and topography: For cell attachment and cell–matrix interactions between grafted and native cartilage.
6. Biodegradation rate: Scaffold serves as a temporary support for the cells and gives way to functional matrix formation.
7. Structural anisotropy: Anisotropic mechanical behavior to influence orientation of cells and extracellular matrix (ECM) deposition.
8. Appropriate mechanical behavior: To allow seamless integration with surrounding cartilage, to withstand in vivo forces, and to avoid stress shielding.
9. Availability: To be reproducible and easily fabricated into a variety of shapes and sizes to adapt to each patient scenario.

◆ Types of Scaffolds

The use of scaffolds for cartilage repair came into practice when monolayered cultured chondrocytes showed progressive dedifferentiation.[9,13] This phenomenon was reverted when they were recultured in 3D media and has led to the development of various scaffolds that can be grouped into four main classes[14]: (1) protein-based scaffolds (e.g., fibrin, collagen, gelatin…), (2) carbohydrate-based scaffolds (e.g., hyaluronan, agarose, alginate, polylactic/polyglycolic acids,…), (3) synthetic or artificial polymer-based scaffolds (e.g., hydroxyapatite, polyethylene glycol,…), and (4) combination of different scaffolds types.

◆ Experience with Currently Available Scaffold Types

Protein-Based Scaffolds

Fibrin

The protein-based scaffolds include fibrin, collagen, and gelatin. Chondrocytes multiply, retain their morphology, and produce matrix as long as they are surrounded by fibrin gel in vitro.[15] Choi et al used a fibrin gel–type autologous chondrocyte (Chondron) implantation for human chondral knee defects during several years without using periosteum or membrane. They concluded that this method appeared to be safe and effective for both decreasing pain and improving knee function.[16] Fibrin glues (FGs) are used extensively to secure other tissue-engineered cartilage in clinical settings. FG has also been used in combination with autologous chondrocytes for the treatment of deep chondral defects in humans. When combined with commercial FG (Tissucol; Baxter, Deerfield, IL), autologous chondrocytes showed better clinical outcomes during the treatment of deep chondral defects compared with abrasive techniques after 1 year in terms of subjective scores.[17] Alternatively, minced cartilage in combination with FG known as DeNovo NT (natural tissue) grafts (Zimmer, Warsaw, IN; ISTO Technologies, St. Louis, MO) is currently under clinical investigation, where cartilage pieces obtained from juvenile allograft donor joint are aseptically minced and mixed intraoperatively with FG and then implanted in the prepared lesions.[18] Juvenile cartilage is chosen depending on the assumption that it would have higher anabolic capability and better expandability. Disease transmission and limited supply are the drawbacks of this technique. A similar product known as DeNovo ET (engineered tissue) graft (ISTO Technologies) is based on the use of juvenile allogeneic cartilage cells. These cells produce in vitro a hyaline-like, scaffold-free, disc-shaped graft that is implantable in a single-stage procedure.[19] The product is implanted into the prepared lesion site and secured via FG. Small cartilage pieces are placed in the chondral defect with the DeNovo NT technique, whereas DeNovo ET uses individual allogenic chondrocytes to form a hyaline-like cartilage disk in vitro, which is then implanted into the cartilage defect.

Collagen

Collagen gels have been evaluated for the treatment of cartilage lesions in animal studies. Promising histological results at 24 weeks have been reported in the treatment of osteochondral defects filled with chondrocytes suspended in a type I collagen gel in the knee joint of full-grown rabbits. However, these gel types showed poor integration with the surrounding host cartilage and a lack of regeneration of a proper subchondral base.[20,21] Recently, predifferentiated MSCs embedded in a collagen I hydrogel were used for the treatment of a chronic osteochondral defect in an ovine stifle joint. This repair strategy showed no signs of degradation after 1 year in vivo. In addition, these gels led to partially superior histological results compared with articular chondrocytes. A similar approach was recently performed in two patients, indicating the feasibility of this approach.[14] Bovine collagen I in combination with bioadhesive is currently used clinically and is known as NeoCart; this material is produced by Histogenics (Waltham, MA). The harvested cartilage samples are processed and chondrocytes grown into collagen; the construct is then applied to the lesion and secured by bioadhesive.[22] Another similar product known as CaReS has been introduced by Arthro Kinetics (Esslingen, Germany) and is also available for clinical application. CaReS utilizes rat tail collagen I, instead of bovine collagen (NeoCart). VeriCart, another product of Histogenics, has been introduced to the market as an adjuvant to microfracture. VeriCart uses a double-structured collagen scaffold.

Ochi et al used autologous chondrocytes, cultured in atelocollagen gel, for the treatment of full-thickness defects of cartilage in human knees.[23,24] Atelocollagen, from which telopeptides have been removed, was chosen because the antigenic determinants on the peptide chains of type I collagen reside mainly in the telopeptide regions.[25,26] In vitro and in vivo experimental results supported the hypothesis that transplanting chondrocytes cultured in atelocollagen gel would be effective in repairing articular cartilage defects, not only in animals but also in humans, by maintaining the chondrocyte phenotype, reducing the risk of leakage, and distributing grafted cells evenly throughout the grafted site.[27–29] It was shown that this technique promoted the restoration of the articular cartilage in the knee.[24]

Gelatin

Allogeneic bone matrix gelatin (BMG) is prepared through the process of defatting, demineralization, and extraction to remove 95% of the noncollagen proteins which would eliminate antigenic materials inside the BMG, rendering it weakly immunogenic and more biocompatible with the host.[30–32] It was suggested that BMG could induce differentiation of mesenchymal cells into chondroblasts in vitro and form hyaline-like cartilage in osteochondral defects in vivo.[33–35] Considering the large supply of banked bone allografts and relatively convenient preparation, allogeneic cancellous BMG should be considered as a promising scaffold for cartilage tissue engineering.[36] Human data concerning the use of gelatin for cartilage repair are currently not available.

Carbohydrate-Based Scaffolds

Agarose

Promising macroscopic and histological results have been reported with the use of agarose as a scaffold for chondrocytes in the treatment of osteochondral defects in 6-week-old rabbits.[37] An increase in proteoglycan and type II collagen synthesis by the transplanted cells was seen, and the implant merged well with the sides of the lesion. The best results were obtained from 18 months on. However, difficulties have been encountered with the use of agarose gels because of their consistency and possible immunological reactions against these matrix molecules. Currently, no human data are available.

Poly-L-Lactic Acid and Polyglycolic Acid

Neocartilage was formed both in vitro and in vivo (subcutaneously in nude mice) when chondrocytes and polyglycolic acid (PGA) or poly-L-lactic acid (PLLA) were used.[38,39] An overall repair frequency of 85% in osteochondral defects in the rabbit was reported, but the cartilage-like quality was variable and none of the specimens appeared normal after 1 year.[40] Osteochondral defects in the trochlea of adult rabbits have also been treated with PGA-scaffolds seeded with allogenic chondrocytes.[41] At 6 months, the total contents of glycosaminoglycans and type II collagen was only one-third of that in parent rabbit cartilage. In contrast, osteochondral defects created in the knee joint of goats with a scaffold (PLLA as a base material and PGA as an additive) seeded with chondrocytes were repaired with hyaline-like cartilage and good underlying bone at 16 weeks.[42] However, its degradation products can eventually lead to the death of the implanted cells, mainly in PGA because of its more rapid degradation when compared with PLLA.[43] These types of scaffolds have not yet been tested in humans in their original forms.

Hyaluronan

Because of its multiple functions in regulating and stabilizing the internal environment of cartilage, hyaluronan is a promising scaffold to promote cartilage repair.[44] HYAFF 11 is the esterified derivative of hyaluronate, and when combined with autologous articular chondrocytes it forms Hyalograft C (Fidia Advanced Biopolymers, Abano Terme, Italy). In terms of the quality of regenerated cartilage and production of chondrocytic markers, trials with Hyalograft C (Fidia) yielded comparable results to that of ACI in animal models and in humans.[44,45] The long-term clinical outcomes of Hyalograft C (Fidia) grafting in humans was similar to that of ACI as indicated by magnetic resonance imaging (MRI) and objective and subjective knee scoring systems. However, one advantage of Hyalograft (Fidia) grafting over ACI is that it can be performed using a less-complicated and minimally invasive surgical procedure.

Alginate

It has been shown that human chondrocytes keep their phenotype in alginate with neosynthesis of an extracellular cartilage matrix and that this chondrocyte/alginate culture setup can be biologically frozen without any impairment in the total of overall aggrecan synthesis rates or its cartilage-specific aggrecan subtypes once thawed.[46,47] A short-term pilot study showed that the alginate-based scaffold containing human mature allogenic chondrocytes is feasible and safe for the treatment of symptomatic cartilage defects of the knee in humans. This technique provided clinical and histologic outcomes that are equal but not superior to those of other cartilage repair techniques.[48,49]

Chitosan and Chitin

Chitosan is the deacetylated derivative of chitin. Chitosan alone or in combination with a wide variety of scaffolds has been used extensively in the tissue engineering of articular cartilage.[50] The seeding of chondrocytes on chitosan-coated coverslips maintained the spherical chondrocyte morphology and promoted expression of collagen II and aggrecan.[51] In addition, chitosan can serve as a cytocompatible space-filling scaffold that can gelate and adhere to cartilage when injected in situ in combination with chondrocytes.[52] In contrast, it has been shown that the rate of cell proliferation and the production of collagen II were lower in the chitosan scaffold compared with chitosan–alginate hybrid after 2 weeks of culturing.[53] These types of scaffolds have not yet been tested in humans in their original forms.

Synthetic or Artificial Polymer-Based Scaffolds

Hydroxyapatite

With the use of hydroxyapatite loaded with chondrocytes in the treatment of osteochondral defects, a layer of fibrous tissue formed around the hydroxyapatite graft.[54] This resulted in a relatively unstable fixation of the hydroxyapatite in the defect with a gradual loss of the newly formed hyaline cartilage-like repair tissue and progressive resorption of the hydroxyapatite. It has been concluded that hydroxyapatite is not a suitable biomaterial to enhance the repair of large articular cartilage defects.

Polyethylene Glycol

Polyethylene glycol (PEG) is a synthetic polymer that has wide biotechnological applications. It is a highly biocompatible material and has been used widely for many medicinal purposes.[55] The neutral and noninteractive nature of PEG facilitates ECM secretion from cells after their encapsulation.[55] Another advantage of the PEG-based scaffold is that it can be laminated by adding a second layer before complete crosslinking of the first layer, which creates zonal patterns resembling the organization of articular cartilage. It has been shown that the mechanical properties of PEG-based scaffolds are dependent on the mesh size. Different mesh sizes can be achieved by changing the PEG molecular weight, concentration, or crosslinking density.[56] In its original form, PEG has not been tested in humans.

Combined Scaffolds

MaioRegen

MaioRegen is an osteochondral nanostructured biomimetic scaffold (Fin-Ceramica Faenza SpA, Faenza, Italy). It has a porous, 3D composite, trilayered structure, to reproduce the cartilaginous layer, the tidemark, and the subchondral bone. The cartilaginous layer, consisting of type I collagen, has a smooth surface. The intermediate layer (tidemark-like) consists of a combination of type I collagen (60%) and hydroxyapatite (40%), whereas the lower layer consists of a mineralized blend of type I collagen (30%) and hydroxyapatite (70%) reproducing the subchondral bone layer.[57] In in vitro and animal studies, this novel biomaterial was tested previously and obtained good results with cartilage and bone tissue formation.[58–60] The same macroscopic, histologic, and radiographic results were observed when implanting scaffolds loaded with autologous chondrocytes or scaffolds alone. The scaffold in the animal model was able to induce an in situ regeneration, through stem cells coming from the surrounding bone marrow. This innovative scaffold was applied without cells to clinical practice.[57] This open one-step procedure was used for the treatment of chondral and osteochondral knee defects. The pilot study highlighted the safety and potential clinical benefit of the graded biomimetic osteochondral scaffold in promoting bone and cartilage tissue restoration by itself and with good clinical and MRI results at the 2-year follow-up.[57]

TruFit

TruFit plugs (Smith & Nephew, Andover, MA), composed of mechanically stable, cylindrical, bilayered PLGA and calcium sulfate, are used clinically nowadays to facilitate the ingrowth of new healing tissue to restore osteochondral defects.[61] The plugs are designed to degrade within a year and are commonly used for back-filling femoral donor site during mosaicplasty.[19] They promote cancellous bone replacement in the subchondral region and formation of fibrocartilage at the surface. Conflicting clinical outcomes have been reported in the recent past about the use of these plugs for cartilage repair in the knee.[62,63] Saithna et al showed promising early clinical results for the repair of small articular cartilage defects within the knee.[62] However, Davidson and Rivenburgh documented some patients who complained of persistent symptoms with an effusion in the knee joint at 6 months after insertion.[63]

Bilayered Collagen I/III Scaffolds

It has been shown that seeding of chondrocytes into collagen I–III scaffolds maintained the cell viability and morphology and promoted the elaboration of articular chondrocytic markers under standard and serum-free conditions.[64,65] A bilayered collagen I–III scaffold in combination with autologous chondrocytes has been used clinically in the last decade and is considered a modification of conventional ACI.[66] The technique is referred to as matrix-associated autologous chondrocyte implantation (MACI), in which autologous chondrocytes are seeded into bilayered porcine collagen before implantation and then secured to the prepared chondral defect using FG.[19,61,66] This technique resulted in clinical improvements comparable to that of ACI or microfracture after 1 year.[19,61] In addition, it showed good to excellent clinical outcomes in 82% of the patients at 4 years of follow-up.[61] The MACI technique was introduced for clinical use by Genzyme (Cambridge, MA). However, other products utilizing the same principle and biomaterial are also available: Carticel produced by Genzyme for application in the United States, and Chondro-Gide produced by Geistlich (Wolhusen, Switzerland) for application in Europe.[22] Chondro-Gide can also be utilized clinically in a single-stage procedure after microfracture, and this is known as autologous matrix-induced chondrogenesis (AMIC). Chondro-Gide membrane in this case provides the scaffold for growth and multiplication of cells released after the microfracture procedure.[22] Satisfactory outcomes have been reported after 2 years of follow-up during the first clinical assessment of 32 patients treated with AMIC in combination with microfracture, in terms of defect filling, functional improvement, pain reduction, and patient satisfaction.[67]

Cartipatch

Cartipatch (Tissue Bank of France, Lyon, France) is a novel agarose–alginate hydrogel scaffold that was developed to improve cell phenotypic stability and ease of surgical handling. A multicenter study on 17 patients was started to investigate the clinical, radiological, arthroscopic, and histological outcome at a minimum follow-up of 2 years after the implantation of autologous chondrocytes embedded in a 3D alginate–agarose hydrogel for the treatment of chondral and osteochondral defects. This technique resulted in a significant clinical improvement at follow-up at 2 years, more so for larger and deeper lesions. The surgical procedure was uncomplicated, and predominantly hyaline cartilage–like repair tissue was observed in eight patients.[68]

Chondrotissue

Chondrotissue is a cell-free sterile matrix chondrotissue (BioTissue AG, Zurich, Switzerland), which consists of an absorbable nonwoven polyglycolic acid textile treated with hyaluronic acid.[69] Hyaluronic acid has been shown to induce mesenchymal progenitor cells from the bone marrow to differentiate along the chondrogenic lineage.[70] This textile scaffold is like a sponge, which may hold the blood clot and progenitor cells within the defect, inducing hemostasis and protecting the underlying tissue.[69] The mechanical stability of the scaffold allows for easy handling and secure fixation in the defect by FG, cartilage or transosseous suture, or resorbable pins.[71,72] In the ovine model of a joint defect, covering a full-thickness cartilage defect with the chondrotissue matrix after microfracture has been shown to improve cartilage repair compared with microfracture alone.[73] Until the present, only case reports have been published concerning this novel technique.[69,74]

Gelrin C

Photopolymerizable PEG-modified fibrinogen, known as PEGylated fibrinogen or Gelrin C (Regentis Biomaterials, Haifa, Israel), crosslinks in situ after exposure to ultraviolet (UV) light. This is now in the stage of clinical trials.[19] The development of PEGylated derivative of fibrin leads to the formation of fibrin gels that have good mechanical strength. In addition, PEG of different molecular weights can be utilized to develop fibrin hydrogels with different mechanical properties.[75,76] Gelrin C is characterized as having innate chondrogenic and osteoconductive potential, while it is nonimmunogenic as indicated from the in vitro study. Implantation of Gelrin C into ovine osteochondral defects showed enhanced production of cartilage-specific markers compared with empty defects that showed scar formation composed mainly of fibrocartilage.[19] Gelrin C is considered an adjunct to microfracture or osteochondral defect filler and its rate of degradation depends on the degree of PEGylation.[19]

Bioseed C

A copolymer of PGA–polylactic acid (PLA) and polydioxanone (PDS), in combination with autologous chondrocytes dispersed into FG, has been introduced into clinical use and is referred to as Bio-Seed-C (Biotissue Technologies, Freiburg, Germany).[66] The expanded chondrocytes are loaded onto the mechanically stable porous matrix using FG for even distribution, and then implanted arthroscopically.[77] Evaluation of midterm results of ACI compared with Bio-Seed-C revealed that they are equally effective as a treatment option for focal degenerative chondral lesions.[78] Good clinical outcomes have been reported with Bio-Seed-C over a period of 4 years.[79]

BST-CarGel

A scaffold composed of chitosan and b-glycerophosphate known as BST-CarGel is used as an adjunct to microfracture to stabilize the blood clot and retain MSCs in the cartilage lesion. BST-CarGel is liquid at room temperature and solidifies at human body temperature. It was introduced for clinical application by Biosyntech (Quebec, Canada).[22] The technique involves mixing fresh autologous blood with BST-CarGel in the operation room, which is then delivered to the holes created by microfracture and the surrounding prepared defect. Interim 6-month data showed better reparative tissue formation in comparison to microfracture alone.[19,61]

ChonDux

A photo-polymerizable hydrogel composed of PEG in combination with chondroitin sulfate (bioadhesive) is now utilized in the clinical setting (ChonDux, introduced to the market by Cartilix [Foster City, CA]). The product is applied in conjunction with microfracture to enhance the cartilage repair process through promotion of chondrogenic differentiation and cartilage tissue formation by bone marrow MSCs. The product is applied in a liquid form that solidifies upon exposure to UVA light.[22]

Cartilage Autograft Implantation System

A new system known as cartilage autograft implantation system (CAIS; DePuy/Mitek, Raynham, MA) utilizes resorbable copolymer of PGA and polycaprolactone reinforced with PDS mesh that is combined with minced cartilage as a source of viable chondrocytes.[19] The system is composed of harvester, disperser, scaffold, and staples. Cartilage pieces are harvested from the non–weightbearing area using the harvester and then minced into small cartilage fragments using the disperser.[18] Minced cartilage is then evenly distributed and stabilized into the PGA–polycaprolactone copolymer using FG and then fixed into the lesion site using the resorbable PDS staples.[18,19] CAIS has been studied in vitro and in animal models but is still in the early phase of clinical trials in humans.[19] Early clinical trials using CAIS showed promising results in comparison to microfracture. However, long-term and randomized human trials have not been conducted yet.[18]

Future Trends

Scaffolds used for osteochondral repair may be either cell- or non-cell-based before implantation in the knee. There is a growing interest in noncell and last-minute cell seeding technologies since they allow for a one-step surgery, eliminating morbidity and the necessity of a previous chondral biopsy. This so-called last-minute cell-seeded scaffold depends on the feasibility to isolate chondrocytes and/or autologous bone marrow MSCs (aBM-MSCs) throughout the same intraoperative setting during knee articular cartilage repair. Currently, a safety and feasibility trial is running combining aBM-MSCs with autologous chondrocytes seeded on a polymer-based scaffold in a single-step surgery for the repair of knee cartilage defects (INSTRUCT therapy, CellCoTec).

Blood-derived MSCs (Bl-MSCs) have also become particularly interesting since these cells may be drawn in an outpatient setting and then differentiated, cultured, and embedded into a scaffold. Regarding MSĆ differentiation, CD90+ and CD105+ MSCs have shown a better phenotype profile and can be aided with endogenous growth factors to obtain adequate collagen II and aggrecan gene expressions. However, there is no defined methodology yet for the best ECM production from well-differentiated chondrocytes via this technique.[80]

The development of scaffolds that maintain well-differentiated cells obtained from MSC technologies should also be emphasized in current tissue-engineering research. It is important to point out that MSCs can be obtained from most types of connective tissue. Concepts that allow generation of 3D, zonally organized, native-like articular cartilage, starting from a single stem cell population, have not yet been reported. The key innovation necessary for this would involve methods to direct a single stem cell population into multiple, spatially distinct phenotypes within a single 3D structure.[81]

Current and clinically available technologies using different types of scaffolds are still miles away from reproducing the exact (or at least similar) structural and biochemical environment found on that of native hyaline cartilage. However, we may finally be close to understanding the multilayered structure and the complex biochemical signaling in cartilage.

Conclusion

Most recent clinical trials regarding microfractures and osteochondral grafting for cartilage lesions in the knee seem to reveal suboptimal outcomes after long-term follow-up. Up to this point, cartilage repair techniques were mainly focused on cells. However, scaffold design and growth factors seem to play a more and more important role in the development of future strategies for osteochondral repair. In this way, acellular scaffold-based techniques are currently being developed and tested. These techniques avoid the cost and the logistics related to cell manipulation and are very interesting for the industry. As we keep obtaining more and more clinical data, scaffold-based technologies may appear to be the better option available so far for surgeons and for patients as well.

Although clinical and histological results from many already available scaffolds seem to be promising, improvements throughout these technologies and the developments of new ones are still necessary to obtain a more efficient biological response as well as to improve the implant's stability. Long after the first reports on human ACI by Brittberg et al in 1994,[7] the development of a so-called perfect technology for osteochondral tissue resuscitation is still one of the most challenging issues in knee surgery. It is also becoming more and more apparent that without support from an intact subchondral bed, any treatment of the surface chondral lesion is likely to fail. Therefore, increased attention is needed for treatment options of the entire osteochondral unit.[10]

References

1. Kim HK, Moran ME, Salter RB. The potential for regeneration of articular cartilage in defects created by chondral shaving and subchondral abrasion. An experimental investigation in rabbits. J Bone Joint Surg Am 1991;73(9):1301–1315
2. Safran MR, Kim H, Zaffagnini S. The use of scaffolds in the management of articular cartilage injury. J Am Acad Orthop Surg 2008;16(6):306–311
3. Cerynik DL, Lewullis GE, Joves BC, Palmer MP, Tom JA. Outcomes of microfracture in professional basketball players. Knee Surg Sports Traumatol Arthrosc 2009;17(9):1135–1139
4. Dorotka R, Windberger U, Macfelda K, Bindreiter U, Toma C, Nehrer S. Repair of articular

cartilage defects treated by microfracture and a three-dimensional collagen matrix. Biomaterials 2005;26(17):3617–3629

5. Steadman JR, Rodkey WG, Singleton SB, et al. Microfracture technique for full-thickness chondral defects: technique and clinical results. Oper Tech Orthop 1997;7:300–304
6. Minas T. Chondrocyte implantation in the repair of chondral lesions of the knee: economics and quality of life. Am J Orthop 1998;27(11):739–744
7. Brittberg M, Lindahl A, Nilsson A, Ohlsson C, Isaksson O, Peterson L. Treatment of deep cartilage defects in the knee with autologous chondrocyte transplantation. N Engl J Med 1994;331(14):889–895
8. Peterson L, Minas T, Brittberg M, Nilsson A, Sjögren-Jansson E, Lindahl A. Two- to 9-year outcome after autologous chondrocyte transplantation of the knee. Clin Orthop Relat Res 2000;374(374):212–234
9. Benya PD, Shaffer JD. Dedifferentiated chondrocytes reexpress the differentiated collagen phenotype when cultured in agarose gels. Cell 1982;30(1):215–224
10. Gomoll AH, Madry H, Knutsen G, et al. The subchondral bone in articular cartilage repair: current problems in the surgical management. Knee Surg Sports Traumatol Arthrosc 2010;18(4):434–447
11. Verdonk PCM, Forsyth RG, Wang J, et al. Characterisation of human knee meniscus cell phenotype. Osteoarthritis Cartilage 2005;13(7):548–560
12. Stella JA, D'Amore A, Wagner WR, Sacks MS. On the biomechanical function of scaffolds for engineering load-bearing soft tissues. Acta Biomater 2010;6(7):2365–2381
13. Bonaventure J, Kadhom N, Cohen-Solal L, et al. Reexpression of cartilage-specific genes by dedifferentiated human articular chondrocytes cultured in alginate beads. Exp Cell Res 1994;212(1):97–104
14. Kasemkijwattana C, Hongeng S, Kesprayura S, Rungsinaporn V, Chaipinyo K, Chansiri K. Autologous bone marrow mesenchymal stem cells implantation for cartilage defects: two cases report. J Med Assoc Thai 2011;94(3):395–400
15. Homminga GN, Buma P, Koot HWJ, van der Kraan PM, van den Berg WB. Chondrocyte behavior in fibrin glue in vitro. Acta Orthop Scand 1993;64(4):441–445
16. Choi NY, Kim BW, Yeo WJ, et al. Gel-type autologous chondrocyte (Chondron) implantation for treatment of articular cartilage defects of the knee. BMC Musculoskelet Disord 2010;11:103
17. Visna P, Pasa L, Cizmár I, Hart R, Hoch J. Treatment of deep cartilage defects of the knee using autologous chondrograft transplantation and by abrasive techniques—a randomized controlled study. Acta Chir Belg 2004;104(6):709–714
18. McCormick F, Yanke A, Provencher MT, Cole BJ. Minced articular cartilage—basic science, surgical technique, and clinical application. Sports Med Arthrosc 2008;16(4):217–220
19. McNickle AG, Provencher MT, Cole BJ. Overview of existing cartilage repair technology. Sports Med Arthrosc 2008;16(4):196–201
20. Wakitani S, Kimura T, Hirooka A, et al. Repair of rabbit articular surfaces with allograft chondrocytes embedded in collagen gel. J Bone Joint Surg Br 1989;71(1):74–80
21. Kawamura S, Wakitani S, Kimura T, et al. Articular cartilage repair. Rabbit experiments with a collagen gel-biomatrix and chondrocytes cultured in it. Acta Orthop Scand 1998;69(1):56–62
22. Cascio BM, Sharma B. The future of cartilage repair. Oper Tech Sport Med 2008;16:221–224
23. Ochi M, Uchio Y, Kawasaki K, Wakitani S, Iwasa J. Transplantation of cartilage-like tissue made by tissue engineering in the treatment of cartilage defects of the knee. J Bone Joint Surg Br 2002;84(4):571–578
24. Tohyama H, Yasuda K, Minami A, et al. Atelocollagen-associated autologous chondrocyte implantation for the repair of chondral defects of the knee: a prospective multicenter clinical trial in Japan. J Orthop Sci 2009;14(5):579–588
25. Pontz B, Meigel W, Rauterberg J, Kühn K. Localization of two species specific antigenic determinants on the peptide chains of calf skin collagen. Eur J Biochem 1970;16(1):50–54
26. Furthmayr H, Timpl R. Immunochemistry of collagens and procollagens. Int Rev Connect Tissue Res 1976;7:61–99
27. Ochi M, Uchio Y, Matsusaki M, Wakitani S, Sumen Y. Cartilage repair: a new surgical procedure of cultured chondrocyte transplantation. In: Chan KM, Fu F, Maffuli N, Rolf C, Kurosaka M, Liu S, eds. Controversies in orthopedic sport medicine. Hong Kong: Williams & Wilkins Asia-Pacific Ltd; 1998:549–563
28. Uchio Y, Ochi M, Matsusaki M, Kurioka H, Katsube K. Human chondrocyte proliferation and matrix synthesis cultured in Atelocollagen gel. J Biomed Mater Res 2000;50(2):138–143
29. Katsube K, Ochi M, Uchio Y, et al. Repair of articular cartilage defects with cultured chondrocytes in Atelocollagen gel. Comparison with cultured chondrocytes in suspension. Arch Orthop Trauma Surg 2000;120(3-4):121–127
30. Badylak SF, Freytes DO, Gilbert TW. Extracellular matrix as a biological scaffold material: Structure and function. Acta Biomater 2009;5(1):1–13
31. Hu X, Yao L, Lu C, Wang S, Chen Y. Experimental and clinical investigations of human insoluble bone matrix gelatin. A report of 24 cases. Clin Orthop Relat Res 1993;293(293):360–365
32. Iwata H, Sakano S, Itoh T, Bauer TW. Demineralized bone matrix and native bone morphogenetic protein in orthopaedic surgery. Clin Orthop Relat Res 2002;395(395):99–109
33. Nogami H, Oohira A, Terashima Y, Urist MR. Ultrastructure of chondrogenetic interactions between bone matrix gelatin and mesenchymal cells. Clin Orthop Relat Res 1978;133(133):238–245
34. Mizutani H, Urist MR. The nature of bone morphogenetic protein (BMP) fractions derived from bovine bone matrix gelatin. Clin Orthop Relat Res 1982;171(171):213–223
35. Dahlberg L, Kreicbergs A. Demineralized allogeneic bone matrix for cartilage repair. J Orthop Res 1991;9(1):11–19
36. Yang B, Yin Z, Cao J, et al. In vitro cartilage tissue engineering using cancellous bone matrix gelatin as a biodegradable scaffold. Biomed Mater 2010;5(4):045003

37. Rahfoth B, Weisser J, Sternkopf F, Aigner T, von der Mark K, Bräuer R. Transplantation of allograft chondrocytes embedded in agarose gel into cartilage defects of rabbits. Osteoarthritis Cartilage 1998;6(1):50–65

38. Freed LE, Marquis JC, Nohria A, Emmanual J, Mikos AG, Langer R. Neocartilage formation in vitro and in vivo using cells cultured on synthetic biodegradable polymers. J Biomed Mater Res 1993;27(1):11–23

39. Freed LE, Hollander AP, Martin I, Barry JR, Langer R, Vunjak-Novakovic G. Chondrogenesis in a cell-polymer-bioreactor system. Exp Cell Res 1998;240(1):58–65

40. Chu CR, Coutts RD, Yoshioka M, Harwood FL, Monosov AZ, Amiel D. Articular cartilage repair using allogeneic perichondrocyte-seeded biodegradable porous polylactic acid (PLA): a tissue-engineering study. J Biomed Mater Res 1995; 29(9):1147–1154

41. Freed LE, Grande DA, Lingbin Z, Emmanual J, Marquis JC, Langer R. Joint resurfacing using allograft chondrocytes and synthetic biodegradable polymer scaffolds. J Biomed Mater Res 1994;28(8):891–899

42. Niederauer GG, Slivka MA, Leatherbury NC, et al. Evaluation of multiphase implants for repair of focal osteochondral defects in goats. Biomaterials 2000;21(24):2561–2574

43. Rotter N, Aigner J, Naumann A, et al. Cartilage reconstruction in head and neck surgery: comparison of resorbable polymer scaffolds for tissue engineering of human septal cartilage. J Biomed Mater Res 1998;42(3):347–356

44. Marcacci M, Berruto M, Brocchetta D, et al. Articular cartilage engineering with Hyalograft C: 3-year clinical results. Clin Orthop Relat Res 2005; 435(435):96–105

45. Grigolo B, Roseti L, Fiorini M, et al. Transplantation of chondrocytes seeded on a hyaluronan derivative (hyaff-11) into cartilage defects in rabbits. Biomaterials 2001;22(17):2417–2424

46. Almqvist KF, Wang L, Broddelez C, Veys EM, Verbruggen G. Biological freezing of human articular chondrocytes. Osteoarthritis Cartilage 2001;9(4):341–350

47. Almqvist KF, Wang L, Wang J, et al. Culture of chondrocytes in alginate surrounded by fibrin gel: characteristics of the cells over a period of eight weeks. Ann Rheum Dis 2001;60(8):781–790

48. Almqvist KF, Dhollander AAM, Verdonk PCM, Forsyth R, Verdonk R, Verbruggen G. Treatment of cartilage defects in the knee using alginate beads containing human mature allogenic chondrocytes. Am J Sports Med 2009;37(10):1920–1929

49. Dhollander AAM, Huysse WCJ, Verdonk PC, et al. MRI evaluation of a new scaffold-based allogenic chondrocyte implantation for cartilage repair. Eur J Radiol 2010;75(1):72–81

50. Di Martino A, Sittinger M, Risbud MV. Chitosan: a versatile biopolymer for orthopaedic tissue-engineering. Biomaterials 2005;26(30):5983–5990

51. Lahiji A, Sohrabi A, Hungerford DS, Frondoza CG. Chitosan supports the expression of extracellular matrix proteins in human osteoblasts and chondrocytes. J Biomed Mater Res 2000;51(4):586–595

52. Hoemann CD, Sun J, Légaré A, McKee MD, Buschmann MD. Tissue engineering of cartilage using an injectable and adhesive chitosan-based cell-delivery vehicle. Osteoarthritis Cartilage 2005;13(4):318–329

53. Li Z, Zhang M. Chitosan-alginate as scaffolding material for cartilage tissue engineering. J Biomed Mater Res A 2005;75(2):485–493

54. van Susante JL, Buma P, Homminga GN, van den Berg WB, Veth RP. Chondrocyte-seeded hydroxyapatite for repair of large articular cartilage defects. A pilot study in the goat. Biomaterials 1998;19(24):2367–2374

55. Barker TH, Fuller GM, Klinger MM, Feldman DS, Hagood JS. Modification of fibrinogen with poly(ethylene glycol) and its effects on fibrin clot characteristics. J Biomed Mater Res 2001;56(4):529–535

56. Bryant SJ, Bender RJ, Durand KL, Anseth KS. Encapsulating chondrocytes in degrading PEG hydrogels with high modulus: engineering gel structural changes to facilitate cartilaginous tissue production. Biotechnol Bioeng 2004;86(7):747–755

57. Kon E, Delcogliano M, Filardo G, Busacca M, Di Martino A, Marcacci M. Novel nano-composite multilayered biomaterial for osteochondral regeneration: a pilot clinical trial. Am J Sports Med 2011;39(6):1180–1190

58. Kon E, Delcogliano M, Filardo G, et al. Orderly osteochondral regeneration in a sheep model using a novel nano-composite multilayered biomaterial. J Orthop Res 2010;28(1):116–124

59. Kon E, Mutini A, Arcangeli E, et al. Novel nanostructured scaffold for osteochondral regeneration: pilot study in horses. J Tissue Eng Regen Med 2010;4(4):300–308

60. Tampieri A, Sandri M, Landi E, et al. Design of graded biomimetic osteochondral composite scaffolds. Biomaterials 2008;29(26):3539–3546

61. Kerker JT, Leo AJ, Sgaglione NA. Cartilage repair: synthetics and scaffolds: basic science, surgical techniques, and clinical outcomes. Sports Med Arthrosc 2008;16(4):208–216

62. Saithna AA, Carey-Smith R, Dhillon M, Thompson P, Spalding T. Synthetic polymer scaffolds for repair of small articular cartilage defects of the knee: Early clinical and radiological results. Presented at the British Orthopaedic Association Conference, Liverpool, England, September 2008.

63. Davidson PA, Rivenburgh DW. Prospective evaluation of osteochondral defects in the knee treated with biodegradable scaffolds (SS-45). Presented at the Arthroscopic Association of North America Meeting, Denver, Colorado, 2007.

64. Chen G, Sato T, Ushida T, et al. The use of a novel PLGA fiber/collagen composite web as a scaffold for engineering of articular cartilage tissue with adjustable thickness. J Biomed Mater Res A 2003;67(4):1170–1180

65. Chen G, Sato T, Ushida T, Ochiai N, Tateishi T. Tissue engineering of cartilage using a hybrid scaffold of synthetic polymer and collagen. Tissue Eng 2004;10(3-4):323–330

66. Iwasa J, Engebretsen L, Shima Y, Ochi M. Clinical application of scaffolds for cartilage tissue engineering. Knee Surg Sports Traumatol Arthrosc 2009;17(6):561–577

67. Steinwachs MR, Guggi T, Kreuz PC. Marrow stimulation techniques. Injury 2008;39(Suppl 1):S26–S31

68. Selmi TA, Verdonk P, Chambat P, et al. Autologous chondrocyte implantation in a novel alginate-agarose hydrogel: outcome at two years. J Bone Joint Surg Br 2008;90(5):597–604
69. Patrascu JM, Freymann U, Kaps C, Poenaru DV. Repair of a post-traumatic cartilage defect with a cell-free polymer-based cartilage implant: a follow-up at two years by MRI and histological review. J Bone Joint Surg Br 2010;92(8):1160–1163
70. Hegewald AA, Ringe J, Bartel J, et al. Hyaluronic acid and autologous synovial fluid induce chondrogenic differentiation of equine mesenchymal stem cells: a preliminary study. Tissue Cell 2004;36(6):431–438
71. Knecht S, Erggelet C, Endres M, Sittinger M, Kaps C, Stüssi E. Mechanical testing of fixation techniques for scaffold-based tissue-engineered grafts. J Biomed Mater Res B Appl Biomater 2007; 83(1):50–57
72. Zelle S, Zantop T, Schanz S, Petersen W. Arthroscopic techniques for the fixation of a three-dimensional scaffold for autologous chondrocyte transplantation: structural properties in an in vitro model. Arthroscopy 2007;23(10):1073–1078
73. Erggelet C, Neumann K, Endres M, Haberstroh K, Sittinger M, Kaps C. Regeneration of ovine articular cartilage defects by cell-free polymer-based implants. Biomaterials 2007;28(36): 5570–5580
74. Zantop T, Petersen W. Arthroscopic implantation of a matrix to cover large chondral defect during microfracture. Arthroscopy 2009;25(11): 1354–1360
75. Barker TH, Fuller GM, Klinger MM, Feldman DS, Hagood JS. Modification of fibrinogen with poly(ethylene glycol) and its effects on fibrin clot characteristics. J Biomed Mater Res 2001;56(4): 529–535
76. Temenoff JS, Athanasiou KA, LeBaron RG, Mikos AG. Effect of poly (ethylene glycol) molecular weight on tensile and swelling properties of oligo(poly(ethylene glycol) fumarate) hydrogels for cartilage tissue engineering. J Biomed Mater Res 2002;59(3):429–437
77. Kon E, Delcogliano M, Filardo G, Montaperto C, Marcacci M. Second generation issues in cartilage repair. Sports Med Arthrosc 2008;16(4):221–229
78. Erggelet C, Kreuz PC, Mrosek EH, et al. Autologous chondrocyte implantation versus ACI using 3D-bioresorbable graft for the treatment of large full-thickness cartilage lesions of the knee. Arch Orthop Trauma Surg 2010;130(8):957–964
79. Kreuz PC, Müller S, Ossendorf C, Kaps C, Erggelet C. Treatment of focal degenerative cartilage defects with polymer-based autologous chondrocyte grafts: four-year clinical results. Arthritis Res Ther 2009;11(2):R33
80. Ahmed TA, Giulivi A, Griffith M, Hincke M. Fibrin glues in combination with mesenchymal stem cells to develop a tissue-engineered cartilage substitute. Tissue Eng Part A 2011;17(3-4): 323–335
81. Nguyen LH, Kudva AK, Saxena NS, Roy K. Engineering articular cartilage with spatially-varying matrix composition and mechanical properties from a single stem cell population using a multi-layered hydrogel. Biomaterials 2011;32(29):6946–6952

15

Toward Engineering a Biological Joint Replacement

Grace D. O'Connell, Eric G. Lima, Liming Bian, Nadeen O. Chahine, Michael B. Albro, James L. Cook, Gerard A. Ateshian, and Clark T. Hung

Over 20% of the adults in the United States (25 years and above) have osteoarthritis (OA) of the hip or knee (> 46 million Americans).[1] OA is ranked as one of the top three causes for disability and contributes more than $185 billion dollars a year in related medical costs.[2–4] More than 580,000 arthroplasty procedures are performed each year in the U.S.[5] While joint replacement generally succeeds in decreasing or eliminating pain and restoring joint function, the lifespan of prostheses is limited due to wear, loosening, infection, and fracture of the implant or surrounding bone.[6–9] Alternative treatments have not yet been successful in providing a viable long-term option for cartilage repair. For example, allografts are limited by donor tissue availability and graft viability,[10,11] while autografts are limited by the availability of healthy tissue and donor site morbidity. Bioengineered repair strategies that circumvent these limitations, while preserving the natural function of the joint and using a procedure less invasive than total joint arthroplasty, may be optimal for treating younger OA patients.

Articular cartilage serves as the load-bearing material of joints and possesses excellent friction, lubrication, and wear characteristics.[12] Successful replacement of damaged or injured articular cartilage will hinge on the ability to recapitulate the mechanical and structural properties of the healthy native tissue before implantation. Over the past two decades, there has been a wide interest in developing functional engineered cartilage. To grow cartilage tissue, cells are cultured within a three-dimensional (3D) scaffold that provides an initial structure for the de novo tissue[13–15]; alternatively, cells may be cultured using scaffold-less techniques.[16,17] For example, autologous chondrocyte implantation (ACI) is a cell-based strategy in which cells are injected directly into focal lesions and covered with a periosteal flap,[18,19] whereas CARTIPATCH[18] (TBF Tissue Bioengineering, Bron, France) uses a 3D agarose hydrogel scaffold to prevent leakage of cells, stabilize the chondrocyte phenotype, and promote a homogeneous distribution of cells.[18] Although these techniques are designed for repair and regeneration of cartilage focal lesions, they can be scaled up to replace an entire articulating layer (**Fig. 15.1**). However, nutrient diffusion through the depth of these large scaffolds represents a major challenge facing the field.

There are two prevailing points of view regarding implantation of engineered cartilage constructs; one approach places the cell–scaffold construct immediately into the defect site and relies on the in situ biological and loading environment to foster construct development (e.g.,[20–27]). Using this approach, poly ε-caprolactone (PCL) scaffolds have been designed to provide sufficient mechanical

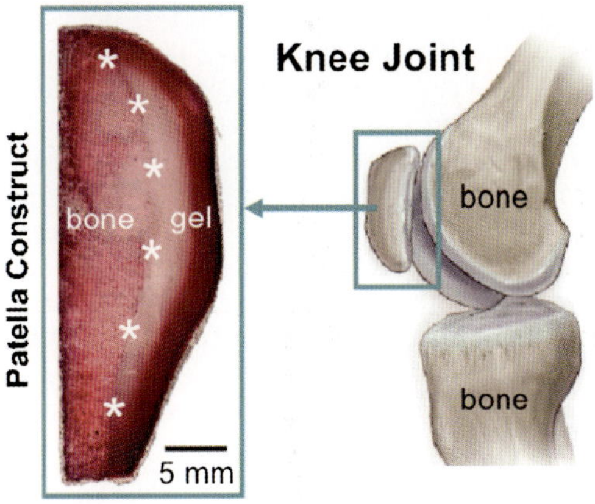

Fig. 15.1 Engineered patella construct showing proteoglycan-rich matrix (red Safranin-O stain) limited to gel periphery indicating diffusion limitations. *Interface between gel–bony substrate. (Adapted from Hung et al).[14]

support upon implantation, while being porous enough to permit de novo tissue development.[26,27] Lee et al demonstrated tissue growth in a full surface repair of a rabbit proximal humeral head following implantation of an acellular PCL scaffold infused with transforming growth factors.[27] Scaffolds that provide sufficient mechanical support at implantation reduce the need for extended culture periods before the repair surgery. The regenerated tissue develops under physiological loading conditions, which may ideally provide better functional tissue.

Another approach is to first precondition the cell–scaffold construct in vitro before implantation into the defect (e.g.,[28–32]). In vitro cultivation provides a controlled nutrient supply and loading environment that may be optimized for matrix synthesis to produce stiff cartilage-like constructs that may ideally sustain physiological loading following implantation. The required mechanical properties of the engineered cartilage will be dictated by the extent of the damaged region and its mechanical demands. In this type of approach, studies have demonstrated that the most robust tissue properties may be achieved by optimizing the media formulation as well as the transport of solutes in the developing tissue. Applied dynamic deformational loading to cell-seeded hydrogel constructs provides physical cues to cells and enhanced solute transport, leading to improved mechanical properties compared with free-swelling (unloaded) control.[14,24,33–39]

Articular cartilage is a highly hydrated soft tissue whose solid organic matrix is composed mostly of collagen fibrils (10 to 20% mass by wet weight) and proteoglycans (5 to 10% mass by wet weight).[40–44] Chondrocytes compose less than 10% of the tissue volume[45] and maintain the tissue by synthesizing and secreting extracellular matrix. Chondrocyte morphology and biochemical and mechanical properties vary through the depth of the tissue. Near the articular surface cells are more elliptical, and the tissue is softer than in the middle and deep zones.[46–49] There have been many improvements in biological replacement strategies for cartilage; however, there are still some limitations and challenges that remain to be addressed for successful repair and regeneration.[50] The purpose of this review is to summarize our advances in engineering cartilage and to identify approaches for scaling up these strategies to engineer large constructs suitable for replacing entire articular surfaces in cases of traumatic injury and advanced joint degeneration.

Cell Sources for Cartilage Tissue Engineering

Previous studies have been successful in cultivating functional engineered cartilage using cells from juvenile bovine and adult canine cartilage.[24,51,52] These studies have reported equilibrium compressive mechanical properties and glycosaminoglycan (GAG) content similar to native values.[25,35,36,53] Cultivating functional tissue with adult chondrocytes shows promise for using these techniques as a clinical repair strategy, since OA is commonly observed in older patients. Autologous chondrocytes from a nearby healthy region of the joint are the ideal cell source for clinical applications of engineered cartilage. However, acquiring enough cells while minimizing donor site morbidity remains a major challenge. Donor cells can be expanded in culture before implantation, but the proliferation rate of human adult cells is generally low and would lengthen the time between surgeries.

Alternatively, stem cells have been investigated as a possible cell source for cartilage regeneration.[26,30,51,54–58] Mesenchymal stem cells (MSCs) from adipose tissue, bone, and synovium express surface markers also expressed by chondrocytes, suggesting that MSCs can potentially differentiate toward a chondrogenic lineage.[55,59–61] Cells can be encouraged toward chondrogenesis by using a cocktail of growth factors to expand the cells in culture.[54,55,57,62] Including growth factors, such as basic fibroblast growth factor (FGF), platelet-derived growth factor (PDGF), and transforming growth factor (TGF) in the expansion culture medium improves the mechanical and biochemical properties of engineered cartilage in 3D culture.[55] Expansion of MSCs in vitro reduces the amount of tissue required to obtain a sufficient number of cells, mitigating damage to healthy tissue. Furthermore, encapsulation of MSCs in hydrogels and woven scaffolds demonstrates that these cells are capable of producing cartilage-like tissue with mechanical and biochemical properties tending toward native cartilage values (**Fig. 15.2**).[26,30,55]

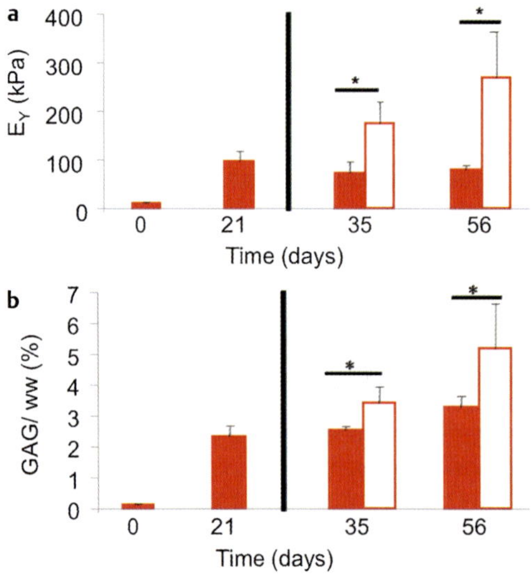

Fig. 15.2 (**a**) Equilibrium modulus (E_Y) and (**b**) glycosaminoglycan content (GAG) normalized to percent wet weight (ww) for synovium-derived stem cell-encapsulated hydrogels cultured with continuous growth factor (*solid bar*) or with transient release of growth factors at day 21 (*white bar*), mean ± SD, $^*p < 0.05$ for continuous versus transient groups for $n = 5$ constructs/group. (Adapted from Sampat et al).[55]

Previous studies that have used bone- or synovium-derived MSCs have demonstrated that these cells can be passaged and expanded in monolayer culture with a medium defined to promote chondrogenesis.[51,55,62,63] In contrast, culturing chondrocytes on a hard surface, such as a glass plate or tissue culture flask, causes the cell to dedifferentiate and become more fibroblast-like. The change in cell behavior may be beneficial for increasing collagen production in vitro. A recent study by Anderson and Athenasiou demonstrated that passaged chondrocytes can self-assemble into engineered cartilage plugs and that passaged cells produce two times more GAG and five times more collagen than primary chondrocytes.[64] However, most of the collagen produced with passaged cells was type I collagen, which is not ideal for recapitulating the native cartilage tissue composition.[64] In contrast, culturing differentiated stem cells in a 3D scaffold that promotes chondrogenesis (i.e., hydrogels) improves collagen type II production.[63] The loading environment experienced by the cell in two-dimensional (2D) culture significantly affects the matrix produced by the cells in 3D culture.[55,64]

Strategies for Improving Nutrient Diffusion: Enhanced Transport of Nutrients

Closer examination of engineered cartilage constructs reveals that the mechanical properties and matrix distribution are spatially heterogeneous. These constructs typically exhibit mechanically stiffer regions and greater matrix deposition near their peripheral boundaries, whereas regions deeper inside the constructs are typically softer, showing less matrix deposition.[65] Nutrient diffusion deep into the construct is a challenge in the field that will be considerably amplified when thicker and wider constructs are cultured. Human articular cartilage can be up to 7 mm thick[66,67]; therefore, thicker constructs will be necessary for clinical applications.

Immature articular cartilage is well vascularized with canals that provide nutrients to the developing tissue and remove waste by-products.[68–71] Experimental studies have

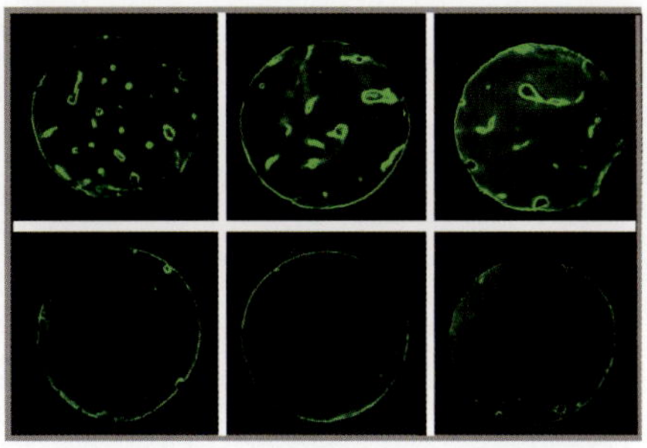

Fig. 15.3 Confocal images of fluorescently labeled 70 kDa dextran solute in cartilage sections after testing under dynamic loading or passive diffusion conditions. These images demonstrate that solute pumping under dynamic loading occurs at cartilage canals as well as the outer boundaries of the explant, as evidenced from the narrow boundary layers of high solute concentrations at 30 seconds and 2 hours. Over time, solute concentration spreads out from the canals into the surrounding matrix (20 hours).

demonstrated that dynamic loading enhances the uptake of nutrients into agarose hydrogels and immature cartilage.[72,73] For smaller macromolecules (i.e., 3 kDa), uptake by cartilage under dynamic loading conditions was threefold higher than uptake by passive diffusion. The effects of dynamic loading were found to be more pronounced with larger macromolecules. A 70 kDa molecule achieved a concentration nine times higher under dynamic loading than under passive diffusion into cartilage. This enhanced uptake of nutrients by the tissue was considerably facilitated by the network of cartilage canals, allowing for nutrients to transport deep into the tissue (**Fig. 15.3**). The increase in nutrient uptake was attributed solely to dynamic loading by demonstrating a return of the solute concentration to passive diffusion levels after termination of loading. Thus, loading provides an active solute-pumping mechanism because the solid matrix of the immature cartilage can impart momentum to the solute at the tissue–bath interface, pulling it into the tissue.[72,73]

Similarly, short-term dynamic loading of engineered cartilage constructs (i.e., for less than 3 hours) has demonstrated improved nutrient diffusion.[33,34,74] Loading of anatomical-size patellar constructs doubled the concentration of large molecules (70 kDa) in the constructs compared with constructs under free-swelling conditions (control; **Fig. 15.4**).[34] Dynamic loading has been shown to significantly improve mechanical properties of engineered cartilage, suggesting that the increased nutrient uptake during the 3 hours of daily loading influences matrix production and deposition.[24,34,36,74,75] The loading type, duration, and frequency can greatly impact the mechanotransduction response of chondrocyte-seeded scaffolds.[74–76] Long-duration loading protocols (6 hours) result in a decrease in the nutrient diffusion into large constructs during loading, primarily due to decreased surface area available for free diffusion.[34] However, these constructs produce stiffer engineered cartilage than constructs cultured under free-swelling conditions.[75] In contrast to the beneficial effects of dynamic loading, static loading significantly decreases the nutrient uptake by engineered cartilage (**Fig. 15.4**) These findings in engineered cartilage are consistent with the observation that dynamic loading produces enhanced uptake of solutes into agarose and cartilage,[72,73] considerably greater than under passive diffusion. However, other mechanotransduction pathways may also be at work when constructs are being loaded dynamically. Although the precise nature of these mechanisms is not completely understood, dynamic loading can be used to improve nutrient transport into large constructs designed to replicate entire articular layers such as the human retropatellar surface (**Figs. 15.1** and **15.4**).

Perfusion, a convective transport method, applies a biomimetic approach to provide nutrients into engineered constructs by mimicking the function of the vascular canals in developing cartilage (**Fig. 15.4b, c**). Currently, there are conflicting findings in the literature for the beneficial effects of perfusion, which may suggest further research is needed to determine the optimal flow rate and duration, and when perfusion should be applied over the tissue maturation period. Raimondi et al demonstrated that perfusion of chondrocyte-seeded constructs can potentially improve cell viability, GAG synthesis, and mechanical properties.[77,78] Grayson et al demonstrated that perfusion of nutrients through the bone region of an osteochondral construct improves matrix production and

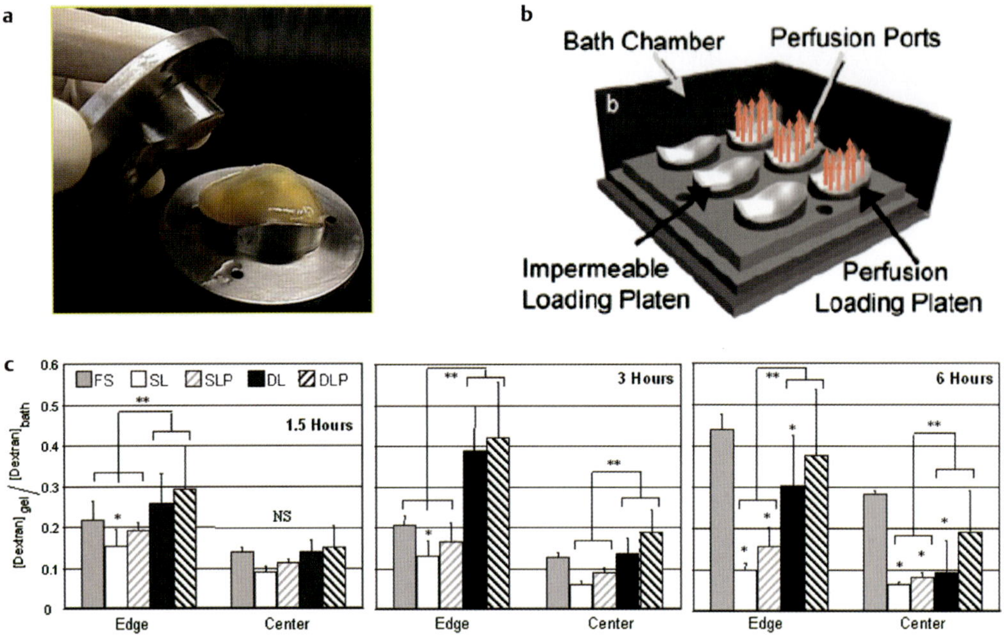

Fig. 15.4 (**a**) Human patella-shaped stainless steel molds for fabrication of patella constructs. (**b**) Schematic of loading platens for applying deformational loading to patella constructs with and without concomitant perfusion. (**c**) Solute transport study of engineered patella constructs using 70 kDa dextran analyzed at the peripheral edge or center region. The experimental groups are FS, free swelling; SL, static loading; SLP, static loading with perfusion; DL, dynamic loading; and DLP, dynamic loading with perfusion. Dextran concentration in engineered cartilage (normalized to bathing concentration) after 1.5, 3, and 6 hours (mean S.D. $N = 3$; $n = 8$–16 per group). $^{*}p < 0.05$ relative to FS; $^{**}p < 0.05$ relative to DL and DLP.

distribution in the engineered cartilage.[79] In contrast, studies that have combined perfusion with dynamic loading have not observed additional nutritional benefits from perfusion.[34,78,80] The findings of these studies suggest some potential benefits of using perfusion in the absence of mechanical loading stimuli to improve the compositional and mechanical properties of immature osteochondral constructs. This will be especially important for large osteochondral constructs because the bone–substrate interface will make limited nutrient diffusion into constructs an even greater challenge to overcome.

◆ Strategies for Improving Nutrient Diffusion: Designing Multiscale Nutrient Pathways

Inspired by the anatomy and physiology of developing native tissue, microscopic and macroscopic channels have been incorporated in engineered cartilage constructs[81–85] to provide pathways for improving nutrient transport. Large vascular-like canals can be incorporated at the macroscopic level by creating one or more channels through the thickness of the scaffold during fabrication. Adding a macroscopic channel (1-mm diameter) in the center of a cylindrical hydrogel construct (4-mm diameter) is a very effective method for decreasing the nutrient path length and improving the depth-dependent mechanical properties over time in culture (**Fig. 15.5**).[81] Over time in culture, chondrocytes located in proximity to the channel deposit extracellular matrix that progressively fills it. Thus, even though channels may improve nutrient supply only initially, they may be most beneficial in the formative stages of large engineered constructs and may not be as critical for maintaining tissue properties following implantation into a joint. This nutrient channel method may be scaled up for larger scaffolds by adding more

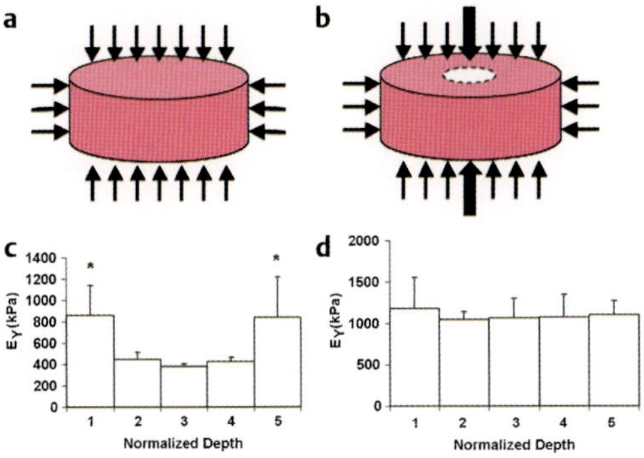

Fig. 15.5 Schematic showing improved nutrient diffusion into (**a**) a hydrogel scaffold by adding a macroscopic channel (*white circle*) to the scaffold (**b**). Representative Young's modulus (E_Y) through the depth of a mature construct for a (**c**) control construct and (**d**) a construct with a channel in the center. (Adapted from Bian et al 2009).[81] Using a microscopy-based material testing device and digital image correlation, the modulus is determined and plotted for each construct, divided into five layers of equal thickness across the depth of the discs. $^{*}p < 0.05$ versus the central region (2–4).

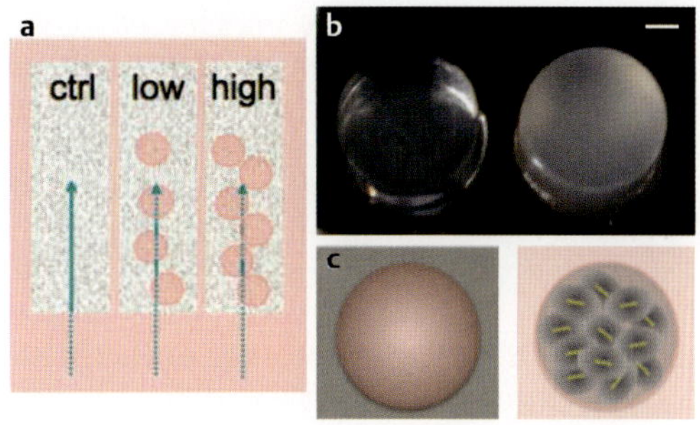

Fig. 15.6 (**a**) A schematic of solute diffusion in agarose hydrogel (ctrl). Microbubbles (*pink circles*) at varying concentrations can be incorporated into the hydrogel to increase the relative porosity of the scaffold upon their dissolution, thereby decreasing the nutrient path length to the center of the construct. Strategies for increasing collagen content of engineered cartilage include controlled enzymatic digestion of proteoglycans. (**b**) Gross image of an acellular hydrogel without (left) and with (right) microtubes. Bar ¼ 1 mm. (**c**) Schematic representing enzyme diffusion into the scaffold from the culture media bath (left) and from scaffold encapsulated lipid microtubes (right). The darker shading represents increased enzyme concentration.

channels. In a study of 10-mm-diameter constructs, the placement of three channels produced tissue with mechanical properties similar to native cartilage.[81] As cartilage tissue engineering moves toward cultivating biological replacements for the entire articular surface, an array of channels may be critical for achieving adequate mechanical and biochemical properties.

Another approach that is being investigated is to incorporate lipid-shelled microbubbles or microtubes as a porogens for hydrogel scaffolds.[85–88] Originally designed for drug delivery,[89–91] these biocompatible porogens are utilized directly with cells during the hydrogel scaffold cross-linking process. This allows engineers to create microlevel porosity in the superstructure of the hydrogel, while maintaining tight, nano-level porosity in the scaffolding directly around the embedded cells. The porous superstructure creates pockets of fluid–fluid nutrient reservoirs that provide less resistance to solute diffusion (**Fig. 15.6a**).[88] Preliminary studies suggest that microporogens improve the homogeneity of mechanical properties through the depth of the construct.[81,88] Constructs with a relatively low concentration of microporogens (0.2% wet by volume or 10% of the agarose hydrogel concentration) are more opaque and are two times stiffer than control constructs without microporogens (**Fig. 15.6b**).[85,88]

The gas-filled lipid microbubbles are incorporated into the scaffold filled with a stable gas. The size of the microbubbles and the gas that is used to create them can be altered to control parameters such as the dispersion rate of the bubbles and the porosity of the hydrogel. Intriguingly, it may be possible to create microbubbles that maintain their gas phase for extended periods of time. Under this scenario, it may be possible to purge the gas (and thereby create a fluid-filled pocket) later in culture. This would allow both spatial and temporal control of hydrogel porosity. It may even allow platen-less dynamic deformational loading as the gas phase of the bubbles is utilized in a hydrostatic pressure chamber.

Alternatively, lipid microtubes provide a hard tubular shell that may act as a nutrient channel on the microscopic scale (diameter = 0.5 μm, length = 40 μm).[92] Similar to the lipid-microbubbles, preliminary data suggest that these porogens can be incorporated into hydrogels to improve nutrient diffusion into engineered constructs. Moreover, the length of the microtubes can be increased to provide larger fluid-filled pockets. This approach combines the decreased nutrient path length provided by channels and the enzyme or nutrient loading capability of microporogens. Since microtubes have a lipid wall between the cells and the open channel, they are not expected to fill with extracellular matrix with time in culture, providing long-term enhanced nutrient diffusion. Although these studies have shown promise for using microtubes to increase the scaffold porosity, nutrient diffusion, extracellular matrix production, and mechanical properties,[85,87,88] future work is needed to confirm that the micropores are maintained with long-term culture and with physiological levels of loading.

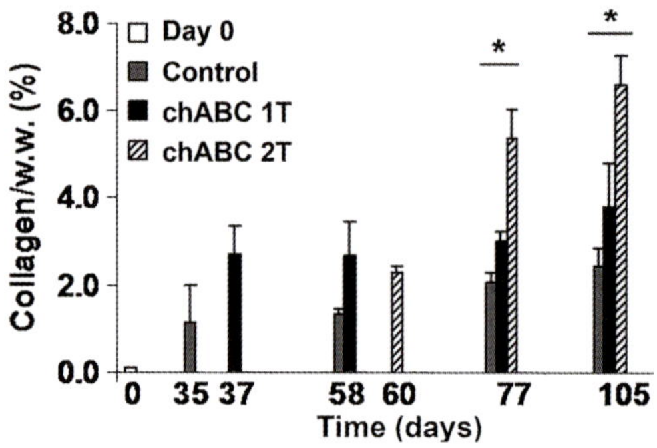

Fig. 15.7 Digestion of mature engineered constructs with chondroitinase ABC (chABC) added to the culture medium increases the collagen content. Multiple applications of chABC added to the medium result in further increases in the collagen content. Control, undigested constructs; chABC 1T, single chABC digestion at day 35; chABC 2T, chABC digestions at days 35 and 58. $^*p < 0.05$. (Adapted from Bian et al).[53]

Strategies for Improving Collagen Production

Collagen type II is a major constituent of the articular cartilage matrix. Although engineered cartilage is capable of producing native levels of GAG, recapitulating the collagen composition and structure remains a significant challenge for the field. Previous studies have described methods to increase in vitro collagen production by differentiating cells in monolayer culture before 3D encapsulation,[64] digesting the deposited GAGs or scaffold.[53,85,93–95]

Digestion of deposited GAGs with chondroitinase ABC (chABC) is a counterintuitive approach but has been a successful strategy for increasing the collagen content, since the depletion of GAGs may provide more space for cells to deposit collagen fibrils. In this strategy, ~ 90% of the GAGs are removed with chABC digestion of mature engineered constructs.[53] During the subsequent culture period, the GAG content recovers to the same level as undigested control samples within 4 weeks.[53,85,95–97] Following GAG recovery, the tensile and compressive mechanical properties of digested constructs are significantly better than the undigested constructs.[53,85,95–97] Multiple digestions with chABC can be applied throughout the culture period with additive increases in the collagen content and mechanical properties (**Fig. 15.7**).[53,96] These intriguing results suggest that additional studies are needed to understand the interactions between GAG and collagen synthesis and deposition in engineered constructs.

An alternative but related approach is to digest the agarose hydrogel scaffold to increase the collagen content.[53,94] In one such study, exposing mature constructs to agarase delivered through the culture medium digested approximately half of the agarose content in 4-mm-diameter constructs. The digestion decreased the compressive modulus by ~ 45%, leaving enough extracellular matrix to maintain construct integrity, while cell viability was unaffected.[94] After the digestion, the collagen content (normalized by wet weight) continued to increase throughout the culture period and achieved levels significantly greater than the undigested control.[53] After 8 weeks of digestion (15 weeks in culture), the collagen content per wet weight was 6 to 7% for the agarose-treated constructs, which was 2.2 times greater than the undigested control. The dynamic and equilibrium moduli and the GAG content of the digested constructs recovered to the undigested control values within 7 weeks. The results of these studies suggest that biodegradable scaffolds designed to degrade within 4 to 6 weeks may provide ideal conditions for cultivating fully bioengineered cartilage replacements.[98]

In previous studies, chABC or agarase was added directly to the culture medium. The beneficial effects of the digestion were limited to the periphery of the construct, due to the diffusion-reaction gradient of enzyme toward the center of the construct (**Fig. 15.6c**).[53,94] As described above, there is a growing interest to encapsulate biomaterials originally designed for drug delivery into hydrogels for cartilage tissue engineering.[85–88] Furthermore, these biomaterials can be modified to encapsulate enzymes or growth factors. Lee et al developed a method for delivering thermostabilized chABC using sugar trehalose and hydrogel-microtubes for applications requiring extended enzyme release (~ 10-day release).[90] Encapsulation of microtubes during the fabrication process of engineered cartilage constructs allows for uniform distribution of the microtubes throughout the scaffold (**Fig. 15.6c**). This concept has been tested in a preliminary study by encapsulating chABC-loaded microtubes in chondrocyte-seeded hydrogels.[85] This study demonstrated that early exposure of immature engineered cartilage to a low continual dose of chABC did not inhibit tissue growth or mechanical properties. As observed in previous studies that added chABC to the culture medium, the collagen content of constructs with chABC-loaded microtubes was greater than control constructs at early culture time points. The improved collagen production during the first 2 weeks in culture resulted in improved compressive properties throughout the 7-week culture period. Furthermore, the uniform distribution of microtubes in the construct resulted in a more homogeneous distribution of GAG and collagen.[85] It is anticipated that chABC-loaded microtubes can improve collagen production in vitro, providing more functional engineered cartilage. The microporogens can be modified to alter the release rate or delay initial release of the enzyme, which would be beneficial in optimizing collagen production using enzymatic digestion.

These studies suggest that digestion of the scaffold or GAGs in mature constructs may provide a viable method for culturing engineered cartilage with near native collagen values. It is important to note that the culture time needed to create a functional engineered cartilage for implantation will increase with digestion, as full recovery of the GAGs may take up to 4 weeks. Clinical application of functional tissue engineering will need to balance the need for sufficient collagen content versus the cost of longer culture time.

Many studies have focused on increasing collagen production, but the collagen produced in vitro tends to be oriented randomly throughout the construct. In native cartilage, the collagen fibrils in the superficial layer are aligned tangential to the articular surface, whereas fibrils in the deep zone are oriented radially.[99] Collagen fibrils provide the tissue with tensile strength to help resist the lateral expansion of the tissue when subjected to elevated compressive and shear loads in situ.[100] Therefore, future work may need to focus on directing collagen fibril orientation during construct growth. Our previous work has suggested that applied deformational loading can influence fiber orientation in engineered cartilage, producing alignment perpendicular to the applied axial loading in unconfined compression of cylindrical constructs.[65] Engineering fibrocartilage tissue constructs may be achieved by using stiff fibers to provide a scaffold that can withstand the higher tensile stresses that these tissues experience under physiological loads. For example, microfibers may be fabricated to produce a prescribed nonlinear stress–strain response with a specific Young's modulus.[101] Encapsulation of these micro- or nanoscaled fibers within a hydrogel scaffold may be important for providing a fiber network template.

◆ Conclusion

Encouraging progress has been made in the field of cartilage functional tissue engineering over the last two decades, demonstrating that it is possible to engineer constructs from a variety of cell sources while achieving native levels of GAG content and equilibrium compressive properties. Due to the competing effects of nutrient transport and consumption, engineering functional constructs is necessarily limited to small sizes relative to the overall dimensions of articular layers in human joints. Therefore, such constructs are currently more suitable for repairing focal defects only, and various strategies are needed to scale up these successful methodologies to produce full-size engineered articular layers. Several promising new strategies were reviewed above, including macro- or microporogens, placement of channels, and dynamic loading, all of which aim to enhance the supply of nutrients to chondrocytes.

Active methods for improving nutrient uptake include bioreactors that apply compression[24,29,30,35,36,75,80] or rotation[102,103] and perfusion devices.[78,80,104,105] As discussed in the previous section, physiological levels of compressive dynamic loading (~ 10% strain, less than 6 hours) may be ideal for enhancing nutrient diffusion into engineered cartilage.[33] Future work will need to combine these techniques to understand whether the effects of increased porosity via micro- or macrochannels can be additive to the improved nutrient diffusion from dynamic loading (**Fig. 15.8**).

Another challenge confronting the field of cartilage tissue engineering is the requirement to produce collagen content levels and tensile mechanical properties that reproduce native values. For this challenge, a consistent observation emerging from recent tissue-engineering studies is the apparent hindrance to higher collagen synthesis caused by the presence of hydrogel and the increasing levels of GAG in the constructs. Enzymatic digestion of either agarose or GAG has been utilized successfully to improve collagen content without sacrificing the mechanical integrity of the mature construct.

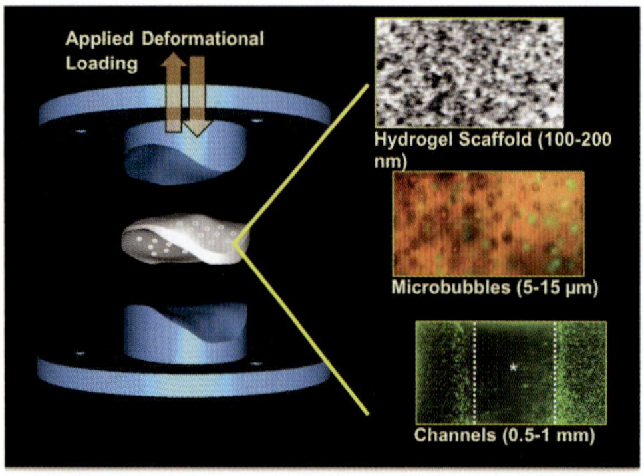

Fig. 15.8 Schematic showing dynamic loading of an engineered construct molded in the shape of a full human patellar surface. The scaffold porosity (AFM image, right column top) can be improved by incorporating microporogens, such as microbubbles (right column—middle; with calcein-labeled chondrocytes in green), or macrochannels (right column—bottom; asterisk and dotted lines indicating channel lumen with calcein-labeled chondrocytes in green).

Although the studies summarized above have demonstrated marked improvements in collagen production and nutrient transport, the resulting constructs do not yet replicate the full complement of functional properties of native cartilage. Consequently, more investigations are needed to combine and refine these new methods, while continuing to develop alternative strategies to produce implantable, full-size articular layers that can withstand the harsh mechanical and biological environment of an osteoarthritic joint.

Acknowledgments

Funding was provided by the NIH (R01AR046568 and R01AR060361), an NIH diversity fellowship, and an FASEB postdoctoral fellowship.

References

1. Helmick CG, Felson DT, Lawrence RC, et al; National Arthritis Data Workgroup. Estimates of the prevalence of arthritis and other rheumatic conditions in the United States. Part I. Arthritis Rheum 2008;58(1):15–25
2. Brault MW, Helmick CG, Theis KA, Armour BS. Prevalence and Most Common Causes of Disability Among Adults—United States, 2005. CDC; 2005.
3. Dibonaventura M, Gupta S, McDonald M, Sadosky A. Evaluating the health and economic impact of osteoarthritis pain in the workforce: results from the National Health and Wellness Survey. BMC Musculoskelet Disord 2011;12:83
4. Kotlarz H, Gunnarsson CL, Fang H, Rizzo JA. Insurer and out-of-pocket costs of osteoarthritis in the US: evidence from national survey data. Arthritis Rheum 2009;60(12):3546–3553
5. AAOS. Total Knee Replacement. 2009; http://orthoinfo.aaos.org/topic.cfm?topic=A00389. Accessed August 9, 2011.
6. Chong DY, Hansen UN, van der Venne R, Verdonschot N, Amis AA. The influence of tibial component fixation techniques on resorption of supporting bone stock after total knee replacement. J Biomech 2011;44(5):948–954
7. Essner A, Herrera L, Hughes P, Kester M. The influence of material and design on total knee replacement wear. J Knee Surg 2011;24(1):9–17

8. Lonner JH, Klotz M, Levitz C, Lotke PA. Changes in bone density after cemented total knee arthroplasty: influence of stem design. J Arthroplasty 2001;16(1):107–111
9. Naudie DD, Ammeen DJ, Engh GA, Rorabeck CH. Wear and osteolysis around total knee arthroplasty. J Am Acad Orthop Surg 2007;15(1):53–64
10. Pallante AL, Bae WC, Chen AC, Görtz S, Bugbee WD, Sah RL. Chondrocyte viability is higher after prolonged storage at 37 degrees C than at 4 degrees C for osteochondral grafts. Am J Sports Med 2009;37(Suppl 1):24S–32S
11. Görtz S, Bugbee WD. Fresh osteochondral allografts: graft processing and clinical applications. J Knee Surg 2006;19(3):231–240
12. Mow VC, Lai M. Biorheology of swelling tissue. Biorheology 1990;27(1):110–119
13. Bian L, Zhai DY, Mauck RL, Burdick JA. Coculture of human mesenchymal stem cells and articular chondrocytes reduces hypertrophy and enhances functional properties of engineered cartilage. Tissue Eng Part A 2011;17(7–8):1137–1145
14. Hung CT, Mauck RL, Wang CC, Lima EG, Ateshian GA. A paradigm for functional tissue engineering of articular cartilage via applied physiologic deformational loading. Ann Biomed Eng 2004; 32(1):35–49
15. Singhal AR, Agrawal CM, Athanasiou KA. Salient Degradation Features of a 50:50 PLA/PGA Scaffold for Tissue Engineering. Tissue Eng 1996;2(3):197–207
16. Elder SH, Cooley AJ Jr, Borazjani A, Sowell BL, To H, Tran SC. Production of hyaline-like cartilage by bone marrow mesenchymal stem cells in a self-assembly model. Tissue Eng Part A 2009;15(10):3025–3036
17. Natoli RM, Skaalure S, Bijlani S, Chen KX, Hu J, Athanasiou KA. Intracellular Na(+) and Ca(2+) modulation increases the tensile properties of developing engineered articular cartilage. Arthritis Rheum 2010;62(4):1097–1107
18. Selmi TA, Verdonk P, Chambat P, et al. Autologous chondrocyte implantation in a novel alginate-agarose hydrogel: outcome at two years. J Bone Joint Surg Br 2008;90(5):597–604
19. Zimmer Holdings and ISTO Technologies Announce Start of Neocartilage Clinical Trial. 2007; http://investor.zimmer.com/releasedetail.cfm?ReleaseID=2300582010
20. Niemeyer P, Köstler W, Salzmann GM, Lenz P, Kreuz PC, Südkamp NP. Autologous chondrocyte implantation for treatment of focal cartilage defects in patients age 40 years and older: A matched-pair analysis with 2-year follow-up. Am J Sports Med 2010;38(12):2410–2416
21. Filardo G, Kon E, Di Martino A, Iacono F, Marcacci M. Arthroscopic second-generation autologous chondrocyte implantation: a prospective 7-year follow-up study. Am J Sports Med 2011;39(10):2153–2160
22. Wegener B, Schrimpf FM, Bergschmidt P, et al. Cartilage regeneration by bone marrow cells-seeded scaffolds. J Biomed Mater Res A 2010; 95(3):735–740
23. Connelly JT, Wilson CG, Levenston ME. Characterization of proteoglycan production and processing by chondrocytes and BMSCs in tissue engineered constructs. Osteoarthritis Cartilage 2008;16(9):1092–1100
24. Mauck RL, Soltz MA, Wang CC, et al. Functional tissue engineering of articular cartilage through dynamic loading of chondrocyte-seeded agarose gels. J Biomech Eng 2000;122(3):252–260
25. Ng KW, Lima EG, Bian L, et al. Passaged adult chondrocytes can form engineered cartilage with functional mechanical properties: a canine model. Tissue Eng Part A 2010;16(3):1041–1051
26. Moutos FT, Guilak F. Functional properties of cell-seeded three-dimensionally woven poly(epsilon-caprolactone) scaffolds for cartilage tissue engineering. Tissue Eng Part A 2010; 16(4):1291–1301
27. Lee CH, Cook JL, Mendelson A, Moioli EK, Yao H, Mao JJ. Regeneration of the articular surface of the rabbit synovial joint by cell homing: a proof of concept study. Lancet 2010;376(9739): 440–448
28. Bichara DA, Zhao X, Bodugoz-Senturk H, et al. Porous poly(vinyl alcohol)-hydrogel matrix-engineered biosynthetic cartilage. Tissue Eng Part A 2011;17(3-4):301–309
29. Laganà K, Moretti M, Dubini G, Raimondi MT. A new bioreactor for the controlled application of complex mechanical stimuli for cartilage tissue engineering. Proc Inst Mech Eng H 2008; 222(5):705–715
30. Huang AH, Farrell MJ, Kim M, Mauck RL. Long-term dynamic loading improves the mechanical properties of chondrogenic mesenchymal stem cell-laden hydrogel. Eur Cell Mater 2010; 19:72–85
31. Qi Y, Zhao T, Xu K, Dai T, Yan W. The restoration of full-thickness cartilage defects with mesenchymal stem cells (MSCs) loaded and cross-linked bilayer collagen scaffolds on rabbit model. Mol Biol Rep 2012;39:1231–1237
32. Jin CZ, Cho JH, Choi BH, et al. The Maturity of Tissue Engineered Cartilage in vitro Affects the Reparability for Osteochondral Defect. Tissue Eng Part A 2012;18:219
33. Chahine NO, Albro MB, Lima EG, et al. Effect of dynamic loading on the transport of solutes into agarose hydrogels. Biophys J 2009; 97(4):968–975
34. Chahine NO, Wei VI, Lima EG, Albro MB, Hung CT, Ateshian GA. Effect of perfusion and dynamic loading on the distribution and concentration of solutes in patellar shaped agarose constructs. Orthopaedic Research Society. 2006; Chicago
35. Lima EG, Bian L, Mauck RL, et al. The effect of applied compressive loading on tissue-engineered cartilage constructs cultured with TGF-beta3. Conf Proc IEEE Eng Med Biol Soc 2006;1:779–782
36. Lima EG, Bian L, Ng KW, et al. The beneficial effect of delayed compressive loading on tissue-engineered cartilage constructs cultured with TGF-beta3. Osteoarthritis Cartilage 2007;15(9): 1025–1033
37. Mauck RL, Hung CT, Ateshian GA. Modeling of neutral solute transport in a dynamically loaded porous permeable gel: implications for articular cartilage biosynthesis and tissue engineering. J Biomech Eng 2003;125(5):602–614
38. Mauck RL, Nicoll SB, Seyhan SL, Ateshian GA, Hung CT. Synergistic action of growth factors and dynamic loading for articular cartilage tissue engineering. Tissue Eng 2003;9(4):597–611

39. Mauck RL, Wang CC, Oswald ES, Ateshian GA, Hung CT. The role of cell seeding density and nutrient supply for articular cartilage tissue engineering with deformational loading. *Osteoarthritis and cartilage/OARS*. Osteoarthritis and Cartilage 2003;11(12):879–890
40. Treppo S, Koepp H, Quan EC, Cole AA, Kuettner KE, Grodzinsky AJ. Comparison of biomechanical and biochemical properties of cartilage from human knee and ankle pairs. J Orthop Res 2000; 18(5):739–748
41. Nguyen AM, Levenston ME. Comparison of osmotic swelling influences on meniscal fibrocartilage and articular cartilage tissue mechanics in compression and shear. J Orthop Res 2012;30:95–102
42. Mow VC, Huiskes R. Basic Orthopaedics Biomechanics and Mechano-biology. Philadelphia, PA: Lippincott Williams & Wilkins; 2005
43. Saarakkala S, Julkunen P, Kiviranta P, Mäkitalo J, Jurvelin JS, Korhonen RK. Depth-wise progression of osteoarthritis in human articular cartilage: investigation of composition, structure and biomechanics. Osteoarthritis Cartilage 2010; 18(1):73–81
44. Temple-Wong MM, Bae WC, Chen MQ, et al. Biomechanical, structural, and biochemical indices of degenerative and osteoarthritic deterioration of adult human articular cartilage of the femoral condyle. Osteoarthritis Cartilage 2009;17(11): 1469–1476
45. Stockwell RS. Biology of Cartilage Cells. Cambridge: Cambridge Press; 1979
46. Choi JB, Youn I, Cao L, et al. Zonal changes in the three-dimensional morphology of the chondron under compression: the relationship among cellular, pericellular, and extracellular deformation in articular cartilage. J Biomech 2007;40(12): 2596–2603
47. Guilak F, Ratcliffe A, Mow VC. Chondrocyte deformation and local tissue strain in articular cartilage: a confocal microscopy study. J Orthop Res 1995;13(3):410–421
48. Schinagl RM, Gurskis D, Chen AC, Sah RL. Depth-dependent confined compression modulus of full-thickness bovine articular cartilage. J Orthop Res 1997;15(4):499–506
49. Wang CC, Deng JM, Ateshian GA, Hung CT. An automated approach for direct measurement of two-dimensional strain distributions within articular cartilage under unconfined compression. J Biomech Eng 2002;124(5):557–567
50. Redman SN, Oldfield SF, Archer CW. Current strategies for articular cartilage repair. Eur Cell Mater 2005;9:23–32, discussion 23–32
51. Huang AH, Yeger-McKeever M, Stein A, Mauck RL. Tensile properties of engineered cartilage formed from chondrocyte- and MSC-laden hydrogels. Osteoarthritis Cartilage 2008;16(9):1074–1082
52. Mauck RL, Seyhan SL, Ateshian GA, Hung CT. Influence of seeding density and dynamic deformational loading on the developing structure/ function relationships of chondrocyte-seeded agarose hydrogels. Ann Biomed Eng 2002;30(8): 1046–1056
53. Bian L, Crivello KM, Ng KW, et al. Influence of temporary chondroitinase ABC-induced glycosaminoglycan suppression on maturation of tissue-engineered cartilage. Tissue Eng Part A 2009; 15(8):2065–2072
54. Pei M, He F, Boyce BM, Kish VL. Repair of full-thickness femoral condyle cartilage defects using allogeneic synovial cell-engineered tissue constructs. Osteoarthritis Cartilage 2009;17(6): 714–722
55. Sampat SR, O'Connell GD, Fong JV, Alegre-Aguarón E, Ateshian GA, Hung CT. Growth factor priming of synovium-derived stem cells for cartilage tissue engineering. Tissue Eng Part A 2011; 17(17-18):2259–2265
56. Gimble JM, Bunnell BA, Chiu ES, Guilak F. Concise review: Adipose-derived stromal vascular fraction cells and stem cells: let's not get lost in translation. Stem Cells 2011;29(5):749–754
57. Sundelacruz S, Kaplan DL. Stem cell- and scaffold-based tissue engineering approaches to osteochondral regenerative medicine. Semin Cell Dev Biol 2009;20(6):646–655
58. Nishimura K, Solchaga LA, Caplan AI, Yoo JU, Goldberg VM, Johnstone B. Chondroprogenitor cells of synovial tissue. Arthritis Rheum 1999; 42(12):2631–2637
59. De Ugarte DA, Alfonso Z, Zuk PA, et al. Differential expression of stem cell mobilization-associated molecules on multi-lineage cells from adipose tissue and bone marrow. Immunol Lett 2003;89(2-3):267–270
60. Noël D, Caton D, Roche S, et al. Cell specific differences between human adipose-derived and mesenchymal-stromal cells despite similar differentiation potentials. Exp Cell Res 2008;314(7):1575–1584
61. Martínez-Lorenzo MJ, Royo-Cañas M, Alegre-Aguarón E, et al. Phenotype and chondrogenic differentiation of mesenchymal cells from adipose tissue of different species. J Orthop Res 2009;27(11):1499–1507
62. Pei M, He F, Vunjak-Novakovic G. Synovium-derived stem cell-based chondrogenesis. Differentiation 2008;76(10):1044–1056
63. Huang AH, Stein A, Tuan RS, Mauck RL. Transient exposure to transforming growth factor beta 3 improves the mechanical properties of mesenchymal stem cell-laden cartilage constructs in a density-dependent manner. Tissue Eng Part A 2009;15(11):3461–3472
64. Anderson DE, Athanasiou KA. A comparison of primary and passaged chondrocytes for use in engineering the temporomandibular joint. Arch Oral Biol 2009;54(2):138–145
65. Kelly TA, Ng KW, Wang CC, Ateshian GA, Hung CT. Spatial and temporal development of chondrocyte-seeded agarose constructs in free-swelling and dynamically loaded cultures. J Biomech 2006;39(8):1489–1497
66. Ateshian GA, Soslowsky LJ, Mow VC. Quantitation of articular surface topography and cartilage thickness in knee joints using stereophotogrammetry. J Biomech 1991;24(8):761–776
67. Eckstein F, Gavazzeni A, Sittek H, et al. Determination of knee joint cartilage thickness using three-dimensional magnetic resonance chondrocrassometry (3D MR-CCM). Magn Reson Med 1996;36(2):256–265
68. Wilsman NJ, Van Sickle DC. Cartilage canals, their morphology and distribution. Anat Rec 1972; 173(1):79–93
69. Stockwell RA. The ultrastructure of cartilage canals and the surrounding cartilage in the sheep fetus. J Anat 1971;109(Pt 3):397–410

70. Hunt CD, Ollerich DA, Nielsen FH. Morphology of the perforating cartilage canals in the proximal tibial growth plate of the chick. Anat Rec 1979;194(1):143–157
71. Ytrehus B, Carlson CS, Lundeheim N, et al. Vascularisation and osteochondrosis of the epiphyseal growth cartilage of the distal femur in pigs—development with age, growth rate, weight and joint shape. Bone 2004;34(3):454–465
72. Albro MB, Banerjee RE, Li R, et al. Dynamic loading of immature epiphyseal cartilage pumps nutrients out of vascular canals. J Biomech 2011;44(9):1654–1659
73. Albro MB, Chahine NO, Li R, Yeager K, Hung CT, Ateshian GA. Dynamic loading of deformable porous media can induce active solute transport. J Biomech 2008;41(15):3152–3157
74. Waldman SD, Spiteri CG, Grynpas MD, Pilliar RM, Kandel RA. Long-term intermittent compressive stimulation improves the composition and mechanical properties of tissue-engineered cartilage. Tissue Eng 2004;10(9-10):1323–1331
75. Ng KW, Mauck RL, Wang CC, et al. Duty Cycle of Deformational Loading Influences the Growth of Engineered Articular Cartilage. Cell Mol Bioeng 2009;2(3):386–394
76. Waldman SD, Spiteri CG, Grynpas MD, Pilliar RM, Hong J, Kandel RA. Effect of biomechanical conditioning on cartilaginous tissue formation in vitro. J Bone Joint Surg Am 2003;85-A(Suppl 2): 101–105
77. Raimondi MT, Boschetti F, Falcone L, Migliavacca F, Remuzzi A, Dubini G. The effect of media perfusion on three-dimensional cultures of human chondrocytes: integration of experimental and computational approaches. Biorheology 2004; 41(3–4):401–410
78. Raimondi MT, Candiani G, Cabras M, et al. Engineered cartilage constructs subject to very low regimens of interstitial perfusion. Biorheology 2008;45(3–4):471–478
79. Grayson WL, Bhumiratana S, Grace Chao PH, Hung CT, Vunjak-Novakovic G. Spatial regulation of human mesenchymal stem cell differentiation in engineered osteochondral constructs: effects of pre-differentiation, soluble factors and medium perfusion. Osteoarthritis Cartilage 2010;18(5):714–723
80. Seidel JO, Pei M, Gray ML, Langer R, Freed LE, Vunjak-Novakovic G. Long-term culture of tissue engineered cartilage in a perfused chamber with mechanical stimulation. Biorheology 2004; 41(3–4):445–458
81. Bian L, Angione SL, Ng KW, et al. Influence of decreasing nutrient path length on the development of engineered cartilage. Osteoarthritis Cartilage 2009;17(5):677–685
82. Choi NW, Cabodi M, Held B, Gleghorn JP, Bonassar LJ, Stroock AD. Microfluidic scaffolds for tissue engineering. Nat Mater 2007; 6(11):908–915
83. Ling Y, Rubin J, Deng Y, et al. A cell-laden microfluidic hydrogel. Lab Chip 2007;7(6): 756–762
84. Golden AP, Tien J. Fabrication of microfluidic hydrogels using molded gelatin as a sacrificial element. Lab Chip 2007;7(6):720–725
85. O'Connell GD, Gollnick C, Ateshian GA, Bellamkonda RV, Hung CT. Beneficial effects of chondroitinase ABC release from lipid microtubes encapsulated in chondrocytes-seeded hydrogel constructs. Paper presented at: ASME2011 Summer Bioengineering Conference 2011; Farmington, PA
86. Wang X, Wenk E, Zhang X, Meinel L, Vunjak-Novakovic G, Kaplan DL. Growth factor gradients via microsphere delivery in biopolymer scaffolds for osteochondral tissue engineering. J Control Release 2009;134(2):81–90
87. Durney KM, Sirsi SR, Nover A, et al. Using Microbubbles to Modulate Hydrogel Scaffold Properties for Cartilage Tissue Engineering. Paper presented at: Orthopaedic Research Society 2010; San Diego, CA.
88. Durney KM, Sirsi SR, Nover A, et al. Microbubbles Improve Depth-Dependent Mechanical Properties of Cartilage Tissue Engineered Constructs. Orthopaedic Research Society 2011; Long Beach, CA
89. Chen CC, Borden MA. The role of poly(ethylene glycol) brush architecture in complement activation on targeted microbubble surfaces. Biomaterials 2011;32(27):6579–6587
90. Lee H, McKeon RJ, Bellamkonda RV. Sustained delivery of thermostabilized chABC enhances axonal sprouting and functional recovery after spinal cord injury. Proc Natl Acad Sci U S A 2010; 107(8):3340–3345
91. Wang X, Yucel T, Lu Q, Hu X, Kaplan DL. Silk nanospheres and microspheres from silk/pva blend films for drug delivery. Biomaterials 2010; 31(6):1025–1035
92. Meilander NJ, Yu X, Ziats NP, Bellamkonda RV. Lipid-based microtubular drug delivery vehicles. J Control Release 2001;71(1):141–152
93. Asanbaeva A, Masuda K, Thonar EJ, Klisch SM, Sah RL. Mechanisms of cartilage growth: modulation of balance between proteoglycan and collagen in vitro using chondroitinase ABC. Arthritis Rheum 2007;56(1):188–198
94. Ng KW, Kugler LE, Doty SB, Ateshian GA, Hung CT. Scaffold degradation elevates the collagen content and dynamic compressive modulus in engineered articular cartilage. Osteoarthritis Cartilage 2009;17(2):220–227
95. Natoli RM, Revell CM, Athanasiou KA. Chondroitinase ABC treatment results in greater tensile properties of self-assembled tissue-engineered articular cartilage. Tissue Eng Part A 2009; 15(10):3119–3128
96. Natoli RM, Responte DJ, Lu BY, Athanasiou KA. Effects of multiple chondroitinase ABC applications on tissue engineered articular cartilage. J Orthop Res 2009;27(7):949–956
97. O'Connell GD, Fong JV, Joffe A, Moy MY, Newman IB, Hung CT. Trimethylamine N-oxide enhances the mechanical and biochemical properties of tissue engineered cartilage. Paper presented at: Orthopaedic Research Society 2011; Long Beach, CA.
98. Ifkovits JL, Devlin JJ, Eng G, Martens TP, Vunjak-Novakovic G, Burdick JA. Biodegradable fibrous scaffolds with tunable properties formed from photo-cross-linkable poly(glycerol sebacate). ACS Appl Mater Interfaces 2009;1(9):1878–1886
99. Wu JP, Kirk TB, Zheng MH. Study of the collagen structure in the superficial zone and physiological state of articular cartilage using a 3D confocal imaging technique. J Orthop Surg 2008;3:29

100. Wong BL, Sah RL. Mechanical asymmetry during articulation of tibial and femoral cartilages: local and overall compressive and shear deformation and properties. J Biomech 2010;43(9):1689–1695

101. Guthold M, Liu W, Sparks EA, et al. A comparison of the mechanical and structural properties of fibrin fibers with other protein fibers. Cell Biochem Biophys 2007;49(3):165–181

102. Cinbiz MN, Tiğli RS, Beşkardeş IG, Gümüşderelioğlu M, Colak U. Computational fluid dynamics modeling of momentum transport in rotating wall perfused bioreactor for cartilage tissue engineering. J Biotechnol 2010;150(3):389–395

103. Vunjak-Novakovic G, Martin I, Obradovic B, et al. Bioreactor cultivation conditions modulate the composition and mechanical properties of tissue-engineered cartilage. J Orthop Res 1999; 17(1):130–138

104. Klein TJ, Sah RL. Modulation of depth-dependent properties in tissue-engineered cartilage with a semi-permeable membrane and perfusion: a continuum model of matrix metabolism and transport. Biomech Model Mechanobiol 2007;6(1-2):21–32

105. Darling EM, Athanasiou KA. Articular cartilage bioreactors and bioprocesses. Tissue Eng 2003; 9(1):9–26

III

Postop Management and Outcomes Assessments

16

Postoperative Management of Patients with Articular Cartilage Repair

Jennifer Yasu Stone and Robert Schaal

The postoperative management of articular cartilage repairs in the knee is key to the continued quality of life for the patient. Because of tissue vulnerability following articular cartilage repair, it is vital that a structured program including early, controlled motion and weight bearing be utilized appropriately to promote an optimal healing environment.[1–6] When used appropriately, the early motion and exercise promoted by postoperative rehabilitation can support maturation of the tissue via the principle of mechanotransduction (process by which the body translates a mechanical load into a cellular response).[7]

Through various animal, cadaver, and human studies it has been determined that mechanotransduction may be thought of as a three-step process: mechanocoupling, cell-to-cell communication, and the effector cell response. Mechanocoupling refers to the process of the physical perturbation to a cell that occurs during a mechanical load and leads to a variety of chemical responses/signals within the cell. On a gross cellular level, the result is cell-to-cell communication, meaning that even those more distant cells that do not receive the direct load will have a response. Finally, the effector cell response refers to the overall result of the mechanotransduction process: cell remodeling and healing. For articular cartilage, it is now well accepted that this creates a healing environment that helps to stimulate matrix production, thus leading to stronger and healthier tissue.[5–7]

It is essential to facilitate healing while avoiding the potentially deleterious effects of overloading or overstraining developing tissues. For any articular cartilage repair, the rate of progression, amount of range of motion (ROM) desired/allowable, and the amount of weight bearing will all depend on the physiological healing process of the repair in question. This chapter will focus on describing the most current evidence-based rehabilitation for patients who have undergone an articular cartilage repair of the knee. We will also address the issues of postoperative pain management, evaluation for readiness to progress, and potential complicating factors.

◆ Rehabilitation for Articular Cartilage Repairs

Articular cartilage lesions (and thus repairs) vary greatly in size and severity. Thus, the rehabilitative process must take into consideration the size and location of the lesion, as well as type of repair performed. Regardless of the type of cartilage repair, there are several general principles that must be kept in mind to maximize the efficacy of therapy.

Most current protocols call for patients to begin passive motion within hours of undergoing surgery and for 6 to 8 hours per day. Some studies indicate that early motion can promote the formation of the smooth, low-friction surface desired for articular cartilage, while other authors opine that it is helpful in preventing the formation of adhesions.[2,3,5,6] Still other authors state that early motion is neither helpful nor harmful.[8,9] If used, it is important that early motion be passive and within a restricted range to prevent shearing forces associated with muscle activation that could compromise the repair.

All patients should be introduced to controlled, early, partial weight bearing. Studies show that controlled loading/unloading nourishes developing cartilage in addition to promoting the formation of a stronger matrix in response to the loads, thus leading to tissue with improved mechanical properties. Unloading technology (unloading treadmills, unloaded squats or leg presses, water therapy) can be helpful in achieving this goal. Attention should also be directed to retraining of the quadriceps and hip flexors, which can be inhibited by postoperative swelling and pain.[1,3,10–12]

Finally, all programs should include a focus on prevention of soft tissue adhesions. Scar tissue and patellar mobilization should occur early and often, and special attention should be paid to muscle length during the early healing phases.[12]

The rehabilitation protocols for various procedures are very similar. While it is important to understand the differences in procedures to help assess stages of tissue healing, the variation of protocol generally relates more to the location of the defect, as opposed to the type of procedure performed. Defects within the trochlea are treated a little differently than defects on the femoral condyles, and the rehabilitation is adjusted accordingly. As with any protocol, these are guidelines to follow, but each patient is progressed to the next phase only when appropriate. For example, poor quad strength may prolong time spent in the brace and delay progression to more aggressive strengthening exercises.

◆ Rehabilitation Protocol for Articular Cartilage Repairs

0 to 6 Weeks

Weight-Bearing Status

Trochlear Defect Patients may begin weight bearing as tolerated with brace locked in full extension for 0 to 6 weeks.

Femoral Condyle Defect Patients may begin non–weightbearing with brace locked in full extension for 0 to 4 weeks, then touch weight-bearing with brace unlocked for weeks 5 to 6.

Bracing

Trochlear Defect Brace locked in full extension except when performing ROM exercises. Brace may also be removed for all exercises if patient has appropriate quad control.

Femoral Condyle Defect Brace locked in full extension for 0 to 4 weeks except when performing ROM exercises. Brace may also be removed for all exercises if patient has appropriate quad control. Brace is unlocked for 5 to 6 weeks.

Range of Motion

Femoral Condyle and Trochlear Defects Treated the Same Patients may advance to 90 degrees of flexion as tolerated for weeks 0 to 4; progress to passive range of motion (PROM) as tolerated for weeks 5 to 6; and no active extension through long arc or short arc.

Therapeutic Exercise

Femoral Condyle and Trochlear Defects Treated the Same Following exercises should be performed in brace if quad control is not adequate:

- Isometric quad strengthening
- Patellar mobilizations
- Straight-leg raises
- Hip abduction/adduction
- Hamstring isometrics

- ROM
- Stationary bike—only if done passively on involved leg
- Core strengthening—not involving postoperative knee
- Modalities as needed for pain control

6 to 12 Weeks

Weight-Bearing Status

Trochlear Defect Patients may continue weight bearing as tolerated and discard crutch as gait normalizes (no antalgic or abnormal gait pattern).

Femoral Condyle Defect Patient may advance to full weight bearing as tolerated and may discard crutch as gait normalizes (no antalgic or abnormal gait pattern).

Bracing

Femoral Condyle and Trochlear Defects Treated the Same Patient may discontinue use of brace upon demonstrating good quad control (can perform straight-leg raise without a lag).

ROM

Femoral Condyle and Trochlear Defects Treated the Same Patient may achieve full active and passive ROM.

Therapeutic Exercise

Femoral Condyle and Trochlear Defects Treated the Same Following exercises should be performed in brace if quad control is not adequate:

- ROM to gain full flexion.
- Advance stationary bike beyond passive ROM, begin light resistance.
- Begin bilateral closed chain activities with resistance less than patient's body weight through pain-free ROM (no unilateral/single leg dynamic weight-bearing activities).
- Initiate proprioception exercises.
- Initiate progressive resistive exercises for hamstrings, hips, and lower legs.
- Initiate gait training to normalize gait pattern if needed.
- Advance core strengthening.

12 Weeks and Beyond

Therapeutic Exercise

Femoral Condyle and Trochlear Defects Treated the Same Patients should be full weight bearing with a normal gait pattern. They should be out of the brace and have full passive and active ROM. The following exercises can be performed:

- More vigorous treadmill walking
- Progression on stationary bike
- Stairmaster and elliptical as tolerated
- Unilateral balance/proprioception activities
- Closed chain activities progressing to resistance greater than the body weight as tolerated
- Unilateral closed chain/dynamic exercises (month 4)
- Jogging, plyometrics, and sport-specific function training (month 6)

Postoperative Pain Management

Pain and swelling are two of the most common complaints after articular cartilage repair. It is well established that pain and/or significant swelling in the knee invariably leads to inhibition of volitional control of the quadriceps muscle.[13–15] Since the quadriceps functions as a major knee stabilizer, it is vital to keep pain and edema controlled to maintain and improve function of this key muscle group.

Ice has long been used as an analgesic and method of controlling edema. A combination of ice, compression, and elevation is the most effective and efficient method of edema control for patients undergoing knee surgery. Several studies have also shown a decrease in the need for pain medication in patients who used a continuous temperature-controlled cuff device such as the CryoCuff (DJO, Vista, CA), though the use of ice alone or ice and elastic wrap seems to provide the same amount of edema control.[14,15]

Other nonpharmaceutical pain relief options include the use of transcutaneous electrical nerve stimulation (TENS), IFC (interferential current), and manual edema drainage by a physical therapist. Patients should also be aware that compliance with their PROM prescription (use of the CPM machine) can help with neuromodulation of pain and prevent fluid buildup in the knee.

◆ Evaluation of Progress

Progression between Stages

The most important criterion used to assess readiness for progression between stages is tissue healing. Most protocols use generic timelines based upon the average person's healing, but therapists, physicians, and patients should understand that each individual person may vary somewhat from these averages. Thus, the decision of when a patient is ready to move to the next stage in a protocol can be guided, but not dictated, by the timelines given. Additionally, a knowledge of which repairs require articular cartilage healing (microfracture, autologous chondrocyte implantation [ACI]) and which ones require primary bone-to-bone healing (osteochondral autograft transfer [OAT]) can be helpful when deciding whether a patient may be a candidate for a more accelerated protocol.

Another helpful criterion for determining readiness to progress is the level of edema. If a patient tries an exercise or series of exercises from a more advanced level of the protocol and then experiences a dramatic increase in edema, this is a good indication that the patient is not yet ready to advance to that phase. Generally speaking, patients are ready for progression to the next stage if they have met all of the goals for the current stage with no increase in pain and swelling, and they are within a time frame where it would be reasonable to assume that the tissue healing is complete enough to permit the activities of the following stage.[4,11,12]

Several recent studies have suggested that it may be beneficial to progress through rehabilitation programs more quickly than has been traditional.[12,16,17] In 2010, Della Villa et al proposed a different type of rehabilitative process for athletes following ACI, based around clinical outcome goals instead of postoperative weeks. Progression to the next stage was permitted when the goals for the previous stage had been achieved without any increase in pain or swelling. They found that most patients achieved their goals quickly, with good outcomes and no report of graft deficits or complications. Though this was a small study in a selected population (31 elite male athletes) and needs to be verified before it can be incorporated into a widely accepted protocol, it does seem to indicate that it may be possible to accelerate rehabilitation to facilitate a faster return to sport, particularly in patients who were very active before injury.[16]

Similarly, other studies found that an accelerated weight-bearing protocol (20% weight bearing for 2 weeks instead of 5, and progression to full weight bearing within the first 8 weeks vs. 11) led to an improvement in walking biomechanics compared with the "traditional rehab" group in patients who underwent an ACI repair.[10,17] These articles did not describe specific rehabilitation programs utilized, but it does seem to indicate that a more accelerated weight-bearing protocol could speed return of normal gait mechanics in some patients.

Assessing Readiness to Return to Sport

Articular cartilage repair has been shown to allow a high rate of return to sport (up to 73% in one systematic review, with as many as 68% returning to a premorbid level of play).[18] One of the challenges clinicians face, then, is how to assess readiness for return to sport. The most important determining factor is healing of the repair, so it is important for the clinician to understand tissue healing times for each type of repair to ensure that an appropriate amount of time has passed. There are currently no studies specifically addressing clinical determination of readiness for return to sport following cartilage repairs; however, there are several studies for this determination following an ACL reconstruction, as well as ways to predict likelihood of injury for the knee (postsurgical or not) which may be extrapolated to this population.

One method of determining readiness for return to high-level/high-impact activities is via a thorough clinical examination. The most common factors used for this determination in the literature are strength testing, lower extremity symmetry, neuromuscular control (i.e., balance and stability when standing on one leg, landing from a jump or hop, and cutting), and ligamentous stability as measured by clinical examination. This method has been used for years by physicians, therapists, and athletic trainers to help determine readiness for return to sport. Unfortunately, these systems rely almost entirely upon subjective tests, which can be difficult to replicate between clinicians.[18–21]

The functional movement screen (FMS) is a method of prescreening developed recently to identify athletes at risk for injury by assessing their performance of several fundamental movements. The authors contend all healthy individuals should be able to perform these movements, which demonstrate good control and symmetry of structures around the joint in question. The FMS provides a measurable and quantifiable/objective way to assess what clinicians have already been using: symmetry, stability, and strength. There are few validation studies available in the literature at this time; the one that exists showed good predictive value (sensitivity 0.91) for injury prediction in professional football players.[22] The FMS is easily learned via taking a course or reading through available literature. Specific screening movements that would be applicable to postoperative knee patients are the deep squat, the hurdle step, and the in-line lunge.[23–25]

Complications

The major complications that may occur after an articular cartilage repair include infection, deep vein thrombosis (DVT), and contracture/scar tissue adhesion. The therapist and physician need to work together to prevent these complications and to catch them early if they do occur. There are very few to no studies assessing the rates of these complications after articular cartilage repair, but studies do exist assessing the rate of complication following ACL repair and following knee arthroscopies in general.

The rate of infection after knee surgery is relatively low. Rates of septic arthritis following ACL surgery are generally estimated to be between 0.14 and 1.7%.[22] A review of the literature shows that most studies report an initial presentation of infection within 1 week of surgery, with a secondary occurrence around 2 months postoperatively for both knee arthroscopy and ACL reconstructions. Patients with an infection typically present with a fever, more edema than would be typical for that stage of recovery, and complaints of knee pain greater than those of patients with the same surgery who do not have an infection. Some patients also have local warmth and a decrease in ROM. Definitive diagnosis can be made via laboratory cultures of aspirated joint fluid. Patients who smoke, use steroids, and have had previous knee surgery are at an increased risk for the development of a deep joint infection.[22,26]

Deep vein thrombosis incidence following articular cartilage repair has not been well studied in the literature. Data suggest that DVT incidence following a major knee surgery such as a total joint replacement may be as high as 84% without prophylaxis, whereas it is somewhat lower after a knee arthroscopy (estimated as 9.9% without prophylaxis, with only 2.1% of those being proximal). Most of the studies found a significant number of silent DVTs in addition to symptomatic ones. While some studies have suggested low-molecular-weight heparin as a prophylaxis, others argue that the overall incidence is too low to give prophylaxis to everyone. Clinicians should be aware of this possible complication, especially in those patients who demonstrate risk factors for DVT (obesity, smoking, female hormone intake, age > 65, venous insufficiency, prior history of venous thromboembolism) and refer them for ultrasonography diagnosis as appropriate.[27–29]

The third major complication after articular cartilage repair is the risk of adhesion, which could limit ROM and, in some scenarios, prevent the patient from returning to the prior level of activity. This risk is higher with allograft, OAT, and ACI due to the size of the incisions. Therapists should mobilize scar tissue and other soft tissue structures as soon as the wounds are adequately healed, and patients should be taught to work on their scar tissue as well. Careful monitoring should occur throughout the rehabilitative process. Modalities such as ultrasound may be used to assist with scar tissue breakup as needed.[12]

◆ Comorbidities That May Alter the Protocol

Most patients with articular cartilage lesions have either a history of another knee injury (commonly a cruciate ligament or meniscal injury) or a concurrent knee comorbidity. No studies currently exist addressing changes to a rehabilitation protocol for an articular cartilage lesion based on a concurrent ligament reconstruction, meniscectomy, meniscal transplant, etc. Generally speaking, clinicians should be aware that the rehabilitative process length will be increased as the number of procedures increases. The general consensus when dealing with multiple reconstructions is that the rehabilitative process should follow the guidelines for the most restrictive protocol.[30]

◆ Conclusion

Rehabilitation and postoperative management following articular cartilage repair are areas in which continued research is needed. However, it is clear that a good rehabilitative program is essential to regaining optimal function, especially in patients who wish to return to sport. It is vital for surgeons and physical therapists to communicate well and work together to progress the rehabilitative process as appropriate, and catch any complications early.

References

1. Lewis PB, McCarty LP III, Kang RW, Cole BJ. Basic science and treatment options for articular cartilage injuries. J Orthop Sports Phys Ther 2006;36(10):717–727
2. Williams JM, Moran M, Thonar EJ, Salter RB. Continuous passive motion stimulates repair of rabbit knee articular cartilage after matrix proteoglycan loss. Clin Orthop Relat Res 1994;304(304):252–262
3. Salter RB. The physiologic basis of continuous passive motion for articular cartilage healing and regeneration. Hand Clin 1994;10(2):211–219
4. Nho SJ, Pensak MJ, Seigerman DA, Cole BJ. Rehabilitation after autologous chondrocyte implantation in athletes. Clin Sports Med 2010;29(2): 267–282, viii
5. Alford JW, Cole BJ. Cartilage restoration, part 1: basic science, historical perspective, patient evaluation, and treatment options. Am J Sports Med 2005;33(2):295–306
6. Alford JW, Cole BJ. Cartilage restoration, part 2: techniques, outcomes, and future directions. Am J Sports Med 2005;33(3):443–460
7. Khan KM, Scott A. Mechanotherapy: how physical therapists' prescription of exercise promotes tissue repair. Br J Sports Med 2009;43(4):247–252
8. Beaupré LA, Davies DM, Jones CA, Cinats JG. Exercise combined with continuous passive motion or slider board therapy compared with exercise only: a randomized controlled trial of patients following total knee arthroplasty. Phys Ther 2001;81(4):1029–1037
9. Bible JE, Simpson AK, Biswas D, Pelker RR, Grauer JN. Actual knee motion during continuous passive motion protocols is less than expected. Clin Orthop Relat Res 2009;467(10):2656–2661
10. Ebert JR, Lloyd DG, Ackland T, Wood DJ. Knee biomechanics during walking gait following matrix-induced autologous chondrocyte implantation. Clin Biomech (Bristol, Avon) 2010;25(10):1011–1017
11. Della Villa S, Kon E, Filardo G, et al. Does intensive rehabilitation permit early return to sport without compromising the clinical outcome after arthroscopic autologous chondrocyte implantation in highly competitive athletes? Am J Sports Med 2010;38(1):68–77
12. Logerstedt DS, Snyder-Mackler L, Ritter RC, Axe MJ; Orthopedic Section of the American Physical Therapy Association. Knee pain and mobility impairments: meniscal and articular cartilage lesions. J Orthop Sports Phys Ther 2010;40(6):A1–A35
13. Spencer JD, Hayes KC, Alexander IJ. Knee joint effusion and quadriceps reflex inhibition in man. Arch Phys Med Rehabil 1984;65(4):171–177
14. Whitelaw GP, DeMuth KA, Demos HA, Schepsis A, Jacques E. The use of the Cryo/Cuff versus ice and elastic wrap in the postoperative care of knee arthroscopy patients. Am J Knee Surg 1995;8(1): 28–30, discussion 30–31
15. Woolf SK, Barfield WR, Merrill KD, McBryde AM Jr. Comparison of a continuous temperature-controlled cryotherapy device to a simple icing regimen following outpatient knee arthroscopy. J Knee Surg 2008;21(1):15–19
16. Della Villa S, Kon E, Filardo G, et al. Does intensive rehabilitation permit early return to sport without compromising the clinical outcome after arthroscopic autologous chondrocyte implantation in highly competitive athletes? Am J Sports Med 2010;38(1):68–77
17. Ebert JR, Robertson WB, Lloyd DG, Zheng MH, Wood DJ, Ackland T. Traditional vs accelerated approaches to post-operative rehabilitation following matrix-induced autologous chondrocyte implantation (MACI): comparison of clinical, biomechanical and radiographic outcomes. Osteoarthritis Cartilage 2008;16(10):1131–1140
18. Barber-Westin SD, Noyes FR. Objective criteria for return to athletics after anterior cruciate ligament reconstruction and subsequent reinjury rates: a systematic review. Phys Sportsmed 2011; 39(3):100–110

19. Mithoefer K, Hambly K, Della Villa S, Silvers H, Mandelbaum BR. Return to sports participation after articular cartilage repair in the knee: scientific evidence. Am J Sports Med 2009;37(Suppl 1): 167S–176S

20. Neeter C, Gustavsson A, Thomeé P, Augustsson J, Thomeé R, Karlsson J. Development of a strength test battery for evaluating leg muscle power after anterior cruciate ligament injury and reconstruction. Knee Surg Sports Traumatol Arthrosc 2006;14(6):571–580

21. Fuller CW, Walker J. Quantifying the functional rehabilitation of injured football players. Br J Sports Med 2006;40(2):151–157, discussion 151–157

22. Wang C, Ao Y, Wang J, Hu Y, Cui G, Yu J. Septic arthritis after arthroscopic anterior cruciate ligament reconstruction: a retrospective analysis of incidence, presentation, treatment, and cause. Arthroscopy 2009;25(3):243–249

23. Cook G, Burton L, Hoogenboom B. Pre-participation screening: the use of fundamental movements as an assessment of function—part 1. N Am J Sports Phys Ther 2006;1(2):62–72

24. Cook G, Burton L, Hoogenboom B. Pre-participation screening: the use of fundamental movements as an assessment of function—part 2. N Am J Sports Phys Ther 2006;1(3):132–139

25. Kiesel K, Plisky PJ, Voight ML. Can serious injury in professional football be predicted by a pre-season functional movement screen? N Am J Sports Phys Ther 2007;2(3):147–158

26. Marmor S, Farman T, Lortat-Jacob A. Joint infection after knee arthroscopy: medicolegal aspects. Orthop Traumatol Surg Res 2009;95(4):278–283

27. Hoppener MR, Ettema HB, Henny CP, Verheyen CC, Büller HR. Low incidence of deep vein thrombosis after knee arthroscopy without thromboprophylaxis: a prospective cohort study of 335 patients. Acta Orthop 2006;77(5):767–771

28. Ilahi O, Reddy J, Ahmad I. Deep venous thrombosis after knee athroscopy: a meta-analysis. Arthroscopy 2005;21(6):727–730

29. Fisher WD. Impact of venous thromboembolism on clinical management and therapy after hip and knee arthroplasty. Can J Surg 2011;54(5):344–351

30. Yenchak AJ, Wilk KE, Arrigo CA, Simpson CD, Andrews JR. Criteria-based management of an acute multistructure knee injury in a professional football player: a case report. J Orthop Sports Phys Ther 2011;41(9):675–686

17

Biomechanical Outcomes of Cartilage Repair of the Knee

Carmen E. Quatman, Joshua D. Harris, and Timothy E. Hewett

Articular cartilage defects of the knee have been documented in over 60% of all patients undergoing knee arthroscopy.[1] In the athletic population, the prevalence of these lesions may be even greater than in the general population.[2] Patients with focal cartilage defects may have major problems with pain and loss of function, equivalent to those scheduled to undergo total knee arthroplasty.[3] Cartilage repair or restoration is a reliable option for symptomatic patients who have failed nonoperative treatment. Clinical outcomes measures are important to aid in the development and optimization of articular cartilage treatment strategies. As techniques in cartilage surgery continue to improve, it is vital that high-level evidence be employed to evaluate patient-reported surgical and functional outcomes. Despite the high levels of interest in cartilage repair and restoration techniques, there are few studies that have evaluated the effects of cartilage surgery on biomechanical and neuromuscular function of the joint and the lower extremity. The focus of the following review is to evaluate the current evidence available on biomechanical functional outcomes after cartilage repair with clinical applications to the rehabilitation process.

As one of the primary load-bearing surfaces of the body, the knee joint surface endures high loads not only during high-impact sports activities but also during activities of daily living. The knee joint may be subjected to loads two to eight times body weight during simple squatting tasks and gait.[4–7] The articular surface of the knee is able to withstand these high loads due to its highly durable, biologically active tissue, with nearly frictionless and elastic material properties that withstand large compressive loads without permanent deformation. The biphasic nature of articular cartilage, composed primarily of water and extracellular matrix, provides unique properties necessary to sustain a healthy knee joint, despite the heavy load demands sustained by the tibiofemoral and patellofemoral surfaces. However, the hypocellular, relatively avascular environment of articular cartilage hinders the healing potential of damaged articular cartilage. In the setting of a neural structure, patients may have large cartilage defects with minimal symptoms and subconscious gait alterations that may help unload damaged areas. Given the poor healing potential of articular cartilage lesions and the subsequent disability, there is a high interest in cartilage repair and tissue-engineering techniques.

The knee joint is subjected to a variety of external loads (ground reaction forces and perturbations) and internal loads (muscle forces, joint reaction forces as a result of the anatomy) during functional activities such as gait, squatting, and activities of daily living.

Muscles that span the knee joint as well as muscles that do not cross the joint can lead to high knee joint loads. Small alterations in walking kinematics, in particular knee flexion angles, may significantly affect the magnitude of knee joint forces.[8–10] Alterations in chondral surfaces and gross cartilage defects alter the load-bearing capabilities of the surrounding tissue.[11,12] The rims of chondral defects bear significantly higher loads than healthy cartilage areas and may lead to defect progression.[12–17]

Knee joint loading profiles in both the tibiofemoral and patellofemoral compartments are influenced by the thickness of the articular cartilage, the joint's radii of curvature, the shape of the menisci, mechanical axis, patellofemoral alignment, and ligamentous stability. In the setting of cartilage repair, achievement of joint surface congruity and defect filling likely decreases the rim stresses of articular cartilage at focal defect areas (**Fig. 17.1**). Theoretically a decrease in rim stresses may prevent or minimize defect progression and may even lead to full integration of cartilage to surrounding normal articular cartilage and underlying subchondral bone. Cartilage repair techniques such as abrasion arthroplasty, drilling, and microfracture penetrate the subchondral bone to induce fibrocartilage formation and demonstrate positive functional and objective outcomes in the short term. However, longer follow-up studies indicate that functional and objective outcomes after these marrow stimulation techniques decline with time, particularly for larger defects.[18–26] In contrast, cartilage restoration techniques (osteochondral autograft, osteochondral allograft, mosaicplasty, and autologous chondrocyte implantation [ACI]) that incorporate more hyaline-like cartilage in focal defects may offer more long-term successful outcomes.[27]

◆ Knee Joint Biomechanics

Two separate layers of calcified and uncalcified cartilage contribute to the biomechanical properties of articular cartilage in the knee. The uncalcified region (includes superficial, transitional, and deep zones) possesses a unique structural composition and provides both tensile strength and resilience to compression and deformation. The calcified layer is deep to the tidemark, overlies the highly vascularized subchondral bone region, and provides an attachment to the underlying subchondral bone.

Tibiofemoral Joint

Knee joint motions between the femoral condyles and the tibial plateau can occur in all three planes (sagittal, frontal, and transverse) with six degrees of freedom that allow 12 directional motions (**Fig. 17.2**).[28,29] The tibiofemoral joint can rotate in the sagittal plane by flexion and extension, in the frontal plane by abduction and adduction, and in the transverse plane by internal and external rotation. Translations of the tibiofemoral joint occur in the sagittal plane anteriorly

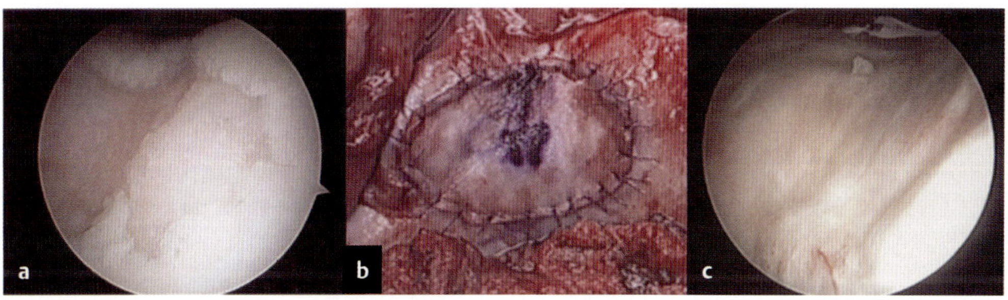

Fig. 17.1 Images of a trochlea articular cartilage defect demonstrating reconstitution/filling of defect and integration of implant after autologous chondrocyte implantation (ACI). (**a**) Trochlea before ACI treatment demonstrating large cartilage defect, (**b**) trochlea during open ACI procedure, and (**c**) the same trochlea at second-look arthroscopy 8 months after ACI.

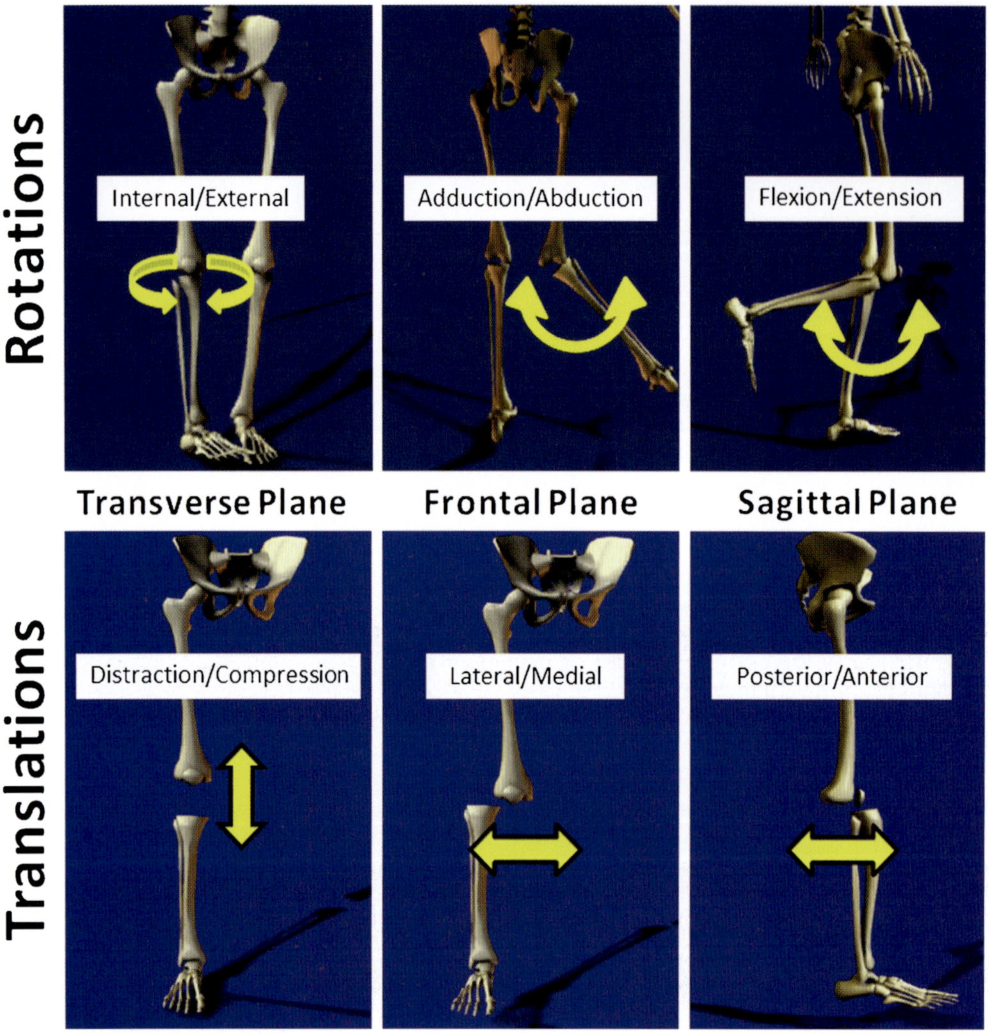

Fig. 17.2 Rotations and translations of the tibiofemoral joint. (Reproduced from Quatman CE, Quatman CC, Hewett TE. A "Plane" Explanation of Anterior Cruciate Ligament Injuries. Sports Med 40(9);729–746, with permission from Adis, a Wolters Kluwer business.

and posteriorly, in the frontal plane medially and laterally, and in the transverse plane by compression and distraction (**Fig. 17.2**). While the knee can move in all 12 of these potential directions, with the exception of sagittal plane rotations (flexion/extension), most of these motions occur through a relatively limited range. In addition, motions in the transverse and frontal planes (as well as sagittal plane translations) are influenced by the amount of knee flexion and the anatomy of the femur, tibia, and surrounding ligamentous structures.

Movements between the articulating surfaces of the tibia and femur play key roles in knee joint function. The menisci also contact both the tibial and the femoral surfaces. The menisci decrease the stress transmission to the articular cartilage by increasing the surface contact area.[30] The contact points of the tibia and femur in the sagittal plane move posteriorly during knee flexion via a combination of both rolling and gliding motions. The femoral condyles glide anteriorly as the femur rolls posteriorly during flexion, reducing the posterior progression of the rolling effect and preventing the subluxation of the femur off the back of the tibia.[30,31] There is a significant asymmetry in the anatomy and kinematics of the medial and lateral tibiofemoral compartments.[32,33] The medial compartment, with the medial tibia possessing a firmly attached medial meniscus on two angled flat surfaces of the medial tibia and articulating with the convex medial femoral condyle, experiences minimal gliding in the anterior-posterior directions and minimal rotations in the transverse plane in unconstrained, nonweightbearing conditions.[34] In contrast, the lateral compartment is composed of a convex-on-convex articulation, due to the shape of the lateral tibial plateau.[35,36] Increased anterior-posterior glide, as well as transverse rotations during unconstrained, non–weightbearing conditions occur in the lateral compartment due to the less adherent lateral meniscus and less bony constraint imposed by the lateral plateau relative to the medial plateau.[37] Thus, the center of rotation of the knee joint during non–weightbearing conditions is likely in the medial compartment.[32,37,38] In contrast to non–weightbearing conditions, knee kinematics are different during gait. Koo and Andriacchi demonstrated that, during gait, the center of rotation of the knee joint is primarily on the lateral side with the medial side demonstrating greater anterior-posterior translation.[39] The convex–concave surface contact area of the medial compartment is much greater than the convex–convex surface contact area of the lateral side during weight bearing. The greater potential surface area that may change with an alteration in gait (i.e., that may occur with pathology such as osteoarthritis) may make the medial compartment more sensitive to small changes in knee joint kinematics.[35]

Patellofemoral Joint

The bony anatomy of the patella and the trochlea of the femur provide inherent stability to the patellofemoral joint, with the medial and lateral facets of the posterior surface of the patella articulating and tracking within the trochlear groove. The superior 75% of the posterior surface of the patella is made up of articular cartilage, and, with thicknesses up to 5 to 6 mm, it has the thickest articular cartilage surface in the human body.[34] Since contact pressures of up to 12 MPa may occur during maximal extensor mechanism contractions, the thick articular surface and smooth congruent surface contact between the trochlea and patella are necessary for optimal performance of the extensor mechanism.[40–42]

The patella undergoes both a rolling and a gliding motion along the articulating surface of the femur as it increases the mechanical advantage of the extensor muscles. The quadriceps muscles (rectus femoris, vastus lateralis, medialis, and intermedius), which are important for knee extension movements, originate from the pelvis (ilium) and proximal femur. The quadriceps muscles wrap around the patella and unite to form the patellar tendon that attaches the patella to the tibial tubercle. The patella serves as a fulcrum for quadriceps muscle moment generation since it displaces the quadriceps tendon anteriorly and increases the internal moment arm of the knee extensor mechanism (the larger the internal moment arm, the greater the internal moment produced per level of force generation).[43] The knee flexion angle influences the patellar–trochlear surface area contact and pressure. The patella engages with the trochlea between 20 and 30 degrees of knee flexion. Near full extension of the knee, the inferior surface of the patella articulates with the superior portion of the trochlea in a single-contact, horizontal fashion. As knee flexion increases, more proximal portions of the patella contact the trochlea and the horizontal contact transitions into separate areas of contact on the lateral and medial facets at ~ 130 degrees of knee flexion.[42] Patellofemoral surface contact increases ~ 24% during flexion with weight-bearing activities compared with non–weightbearing conditions.[44] The highest contact pressures in the patellofemoral joint occur between 60 and 90 degrees of knee flexion.[40]

◆ Biomechanics of Cartilage Rehabilitation

The rehabilitation program after cartilage repair is critical to optimize surgical outcomes. Physiologic healing and chondrocyte protection are paramount to the success of the surgical procedure. However, this healing and protection must be balanced with optimization of the rehabilitation program. The goals of the rehabilitation process should be to minimize disability, immobilization, and overall morbidity accounting for the complications that may occur with small arthroscopic procedures and/or large open procedures combined with osteotomies.

Due to the high variability in lesion location, nature of the defect, and surgical procedures performed, personalized rehabilitation strategies are necessary. Awareness of the location of the defect, size of the defect, concomitant procedures, type of healing response anticipated for the surgical procedure performed, and goals of the patient for return to both activities of daily living and high-level activities such as sports are important considerations during the rehabilitation stage.[45] Currently, there is minimal evidence in the literature regarding rehabilitation programs following cartilage repair. Most of the available literature about rehabilitation is based on biomechanical theories and expert or nonexpert opinion.

Although protection of the knee joint after a surgical intervention is important, joint immobilization has known detrimental effects on muscle and articular cartilage.[46] Even short-term immobilization can induce articular cartilage atrophy, and, although reversible, full restoration of cartilage integrity may not occur.[47–49] Animal studies demonstrate that cartilage repair techniques supplemented with early mobilization show superior tissue regeneration and integration.[50–52] Successful cartilage repair is predicated on chondrocyte viability, integration of the new repair tissue with the surrounding healthy tissue, and the ability of the repair to withstand high mechanical stresses. In the early postoperative period after cartilage repair, continuous passive motion (CPM) is often used within the first few hours after surgery to counter the immobilization effects, to enhance metabolic and nutritional activity of the cartilage, and to prevent joint stiffness.[52,53] However, CPM recommendations are largely based on basic science and animal model studies with few in vivo clinical studies available to support this accepted clinical practice.[54] A recent systematic review revealed only four human in vivo studies that evaluated the use of CPM after cartilage repair.[53,55–58] No randomized controlled studies were available to support the use of CPM, and all studies varied significantly in the recommendation of time and duration of use (from 6 to 8 hours per day for 6 days up to 8 weeks of use).[53,55–58] Rodrigo et al showed improvement of the defects on second-look arthroscopy in patients who used CPM compared with the control group; however, no correlation to functional outcome was reported.[58] Although available only on a small group of subjects, histological outcomes on second-look arthroscopy do not show significant differences between patients who use CPM and those who perform only active range of motion.[57,58] In general, rehabilitation programs encourage early, interval increases in joint motion starting with passive range-of-motion activities. Active joint range of motion may have higher joint-contact pressures than passive range of motion; thus, most programs begin with passive range of motion with slow progression into active range-of-motion and strengthening activities.[59]

Weight-bearing guidelines after cartilage repair vary widely in the literature.[59] In general, most recommendations use separate guidelines for femoral/tibial defects compared with patellofemoral defects. The mechanics of weight bearing and walking may increase the risk of articular cartilage breakdown when the normal mechanical environment and balance between loading and biological maintenance of cartilage are compromised.[60,61] During the early postoperative weeks following cartilage repair, the cells contained in the defect start differentiating and develop into a soft repair tissue that is highly vulnerable to, but also may be stimulated by, mechanical loads.[62] It may take up to 2 years for the repaired lesion to develop into the optimal cartilage phenotype.[63,64] Small mechanical loads may promote cartilage metabolic activity and nutrition, similar to CPM effects. Thus, the challenge in the early postoperative period is to provide small bouts of loading stimulus to the healing articular lesion, without subjecting the cartilage repair site to significant, potentially damaging mechanical overload.[65] The restriction of weight bearing in the early postoperative phase, particularly in the tibiofemoral joint, may theoretically protect the cartilage repair and decrease the mechanical stress that the repair site is subjected to during the most vulnerable time period.

Often patients with cartilage repair in the tibiofemoral compartment are restricted to non–weightbearing in the first few weeks after restoration, with slow progression over 2 to 12 weeks to full weight-bearing conditions.[45] Weight-bearing movements result in increased pressure on the tibiofemoral articulation throughout the entire arc of knee motion, and contact forces can reach four times body weight at 90 degrees of knee flexion.[66,67] In addition to restriction of the amount of weight exerted on the limb after cartilage repair, clinicians should consider other external loads during gait that may subject the knee to mechanical stresses. Ebert et al found that in ACI patients the contribution of ground reaction force at different percentages of weight-bearing status demonstrates a large variability in kinematic, spatiotemporal parameters and external knee moments.[68] Thus, in addition to the ground reaction force, other variables may affect the external loads experienced by the knee joint during gait. Also, from a patient perspective, despite strict instruction and practice of partial weight-bearing restrictions on bathroom-type weight scales it is difficult to accurately replicate "partial weight-bearing" restrictions.[69] To prevent mechanical shear and high-contact pressures during the most vulnerable healing time after cartilage repair, weight-bearing restrictions have often been strict, with slow progressions. However, there is limited good evidence regarding how weight-bearing restrictions affect functional outcomes.

An assessment of ACI biopsies taken on 11 patients (12 graft sites on femoral condyles) at second-look arthroscopy at 3 to 7 years after the initial procedure demonstrated stiffness measures of 90% or more of the value of controls in 8 of the 12 biopsies and good to excellent grading of integration of graft to the surrounding cartilage in 10 of the 11 subjects. All patients followed a weight-bearing protocol that started with gradual weight-bearing for 8 weeks with progression to full weight-bearing at 10 weeks.[70] A randomized controlled trial that evaluated clinical outcomes in "accelerated" versus "delayed" weight bearing for matrix-associated ACI of the femoral condyle demonstrated no differences in clinical outcome 2 years postoperatively. The accelerated group was allowed 20% weight bearing for the first 2 weeks, 50% weight bearing between 2 and 4 weeks, and progression to full weight bearing after 6 weeks. The delayed group was allowed toe-touch weight bearing only until 4 weeks postoperatively, followed by 20% weight bearing at 4 to 6 weeks, 50% weight bearing at 6 to 8 weeks, with gradual increase to full weight bearing at postoperative week 10.[71] Ebert et al compared an accelerated rehabilitation that allowed full weight bearing at 8 weeks compared with delayed full weight bearing until 11 weeks postoperatively after matrix-induced autologous chondrocyte implantation (MACI) in the femoral condyle resulted in the accelerated group demonstrating improved function, reduced knee pain, and no significant changes on MRI at 3 months postoperatively compared with the delayed group.[72] Thus, early pilot studies demonstrate that it may be possible to accelerate weight-bearing loads in femoral articular cartilage repair without compromising the repair.

In contrast to tibiofemoral cartilage repair rehabilitation, most patellofemoral cartilage repair protocols allow for immediate weight bearing with the knee in full extension or with restricted range of motion to 0 to 30 degrees of knee flexion. This is theoretically a "safe" range of motion for patellofemoral patients because the patella does not engage with the trochlea until ~ 20 degrees of knee flexion and the patellofemoral joint reaction force is 50% of the quadriceps force in full knee extension.[40,73] In contrast to tibiofemoral cartilage repair, patients with patellofemoral pain or patellofemoral cartilage restoration likely benefit from weight-bearing restrictions and exercises that minimize the patellofemoral joint reaction forces but maximize quadriceps forces. Van Eijden et al demonstrated that between 70 and 100 degrees of knee flexion the patellofemoral joint reaction force is equal to the quadriceps force.[73] In contrast, from full knee extension to 40 degrees of knee flexion the patellofemoral joint reaction force is 50 to 90%, respectively.[73] During squatting activities, the highest patellofemoral joint forces and stresses occur at ~ 90 degrees of knee flexion. Increased external loads (increased weight) significantly increase the patellofemoral joint stresses.[74] Targeted rehabilitation exercises that avoid larger knee flexion angles may allow for quadriceps strengthening exercises that reduce patellofemoral joint stresses. However, timing of implementation of quadriceps activation and maximal muscle force recruitment may need to be modified and taken into consideration for cartilage procedures that are combined with tibial tubercle osteotomies.

◆ Biomechanical Outcomes

Currently, there is a paucity of literature that evaluates functional outcomes after cartilage restoration procedures. The majority of outcome studies utilize patient-reported outcomes via surveys to "assess" function, return to activities of daily living and sport, and overall patient satisfaction. In 2009, Mithoefer et al identified 20 studies that described return to sports of 1,363 patients after cartilage repair. Almost 75% of athletes were able to return to sports within 7 to 18 months of cartilage repair. However, of the patients who returned to sports, only 65% returned to the preinjury level of participation. The ability to return to preinjury level of sport after 3 to 5 years was significantly better after autologous chondrocyte procedures than for microfracture or osteochondral transfer patients.[19] While return to sport indirectly evaluates overall functional outcome after cartilage procedures, the specific functional outcomes such as strength and neuromuscular performance after cartilage repair are not well known.

A rigorous systematic review of the literature on March 1, 2012, identified only two peer-reviewed studies that evaluated the functional outcome of gait after cartilage restoration,[72,75] four peer-reviewed articles that evaluated strength measures,[76–79] and two studies that evaluated overall function with a collective functional index score (hop tests and strength measures).[78,79] All gait studies were performed in patients who had undergone MACI. Functional measures of gait have not been directly measured after cartilage repair for any other cartilage repair techniques. Ebert et al conducted a randomized controlled trial that invested gait differences between MACI patients (medial or lateral femoral condyle lesions) who underwent an accelerated weight-bearing protocol (full weight bearing by 8 weeks) versus a traditional weight bearing protocol (full weight bearing by 11 weeks) after surgery with gait analyses at 3, 6, and 12 months postsurgery.[72,75] Both groups had similar ground reaction forces and spatiotemporal gait parameters during gait postoperatively. However, compared with healthy controls, both the accelerated rehabilitation group and the traditional rehabilitation group had significantly reduced knee extension moments, which persisted 12 months postsurgery. The traditional group also demonstrated significantly reduced peak knee adduction and flexion moment compared with healthy controls that was not found in the accelerated rehabilitation group.[72,75]

Muscle atrophy and weakness are common clinical manifestations after knee surgery.[77,80–82] Quadriceps muscles weakness alters the mechanical loading of the articular cartilage and is necessary for resisting external ground reaction forces and external perturbations.[83] Poor knee joint control and abnormal joint loading may play a role in the development of osteoarthritis.[60] Studies by Ebert et al, Løken et al, Kreuz et al, and Van Assche et al indicate that significant strength deficits may occur after cartilage repair and may persist even 4 to 5 years postoperatively.[76–78,84] Ebert et al found significantly reduced knee extensor torques in the operative limb compared with the contralateral limb 5 years following MACI at all isokinetic speeds tested.[76] Kreuz et al demonstrated that isokinetic strength in patients 4 years after cartilage repair with scaffold-assisted ACI was significantly lower in the operative limb compared with the healthy control limb, with extension peak torque deficits greater than flexor peak torque deficits.[84] Løken et al demonstrated that, compared with the contralateral limb, patients had a 10 to 20% extensor strength deficit in the leg that had cartilage repair from 2 to 7.4 years after the surgery.[77] Van Assche et al evaluated knee extension deficits in ACI patients compared with microfracture patients presurgery and at 6 months and 2-year follow-up after surgery.[78] Both the microfracture and ACI groups experienced a decrease in strength after surgery at the 6-month time point. At 2 years after cartilage repair, 26% of the patients had significant extension strength deficits. Adequate muscle control after cartilage repair is critical to function during activities of daily living. Despite supervised rehabilitation many patients exhibit significant strength deficits that persist for years after cartilage repair.

Lower-extremity asymmetry in strength and functional abilities has been linked to increased risk for lower-extremity injury. Persistence of lower-extremity functional asymmetries may pose a significant risk to injury in both the rehabilitated and the healthy contralateral limb.[85–88] With a large number of cartilage repair patients returning to moderate- to high-level activities, evaluation of neuromuscular strength and control symmetries is important to minimize the risk for injury upon return to sport. However, there is minimal evidence in the literature about the effects of cartilage repair on lower-extremity functional asymmetries. There is evidence of significant strength asymmetries in limbs after cartilage repair, particularly for quadriceps function, and these may persist up to 5 years after cartilage repair.[76–78,84] Van Assche et al found that at 2 years after ACI or microfracture, 30% of the patients had asymmetries greater than 85% compared with their contralateral limb for strength and functional hop tests.[78,79] At 2 years postoperatively, microfracture patients and ACI patients had similar functional outcomes.[79] Persistent abnormal neuromuscular and biomechanical risk factors, such as lower-limb motion and loading asymmetries, may increase risk for future injury and indicate that cartilage repair patients may require more targeted interventions before return to moderate- to high-level sports activities.

Concomitant Procedures: Osteotomy Effects on Biomechanics

Osteotomies have been utilized independently as well as in combination with cartilage repair techniques to improve symptoms of cartilage pathology and patellofemoral symptoms. In the tibiofemoral joint, both opening- and closing-wedge, valgus-producing, high (proximal) tibial osteotomies have gained wide acceptance as a technique to off-load damaged cartilage areas in an attempt to provide symptom relief and prolong time to total knee arthroplasty.[89] For patients with unicompartmental knee pathology, this technique redistributes the loads within the tibiofemoral joint. In the setting of cartilage repair, a concomitant osteotomy can be used to off-load the repair site and perhaps help with the healing process.[89,90] In general, most surgeons aim for over-correction of the varus malaligned medial compartment osteoarthritic patient versus neutral correction in the cartilage repair or restoration patient. In the patellofemoral joint, it is common to find patients with significant patellar maltracking, instability, and/or malalignment, which are often addressed at the time of cartilage repair. Tibial tubercle anteriorization (**Fig. 17.3**) or anteromedialization can significantly reduce patellofemoral contact stresses.[91–93] Maquet demonstrated that elevation of the tibial tubercle by 2 cm reduces patellofemoral joint reaction forces by up to 50% at 45 degrees of knee flexion.[94] There is increased evidence that patellofemoral cartilage repair, in particular ACI techniques, have improved outcomes for patients who have concomitant tibial tubercle osteotomies with either anteromedialization or anteriorization rather than ACI alone.[95–97] Bauer et al

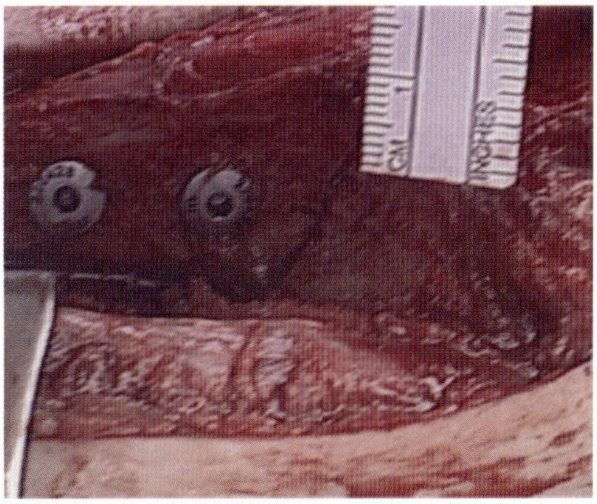

Fig. 17.3 Tibial tubercle osteotomy with anteriorization to reduce patellofemoral contact stresses.

demonstrated significant improvements in patient-reported outcomes after MACI combined with high tibial osteotomy; however, MRI results demonstrated poor filling of cartilage defects postoperatively.[90] Gigante et al also demonstrated significant improvement in patient-reported outcomes after combined tibial tubercle osteotomy and ACI in the patellofemoral joint 36 months after surgery.[98] Unfortunately, no current studies are available that report measured neuromuscular performance outcomes in patients with concomitant cartilage repair and osteotomy procedures. Overall, patients with significant anatomical malalignment may benefit from concomitant procedures that address both the alignment and the cartilage pathology, but there is a significant need for research in this area.[90,99]

Limitations in Current Literature Regarding Biomechanics after Cartilage Repair

Cartilage repair and restoration techniques are increasingly recognized as surgical interventions that may significantly alter patient symptoms and perhaps even modify progression of articular cartilage pathology. However, as the field of cartilage repair has grown, there has been a gap in the literature for evidence on functional outcomes. Even more challenging is the heterogeneity of patients, location of lesions, size of lesions, concomitant injuries, and types of procedures that often become grouped into mean, pooled data to meet statistical power.[100] This poses a significant challenge to clinicians for evaluation of outcomes after cartilage repair. In addition, very few studies exist evaluating functional, biomechanical outcomes of cartilage repair, and the current literature base is limited to short- to moderate-term follow-up studies. Long-term studies, in terms of functional outcome, are currently unavailable.

◆ Conclusion

Clinical outcomes measures are critical to the development and optimization of evidence-based treatment strategies. With the advancements in cartilage restoration, it is imperative that patient-reported surgical and functional outcomes be evaluated with high-level, thorough, systematic research methods. Despite the high level of interest in cartilage restoration techniques and outcomes, there are few studies that evaluate the effects of cartilage restoration on biomechanical and neuromuscular function. The limited studies available indicate that significant strength deficits and gait alterations may occur after cartilage restoration procedures and persist up to 5 years or more postoperatively. Future work should focus on determining functional outcomes such as strength, neuromuscular function, and return to activities of daily living and sport. In addition, development of clinical assessment tools that help evaluate and improve functional deficits during the rehabilitation phases of cartilage restoration may aid in the assessment of outcomes as well as improve rehabilitation strategies.

Acknowledgment

The authors acknowledge support from the National Institutes of Health/National Institute of Arthritis Musculoskeletal and Skin Diseases, grants RO1-AR049735 and RO1-AR05625.

References

1. Curl WW, Krome J, Gordon ES, Rushing J, Smith BP, Poehling GG. Cartilage injuries: a review of 31,516 knee arthroscopies. Arthroscopy 1997;13(4): 456–460
2. Flanigan DC, Harris JD, Trinh TQ, Siston RA, Brophy RH. Prevalence of chondral defects in athletes' knees: a systematic review. Med Sci Sports Exerc 2010;42(10):1795–1801
3. Heir S, Nerhus TK, Røtterud JH, et al. Focal cartilage defects in the knee impair quality of life as much as severe osteoarthritis: a comparison of knee injury and osteoarthritis outcome score in 4 patient categories scheduled for knee surgery. Am J Sports Med 2010;38(2):231–237
4. Fregly BJ, Besier TF, Lloyd DG, et al. Grand challenge competition to predict in vivo knee loads. J Orthop Res 2012;30(4):503–513
5. D'Lima DD, Patil S, Steklov N, Chien S, Colwell CW Jr. In vivo knee moments and shear after total knee arthroplasty. J Biomech 2007;40(Suppl 1): S11–S17
6. Heinlein B, Kutzner I, Graichen F, et al. ESB Clinical Biomechanics Award 2008: Complete data of total

knee replacement loading for level walking and stair climbing measured in vivo with a follow-up of 6-10 months. Clin Biomech (Bristol, Avon) 2009;24(4):315–326

7. Reilly DT, Martens M. Experimental analysis of the quadriceps muscle force and patello-femoral joint reaction force for various activities. Acta Orthop Scand 1972;43(2):126–137
8. Herzog W, Longino D, Clark A. The role of muscles in joint adaptation and degeneration. Langenbecks Arch Surg 2003;388(5):305–315
9. Schipplein OD, Andriacchi TP. Interaction between active and passive knee stabilizers during level walking. J Orthop Res 1991;9(1):113–119
10. Shelburne KB, Torry MR, Pandy MG. Contributions of muscles, ligaments, and the ground-reaction force to tibiofemoral joint loading during normal gait. J Orthop Res 2006;24(10):1983–1990
11. Sasaki K, Neptune RR. Individual muscle contributions to the axial knee joint contact force during normal walking. J Biomech 2010;43(14): 2780–2784
12. Guettler JH, Demetropoulos CK, Yang KH, Jurist KA. Osteochondral defects in the human knee: influence of defect size on cartilage rim stress and load redistribution to surrounding cartilage. Am J Sports Med 2004;32(6):1451–1458
13. Lefkoe TP, Trafton PG, Ehrlich MG, et al. An experimental model of femoral condylar defect leading to osteoarthrosis. J Orthop Trauma 1993;7(5): 458–467
14. Flanigan DC, Harris JD, Brockmeier PM, Siston RA. The effects of lesion size and location on subchondral bone contact in experimental knee articular cartilage defects in a bovine model. Arthroscopy 2010;26(12):1655–1661
15. Minas T, Nehrer S. Current concepts in the treatment of articular cartilage defects. Orthopedics 1997;20(6):525–538
16. Henderson IJ, La Valette DP. Subchondral bone overgrowth in the presence of full-thickness cartilage defects in the knee. Knee 2005;12(6):435–440
17. Radin EL, Rose RM. Role of subchondral bone in the initiation and progression of cartilage damage. Clin Orthop Relat Res 1986; (213):34–40
18. Gudas R, Stankevicius E, Monastyreckiene E, Pranys D, Kalesinskas RJ. Osteochondral autologous transplantation versus microfracture for the treatment of articular cartilage defects in the knee joint in athletes. Knee Surg Sports Traumatol Arthrosc 2006;14(9):834–842
19. Mithoefer K, Williams RJ III, Warren RF, Wickiewicz TL, Marx RG. High-impact athletics after knee articular cartilage repair: a prospective evaluation of the microfracture technique. Am J Sports Med 2006;34(9):1413–1418
20. Gobbi A, Nunag P, Malinowski K. Treatment of full thickness chondral lesions of the knee with microfracture in a group of athletes. Knee Surg Sports Traumatol Arthrosc 2005;13(3):213–221
21. Kreuz PC, Steinwachs MR, Erggelet C, et al. Results after microfracture of full-thickness chondral defects in different compartments in the knee. Osteoarthritis Cartilage 2006;14(11):1119–1125
22. Blevins FT, Steadman JR, Rodrigo JJ, Silliman J. Treatment of articular cartilage defects in athletes: an analysis of functional outcome and lesion appearance. Orthopedics 1998;21(7):761–767, discussion 767–768
23. Nehrer S, Spector M, Minas T. Histologic analysis of tissue after failed cartilage repair procedures. Clin Orthop Relat Res 1999; (365):149–162
24. Oneto JM, Ellermann J, LaPrade RF. Longitudinal evaluation of cartilage repair tissue after microfracture using T2-mapping: a case report with arthroscopic and MRI correlation. Knee Surg Sports Traumatol Arthrosc 2010;18(11):1545–1550
25. Magnussen RA, Dunn WR, Carey JL, Spindler KP. Treatment of focal articular cartilage defects in the knee: a systematic review. Clin Orthop Relat Res 2008;466(4):952–962
26. Saris DB, Vanlauwe J, Victor J, et al; TIG/ACT/01/2000&EXT Study Group. Treatment of symptomatic cartilage defects of the knee: characterized chondrocyte implantation results in better clinical outcome at 36 months in a randomized trial compared to microfracture. Am J Sports Med 2009;37(Suppl 1):10S–19S
27. Harris JD, Siston RA, Pan X, Flanigan DC. Autologous chondrocyte implantation: a systematic review. J Bone Joint Surg Am 2010;92(12):2220–2233
28. Quatman CE, Quatman-Yates CC, Hewett TEA. A 'plane' explanation of anterior cruciate ligament injury mechanisms: a systematic review. Sports Med 2010;40(9):729–746
29. Woo SL, Abramowitch SD, Kilger R, Liang R. Biomechanics of knee ligaments: injury, healing, and repair. J Biomech 2006;39(1):1–20
30. Buckwalter JA, Einhorn TA, O'Keefe RJ; American Academy of Orthopaedic Surgeons. Orthopaedic basic science: foundations of clinical practice. 3rd ed. Rosemont, IL: American Academy of Orthopaedic Surgeons; 2007:xiv, 465 pp.
31. Pope MH, Crowninshield R, Miller R, Johnson R. The static and dynamic behavior of the human knee in vivo. J Biomech 1976;9(7):449–452
32. Iwaki H, Pinskerova V, Freeman MA. Tibiofemoral movement 1: the shapes and relative movements of the femur and tibia in the unloaded cadaver knee. J Bone Joint Surg Br 2000;82(8):1189–1195
33. Pinskerova V, Iwaki H, Freeman MA. The shapes and relative movements of the femur and tibia at the knee. Orthopade 2000;29(Suppl 1):S3–S5
34. Scott WN. Surgery of the Knee. 4th ed. Philadelphia: Churchill Livingstone/Elsevier; 2006
35. Koo S, Rylander JH, Andriacchi TP. Knee joint kinematics during walking influences the spatial cartilage thickness distribution in the knee. J Biomech 2011;44(7):1405–1409
36. Koo S, Andriacchi TP. A comparison of the influence of global functional loads vs. local contact anatomy on articular cartilage thickness at the knee. J Biomech 2007;40(13):2961–2966
37. Blaha JD, Mancinelli CA, Simons WH, Kish VL, Thyagarajan G. Kinematics of the human knee using an open chain cadaver model. Clin Orthop Relat Res 2003;410(410):25–34
38. Freeman MA, Pinskerova V. The movement of the normal tibio-femoral joint. J Biomech 2005; 38(2):197–208
39. Koo S, Andriacchi TP. The knee joint center of rotation is predominantly on the lateral side during normal walking. J Biomech 2008;41(6):1269–1273
40. Huberti HH, Hayes WC. Patellofemoral contact pressures. The influence of q-angle and tendofemoral contact. J Bone Joint Surg Am 1984;66(5): 715–724

41. Miller RK, Murray DW, Gill HS, O'Connor JJ, Goodfellow JW. In vitro patellofemoral joint force determined by a non-invasive technique. Clin Biomech (Bristol, Avon) 1997;12(1):1–7
42. Goodfellow J, Hungerford DS, Zindel M. Patellofemoral joint mechanics and pathology. 1. Functional anatomy of the patello-femoral joint. J Bone Joint Surg Br 1976;58(3):287–290
43. Neumann DA. Kinesiology of the musculoskeletal system: foundations for rehabilitation. 2nd ed. St. Louis, MO: Mosby/Elsevier; 2010:xx, 725 p.
44. Besier TF, Draper CE, Gold GE, Beaupré GS, Delp SL. Patellofemoral joint contact area increases with knee flexion and weight-bearing. J Orthop Res 2005;23(2):345–350
45. Mithoefer K, Hambly K, Logerstedt DS, Ricci M, Silvers H, Della Villa S. Current concepts for rehabilitation and return to sport after knee articular cartilage repair in the athlete. J Orthop Sports Phys Ther 2012;42(3):254–273
46. Brandt KD. Response of joint structures to inactivity and to reloading after immobilization. Arthritis Rheum 2003;49(2):267–271
47. Haapala J, Arokoski J, Pirttimäki J, et al. Incomplete restoration of immobilization induced softening of young beagle knee articular cartilage after 50-week remobilization. Int J Sports Med 2000;21(1):76–81
48. Haapala J, Arokoski JP, Hyttinen MM, et al. Remobilization does not fully restore immobilization induced articular cartilage atrophy. Clin Orthop Relat Res 1999;(362):218–229
49. Hagiwara Y, Ando A, Chimoto E, Saijo Y, Ohmori-Matsuda K, Itoi E. Changes of articular cartilage after immobilization in a rat knee contracture model. J Orthop Res 2009;27(2):236–242
50. O'Driscoll SW, Keeley FW, Salter RB. The chondrogenic potential of free autogenous periosteal grafts for biological resurfacing of major full-thickness defects in joint surfaces under the influence of continuous passive motion. An experimental investigation in the rabbit. J Bone Joint Surg Am 1986; 68(7):1017–1035
51. O'Driscoll SW, Salter RB. The repair of major osteochondral defects in joint surfaces by neochondrogenesis with autogenous osteoperiosteal grafts stimulated by continuous passive motion. An experimental investigation in the rabbit. Clin Orthop Relat Res 1986;(208):131–140
52. Salter RB, Simmonds DF, Malcolm BW, Rumble EJ, MacMichael D, Clements ND. The biological effect of continuous passive motion on the healing of full-thickness defects in articular cartilage. An experimental investigation in the rabbit. J Bone Joint Surg Am 1980;62(8):1232–1251
53. Fazalare JA, Griesser MJ, Siston RA, Flanigan DC. The use of continuous passive motion following knee cartilage defect surgery: a systematic review. Orthopedics 2010;33(12):878
54. Howard JS, Mattacola CG, Romine SE, Lattermann C. Continuous Passive Motion, Early Weight Bearing, and Active Motion following Knee Articular Cartilage Repair Evidence for Clinical Practice. Cartilage 2010;1(4):276–286
55. Alfredson H, Lorentzon R. Superior results with continuous passive motion compared to active motion after periosteal transplantation. A retrospective study of human patella cartilage defect treatment. Knee Surg Sports Traumatol Arthrosc 1999;7(4):232–238
56. Marder RA, Hopkins G Jr, Timmerman LA. Arthroscopic microfracture of chondral defects of the knee: a comparison of two postoperative treatments. Arthroscopy 2005;21(2):152–158
57. Schultz W, Göbel D. Articular cartilage regeneration of the knee joint after proximal tibial valgus osteotomy: a prospective study of different intra- and extra-articular operative techniques. Knee Surg Sports Traumatol Arthrosc 1999;7(1):29–36
58. Rodrigo JJ, Steadman JR, Silliman JF, Fulstone HA. Improvement of full-thickness chondral defect healing in the human knee after debridement and microfracture using continuous passive motion. Am J Knee Surg 1994;7(3):109–116
59. Hambly K, Bobic V, Wondrasch B, Van Assche D, Marlovits S. Autologous chondrocyte implantation postoperative care and rehabilitation: science and practice. Am J Sports Med 2006;34(6): 1020–1038
60. Andriacchi TP, Koo S, Scanlan SF. Gait mechanics influence healthy cartilage morphology and osteoarthritis of the knee. J Bone Joint Surg Am 2009;91(Suppl):(1):95–101
61. Chaudhari AM, Briant PL, Bevill SL, Koo S, Andriacchi TP. Knee kinematics, cartilage morphology, and osteoarthritis after ACL injury. Med Sci Sports Exerc 2008;40(2):215–222
62. Shapiro F, Koide S, Glimcher MJ. Cell origin and differentiation in the repair of full-thickness defects of articular cartilage. J Bone Joint Surg Am 1993;75(4):532–553
63. Gikas PD, Morris T, Carrington R, Skinner J, Bentley G, Briggs T. A correlation between the timing of biopsy after autologous chondrocyte implantation and the histological appearance. J Bone Joint Surg Br 2009;91(9):1172–1177
64. Gikas PD, Bayliss L, Bentley G, Briggs TW. An overview of autologous chondrocyte implantation. J Bone Joint Surg Br 2009;91(8):997–1006
65. Arokoski JP, Jurvelin JS, Väätäinen U, Helminen HJ. Normal and pathological adaptations of articular cartilage to joint loading. Scand J Med Sci Sports 2000;10(4):186–198
66. Adouni M, Shirazi-Adl A. Knee joint biomechanics in closed-kinetic-chain exercises. Comput Methods Biomech Biomed Engin 2009;12(6):661–670
67. Escamilla RF. Knee biomechanics of the dynamic squat exercise. Med Sci Sports Exerc 2001;33(1): 127–141
68. Ebert JR, Lloyd DG, Smith A, Ackland T, Wood DJ. The association between external-ground-reaction force and knee-joint kinetics during partial- and full-weight-bearing gait. Clin Biomech (Bristol, Avon) 2010;25(4):359–364
69. Ebert JR, Ackland TR, Lloyd DG, Wood DJ. Accuracy of partial weight bearing after autologous chondrocyte implantation. Arch Phys Med Rehabil 2008;89(8):1528–1534
70. Peterson L, Brittberg M, Kiviranta I, Akerlund EL, Lindahl A. Autologous chondrocyte transplantation. Biomechanics and long-term durability. Am J Sports Med 2002;30(1):2–12
71. Wondrasch B, Zak L, Welsch GH, Marlovits S. Effect of accelerated weightbearing after matrix-associated autologous chondrocyte implantation on the femoral condyle on radiographic and clinical outcome after 2 years: a prospective, randomized controlled pilot study. Am J Sports Med 2009;37(Suppl 1):88S–96S

72. Ebert JR, Robertson WB, Lloyd DG, Zheng MH, Wood DJ, Ackland T. Traditional vs accelerated approaches to post-operative rehabilitation following matrix-induced autologous chondrocyte implantation (MACI): comparison of clinical, biomechanical and radiographic outcomes. Osteoarthritis Cartilage 2008;16(10):1131–1140
73. van Eijden TM, Kouwenhoven E, Verburg J, Weijs WA. A mathematical model of the patellofemoral joint. J Biomech 1986;19(3):219–229
74. Wallace DA, Salem GJ, Salinas R, Powers CM. Patellofemoral joint kinetics while squatting with and without an external load. J Orthop Sports Phys Ther 2002;32(4):141–148
75. Ebert JR, Lloyd DG, Ackland T, Wood DJ. Knee biomechanics during walking gait following matrix-induced autologous chondrocyte implantation. Clin Biomech (Bristol, Avon) 2010;25(10): 1011–1017
76. Ebert JR, Lloyd DG, Wood DJ, Ackland TR. Isokinetic knee extensor strength deficit following matrix-induced autologous chondrocyte implantation. Clin Biomech (Bristol, Avon) 2012 Feb 14 [Epub ahead of print]
77. Løken S, Ludvigsen TC, Høysveen T, Holm I, Engebretsen L, Reinholt FP. Autologous chondrocyte implantation to repair knee cartilage injury: ultrastructural evaluation at 2 years and long-term follow-up including muscle strength measurements. Knee Surg Sports Traumatol Arthrosc 2009;17(11):1278–1288
78. Van Assche D, Staes F, Van Caspel D, et al. Autologous chondrocyte implantation versus microfracture for knee cartilage injury: a prospective randomized trial, with 2-year follow-up. Knee Surg Sports Traumatol Arthrosc 2010;18(4):486–495
79. Van Assche D, Van Caspel D, Vanlauwe J, et al. Physical activity levels after characterized chondrocyte implantation versus microfracture in the knee and the relationship to objective functional outcome with 2-year follow-up. Am J Sports Med 2009;37(Suppl 1):42S–49S
80. Ericsson YB, Roos EM, Dahlberg L. Muscle strength, functional performance, and self-reported outcomes four years after arthroscopic partial meniscectomy in middle-aged patients. Arthritis Rheum 2006;55(6):946–952
81. Gapeyeva H, Pääsuke M, Ereline J, Pintsaar A, Eller A. Isokinetic torque deficit of the knee extensor muscles after arthroscopic partial meniscectomy. Knee Surg Sports Traumatol Arthrosc 2000;8(5): 301–304
82. Lautamies R, Harilainen A, Kettunen J, Sandelin J, Kujala UM. Isokinetic quadriceps and hamstring muscle strength and knee function 5 years after anterior cruciate ligament reconstruction: comparison between bone-patellar tendon-bone and hamstring tendon autografts. Knee Surg Sports Traumatol Arthrosc 2008;16(11):1009–1016
83. Jefferson RJ, Collins JJ, Whittle MW, Radin EL, O'Connor JJ. The role of the quadriceps in controlling impulsive forces around heel strike. Proc Inst Mech Eng H 1990;204(1):21–28
84. Kreuz PC, Müller S, Freymann U, et al. Repair of focal cartilage defects with scaffold-assisted autologous chondrocyte grafts: clinical and biomechanical results 48 months after transplantation. Am J Sports Med 2011;39(8):1697–1705
85. Hewett TE, Myer GD, Ford KR, et al. Biomechanical measures of neuromuscular control and valgus loading of the knee predict anterior cruciate ligament injury risk in female athletes: a prospective study. Am J Sports Med 2005;33(4):492–501
86. Myer GD, Paterno MV, Ford KR, Quatman CE, Hewett TE. Rehabilitation after anterior cruciate ligament reconstruction: criteria-based progression through the return-to-sport phase. J Orthop Sports Phys Ther 2006;36(6):385–402
87. Klein KK. Asymmetries in the pelvis and legs and their implications in knee injury. Am Correct Ther J 1970;24(3):93–95
88. Noyes FR, Barber SD, Mangine RE. Abnormal lower limb symmetry determined by function hop tests after anterior cruciate ligament rupture. Am J Sports Med 1991;19(5):513–518
89. Tang WC, Henderson IJ. High tibial osteotomy: long term survival analysis and patients' perspective. Knee 2005;12(6):410–413
90. Bauer S, Khan RJ, Ebert JR, et al. Knee joint preservation with combined neutralising High Tibial Osteotomy (HTO) and Matrix-induced Autologous Chondrocyte Implantation (MACI) in younger patients with medial knee osteoarthritis: A case series with prospective clinical and MRI follow-up over 5 years. Knee 2012;19(4):431–439
91. Rue JP, Colton A, Zare SM, et al. Trochlear contact pressures after straight anteriorization of the tibial tuberosity. Am J Sports Med 2008;36(10): 1953–1959
92. Beck PR, Thomas AL, Farr J, Lewis PB, Cole BJ. Trochlear contact pressures after anteromedialization of the tibial tubercle. Am J Sports Med 2005;33(11):1710–1715
93. Shirazi-Adl A, Mesfar W. Effect of tibial tubercle elevation on biomechanics of the entire knee joint under muscle loads. Clin Biomech (Bristol, Avon) 2007;22(3):344–351
94. Maquet P. The biomechanics of the knee and surgical possibilities of healing osteoarthritic knee joints. Clin Orthop Relat Res 1980;(146):102–110
95. Pascual-Garrido C, Slabaugh MA, L'Heureux DR, Friel NA, Cole BJ. Recommendations and treatment outcomes for patellofemoral articular cartilage defects with autologous chondrocyte implantation: prospective evaluation at average 4-year follow-up. Am J Sports Med 2009; 37(Suppl 1):33S–41S
96. Farr J. Autologous chondrocyte implantation improves patellofemoral cartilage treatment outcomes. Clin Orthop Relat Res 2007;463:187–194
97. Henderson IJ, Lavigne P. Periosteal autologous chondrocyte implantation for patellar chondral defect in patients with normal and abnormal patellar tracking. Knee 2006;13(4):274–279
98. Gigante A, Enea D, Greco F, et al. Distal realignment and patellar autologous chondrocyte implantation: mid-term results in a selected population. Knee Surg Sports Traumatol Arthrosc 2009; 17(1):2–10
99. Sterett WI, Steadman JR, Huang MJ, Matheny LM, Briggs KK. Chondral resurfacing and high tibial osteotomy in the varus knee: survivorship analysis. Am J Sports Med 2010;38(7):1420–1424
100. Jakobsen RB, Engebretsen L, Slauterbeck JR. An analysis of the quality of cartilage repair studies. J Bone Joint Surg Am 2005;87(10):2232–2239

18

Clinical Outcomes Assessment for Articular Cartilage Restoration

Kai Mithoefer and Marco Acuna

Articular cartilage injuries of the knee have received increasing attention because of their lack of spontaneous repair, high clinical incidence, and frequent associated symptoms. Arøen et al[1] demonstrated 66% focal articular cartilage defects in 993 consecutive knee arthroscopies, with 11% of these injuries requiring repair. Currently, several established options exist for articular cartilage repair in the knee, including microfracture, osteochondral allograft and autograft transfer, and autologous chondrocyte implantation (ACI). Novel developments in tissue engineering and biomaterials research have been promoting technical modifications and enhancements of the existing techniques, with resultant second- and third-generation technologies. With the continued evolution of cartilage repair technology, clinical instruments for outcome evaluation and comparison are becoming increasingly relevant for scientists, clinicians, and patients. Recent reviews have noticed the limited quality of reported evidence and outcome measures in the orthopedic literature. A meta-analysis of 2,468 randomized trials published in the *Journal of Bone and Joint Surgery* from 1988 to 2000 identified only 72 (2.9%) that met all the authors' criteria for randomized control trials (RCTs).[2] Similarly, other authors[3] found that, of 61 publications on cartilage repair in the knee, only 4 were RCTs and had limited Coleman Methodology Scores (average score: 43.5; range: 0 to 100). These authors also pointed out that 27 different outcome measures had been used in these publications, many of them not validated for cartilage repair.[4] Outcome instruments that have been used in the past include the Lysholm Score,[5,6] International Knee Documentation Committee (IKDC) scores,[7] Hospital for Special Surgery knee scale,[8,9] Knee Society knee scale,[10] Tegner Activity Scale,[5,11,12] Cincinnati knee scale,[12–14] Western Ontario and McMaster Universities Osteoarthritis Index (WOMAC),[10,11,15] and Knee injury Osteoarthritis Outcome Score (KOOS).[16] In addition, instruments to measure health-related quality of life (QOL) have been increasingly used in evidence-based medicine.[17] Besides knee function scores, information about the quality and quantity of the repaired cartilage provides important outcome evaluation. Recent animal and human studies have concluded that currently there is no suitable substitute for histological analyses for assessment of tissue quality.[18–20] However, histological assessment may not always be feasible, and less invasive assessment of repair cartilage by magnetic resonance imaging (MRI) has evolved as a useful and practical alternative for evaluation of articular cartilage repair in the knee. Therefore, the purpose of this chapter is to outline the available options for outcome assessment in knee articular cartilage repair and to present appropriate instruments and knee function measures that are validated for this specific type of procedure or provide clinically relevant functional outcome information.

◆ Functional Outcome Evaluation

Functional outcome measures in the knee such as the Knee Society Score, Hospital for Special Surgery Score, IKDC knee examination form, and Lysholm score have originally been developed and completed by clinicians and did not reflect the patient's perspective. More recently, patient-reported outcome (PRO) measures have turned the focus toward the patient's outcome perspective and have been helpful in improving standardization and decreasing assessment bias by the surgeon. Surgeon-based scores often provide only a single aggregate score. Newer PROs have been designed in separate subscales to help with interpretation of the number of meaningful and cartilage repair–oriented outcomes of clinical studies over time. Subscales include specific and critical domains such as pain, joint function, and activities of daily living (ADL). Accordingly, regulatory institutions such as the Food and Drug Administration have designated pain and physical function as the primary clinically meaningful end points that provide the best evidence of efficacy.[21] When possible, blinding should be used to reduce assessment bias and improve outcome objectivity. Patients in the control group may expect less benefit from the treatment than patients in the experimental group, and surgeons may be less likely to identify treatment responses in the control group. The resultant differences in assessment and interpretation of treatment responses are likely to bias the outcome results. However, effective blinding in cartilage repair studies can be challenging because the evaluated techniques or product may have differences in invasiveness (arthroscopy vs. mini-open vs. arthrotomy), treatment process (two-stage interventions vs. a single surgery), or postsurgical follow-up requirements such as variations in rehabilitation. If it is impossible to blind patients and surgeons based on the compared technologies—for example, when comparing ACI with microfracture—blinded outcome assessment should be performed by independent evaluators—that is, with blinded histological or MRI analysis—to reduce assessment bias and optimize objectivity of the study results.

General Health–Related Quality-of-Life Outcomes

Outcome evaluation traditionally uses both general health–related QOL measures that reflect health-related QOL in different diseases and populations such as the Short-Form 36 items and 12 items of the Medical Outcome Study (SF-36 and SF-12) and EuroQoL-5 Dimensions (EQ-5D). These general scores are different from specific outcome measures that include data related to specific diseases or organs, such as the WOMAC, IKDC, Lysholm, Cincinnati knee scale, or KOOS scores.[22–24] The SF-12, SF-36, and EQ-5D are PRO scales that have been used for evaluation of cartilage defect repair. It is recommended that at least one knee- or cartilage-specific outcome instrument and one general health-related QOL measure be included in cartilage repair evaluations. Outcome measures should be reliable, responsive, and validated for cartilage repair in the knee. ***Reliability*** refers to the reproducibility of the measure, either between subjects (test–retest reliability) or between observers (interobserver reliability). A reliability coefficient of 0.8 to 0.9 is considered adequate for a patient's group study.[25] *Validity* questions whether an outcome instrument actually measures what it is intended to measure. Components of validity include content validity (floor and ceiling effects), criterion validity (how an instrument compares with an accepted gold standard instrument), and construct validity. Responsiveness assesses changes in the instrument value over time or treatment. Specific assessment scales that have been validated for the study of patients with chondral repair techniques include the Lysholm score,[4] IKDC score,[23] and KOOS.[24]

Disease-/Joint-Specific Outcome Measures

IKDC Subjective Knee Form

This PRO instrument assesses daily activity, symptoms, and sports function, and it has been used in different kinds of disorders of the knee (meniscal injuries, patellofemoral syndrome, osteoarthritis, and, recently, chondral lesions). The original instrument was developed in 1987 and the Subjective Knee Form was added in 2000.[26] It consists of 18 items that are summed and expressed as a percentage from 0 to 100, with 100 representing an absence of symptoms and higher levels of functioning. It is part of the International Cartilage Repair Society (ICRS) cartilage injury evaluation package, and it is also in the ICRS Socrates software program. The user manual and the Excel file scoring are available from the American Orthopedic Society for Sports Medicine (www.sportsmed.org/research/IKDC_Forms). The IKDC has been shown to have an internal consistency of 0.92 and a test–retest correlation of 0.94.[27] Women have been found to exhibit lower mean scores than men. It is recommended in studies of patients younger than 18 years and those age 35 years or older to adjust the Subjective Knee Form scores for age differences in both men and women.[28] Hambly and Griva[23] recently determined that the IKDC contains the most relevant items and is more effective than KOOS for monitoring the short- and medium-term outcomes given that KOOS focuses on longer-term consequences.

Knee Injury and Osteoarthritis Outcome Score

The KOOS is a knee-specific instrument created to evaluate the symptoms and function in subjects with a variety of knee injuries that could possibly result in osteoarthritis. It is based on an extension of the WOMAC. It comprises 42 items containing five separately scored subscales: pain (9), other symptoms (7), ADL (17), function in sport and recreation (Sport/Rec) (5), and knee-related QOL (4). Each subscale is scored from 0 to 100 on a worst-to-best scale. The original 24 items of the WOMAC were included in the KOOS, but the Sport/Rec and QOL subscales are unique to the KOOS. Psychometric properties were tested (reliability, construct validity, and responsiveness) for use in clinical studies of patients requiring several orthopedic interventions, including cartilage defect surgery.[24] Content validity testing has demonstrated excellent internal consistency and reliability (reliability coefficient: 0.74 to 0.97) and test–retest reliability (reliability coefficient: 0.78 to 0.82) for the KOOS subscales.[29] The KOOS is also included in the ICRS Socrates outcomes software package and can be downloaded from http://koos.nu.

Lysholm Scoring Scale

This scale was originally designed in 1982 to assess ligament injuries of the knee and modified for patient self-completion in cartilage damage evaluation in 2004.[4] It is a condition-specific outcome measure that contains eight domains evaluated in points: limp (5), locking (15), pain (25), stair climbing (10), support (5), instability (25), swelling (10), and squatting (5), for a total score of 0 to 100, from worst to best. The overall scale has demonstrated acceptable test–retest reliability (average coefficient 0.91) and internal consistency (coefficient 0.71).[4] Smith et al[30] evaluated the validity of the Lysholm score for articular cartilage injury with the Rasch model (a measurement model that sets strict standards for the quality of measurement derived from the scale). They concluded that the Lysholm score demonstrated acceptable psychometric parameters to justify its use in outcomes assessment for chondral disorders.

Activity-Related Outcome Evaluation

The evaluation of postintervention activities presents a useful and practical outcome measure after articular cartilage repair in the knee in athletically active patients. Activity rating scales that have been used to provide quantitative analysis after cartilage repair procedures are the Tegner Activity Scale and Marx Activity Rating Scale. Activity rating instruments are used to compare the activity levels before injury, before surgery, and after surgery. Besides these activity scales, return to sports activity presents an outcome parameter that is often considered particularly relevant for the subjective measurement of success in athletes undergoing cartilage repair procedures.

Tegner Activity Scale

The Tegner Activity Scale was constructed by grading sports activities according to their difficulty and was originally developed for patients with anterior cruciate ligament (ACL) surgery but has been widely used for other knee problems, including articular cartilage injury. The Tegner score uses a numerical scale ranging from 0 to 10.[31] Rating can be performed by the patient or clinician, with each numerical value indicating the ability to participate in a specific sports activity. The score differentiates between recreational and competitive sports participation and includes up to 170 different athletic activities that can be rated. An activity level of 10 corresponds to participation in competitive sports, including soccer, football, and rugby at a professional level. An activity level of 6 points equals participation in recreational sports and has been shown to be the median activity level in a normal population. A score of 0 is assigned to a person who is disabled from sports participation. The Tegner score has been evaluated for validity and reliability only after ACL injury,[32] but its responsiveness for articular cartilage repair was successfully demonstrated with an effect size of 0.67.[24] Tegner Activity Scale scores have been found to decrease with age.

Marx Activity Rating Scale

The Marx Activity Rating Scale uses four questions to measure the frequency with which patients pivot, run, decelerate, and cut. It was developed to provide a standardized measure of athletic activity levels and is independent of the particular sport played by the athlete. It uses scores between 0 and 16 (a score of 0 refers to patients who do these activities less than once a month, and a score of 16 indicates those who do each activity four or more times a week). It has not been formally validated for cartilage injury in the knee, but its responsiveness after cartilage repair procedures has been established.[33] Similar to the Tegner score, it is correlated inversely with the age of the athlete.

Return to Sports Activity

For most athletes, the ability to return to their sport is the most important measure of a successful treatment outcome. Providing an athlete with accurate information about the ability to engage in a particular sport allows for appropriate expectations of the patient and presents a critical component of the pretreatment counseling. The goals and expectations are often variable between individual athletes and sports, but especially between recreational and competitive or even professional athletes.[34] The higher prevalence (35%) of articular cartilage lesions among recreational and professional athletes, compared with the general population (5 to 11%) emphasizes the relevance of this aspect of outcome evaluation.[35] Average rates for return to sports activity in the athletic population have been reported after autologous chondrocyte transplantation (ACT) (74%),[36–38] microfracture (68%),[39] osteochondral autologous transfer[40] (91%), and osteochondral allograft transplantation[40] (88%). A recent systematic review of cartilage repair techniques demonstrated that 65% of athletes returned to the preinjury level after cartilage repair without significant difference between individual techniques.[41,42] Several next-generation techniques have been developed, including matrix-associated chondrocyte implantation (MACI)[43] and scaffold-enhanced microfracture,[44] with similar rates for return to sports compared with the first-generation techniques.[36,42,45] Besides the ability to return to sports, the ability to continue to play presents another important outcome parameter. Although durability of athletic activity was observed in 87% of athletes treated with ACT after 52 months, continued sports activity was observed in only 53% after treatment with microfracture.[36,42,45] Cartilage injury in athletes is often associated with other injuries such as ACL rupture,[46] and several factors have been shown to affect the ability return to sports after cartilage repair. Delay in surgical treatment of more than 12 months,[45,47] lesion size of more than 2 cm,[2,40,45] and patient age younger than 25 years are associated with better return-to-sports activity after cartilage repair.[36]

◆ Structural Outcome Evaluation

In addition to measurement of joint pain and function evaluation of the structural quantity and quality of the repair cartilage tissue filling, the defect presents an important outcome parameter. Several qualitative and quantitative parameters are clinically relevant, including macroscopic and microscopic repair cartilage evaluation as well as radiographic assessment. Although they are not all validated for articular cartilage repair, they can be useful as secondary outcome assessment measures. Both the macroscopic and histological structure and appearance of the repair tissue have been correlated with functional outcome scores after cartilage repair.[40,48,49] To date, there is still an incomplete understanding of the role of structural outcome parameters on long-term functional outcome after cartilage injury and repair in the knee. However, structural outcome parameters continue to assume an increasingly important role as evaluation tools in cartilage research.

Macroscopic Evaluation

Macroscopic assessment of the repaired cartilage tissue can be performed arthroscopically with a graded quantitative and qualitative analysis of the defect repair. Macroscopic assessment of the repair cartilage tissue filling the defect has been validated for articular cartilage repair using the ICRS and Oswestry macroscopic cartilage evaluation scores[50] (**Table 18.1**). The macroscopic evaluation includes criteria such as defect fill grade, peripheral integration, and surface appearance. However, this macroscopic system does not provide mechanical graft information such as graft firmness or information regarding graft overgrowth. Although macroscopic grading provides valuable morphological information, it requires scheduled and standardized second-look arthroscopy. Because of the ethical considerations regarding mandatory second-look arthroscopy and its associated surgical and anesthesia risks, it may not always be feasible. Voluntary second-look arthroscopy or second-look arthroscopy only in symptomatic patients may introduce selection bias but still may provide useful information for outcome evaluation after cartilage repair.

Table 18.1 International Cartilage Repair Society macroscopic cartilage assessment score

Criteria and appearance	Points
1. A	
Level with surrounding cartilage	4
75% repair of defect	3
50% repair of defect	2
25% repair of defect	1
0% repair of defect	0
1. B	
100% survival of initially grafted surface	4
75% survival of initially grafted surface	3
50% survival of initially grafted surface	2
25% survival of initially grafted surface	1
0% survival of initially grafted surface	0
2. Integration	
Complete integration with surrounding cartilage	4
Demarcation border < 1 mm	3
75% integrated, 25% with notable border > 1 mm	2
50% integrated, 50% with notable border > 1 mm	1
0–25% integrated	0
3. Appearance	
Intact, smooth surface	4
Fibrillated surface	3
Small, scattered fissures and cracks	2
Small and large fissures	1
Complete degeneration of graft area	0
4. Overall assessment and score	
Grade 1: normal	12
Grade 2: nearly normal	8–11
Grade 3: abnormal	4–7
Grade 4: severely abnormal	0–3

Histological Evaluation

Histological evaluation provides important information about the structural quality of the cartilage repair tissue and should be done using the ICRS Histology Endpoint Committee guidelines. These guidelines include quantitative scoring systems (ICRS I and ICRS II) that include assessment of tissue surface integrity, matrix organization and characterization, cell distribution and viability, calcified cartilage appearance, subchondral bone morphology, and additional tissue characteristics (**Table 18.2**).[51,52] Tissue samples should be obtained at predetermined intervals after surgery to avoid the influence of repair cartilage tissue maturation on the histological parameters. Histology evaluations may include potential sampling bias due to inhomogeneity of the cartilage repair tissue, and biopsies should consistently be obtained in a predefined location within the repaired cartilage defect. Besides standard hematoxylin and eosin and safranin-O staining, special immunohistochemical evaluation such as for collagen-type and noncollagenous matrix protein expression provides important qualitative information about the repair cartilage tissue. Although some prospective randomized studies have found preliminary evidence for better functional outcome with improved tissue quality, further scientific evaluation is still required to establish a valid correlation between structural and clinical outcome parameters.[52–54]

Table 18.2 International Cartilage Repair Society II visual histological assessment scale

Histological parameter	VAS score
1. Tissue morphology (polarized light)	0–100
2. Matrix staining (metachromasia)	0–100
3. Cell morphology	0–100
4. Chondrocyte clustering	0–100
5. Surface architecture	0–100
6. Basal integration	0–100
7. Tidemark formation	0–100
8. Subchondral bone abnormalities/ marrow fibrosis	0–100
9. Inflammation	0–100
10. Abnormal calcification/ ossification	0–100
11. Vascularization in repair tissue	0–100
12. Surface/superficial assessment	0–100
13. Midzone/deep zone assessment	0–100
14. Overall assessment	0–100

Abbreviation: VAS, visual analog scale.

Magnetic Resonance Imaging Evaluation

Structural analysis of articular cartilage repair tissue using cartilage-sensitive MRI sequences offers both quantitative and qualitative outcome information. The most widely used MRI techniques for cartilage imaging are intermediate-weighted fast-spin-echo and three-dimensional (3D) fat-suppressed T1-weighted gradient-echo techniques. One of the significant advantages of MRI for structural repair assessment is its ability to assess noninvasively the cartilage repair area and surrounding tissue in multiple planes. It can also be helpful for noninvasive longitudinal follow-up and comparison with preoperative appearance (**Fig. 18.1**).

Magnetic Resonance Scoring

Marlovits et al[55] have developed an instrument for quantitative MRI outcome evaluation of cartilage repair tissue (MOCART). The MOCART score combines a descriptive part and 11 variables, including defect fill, peripheral cartilage interface, subchondral bone interface, surface of the repair tissue, structure of the repair tissue, signal intensity of the repair tissue, subchondral lamina, intralesional osteophyte formation, bone marrow edema, subchondral bone, and effusion, and it sums up the quantitative MRI observations of these individual variables into a single overall score (**Table 18.3**). One of the most clinically relevant variables is the defect fill after cartilage repair. The MOCART variable has been shown to correlate with clinical symptoms following cartilage repair. Clinical correlation testing of the MOCART score

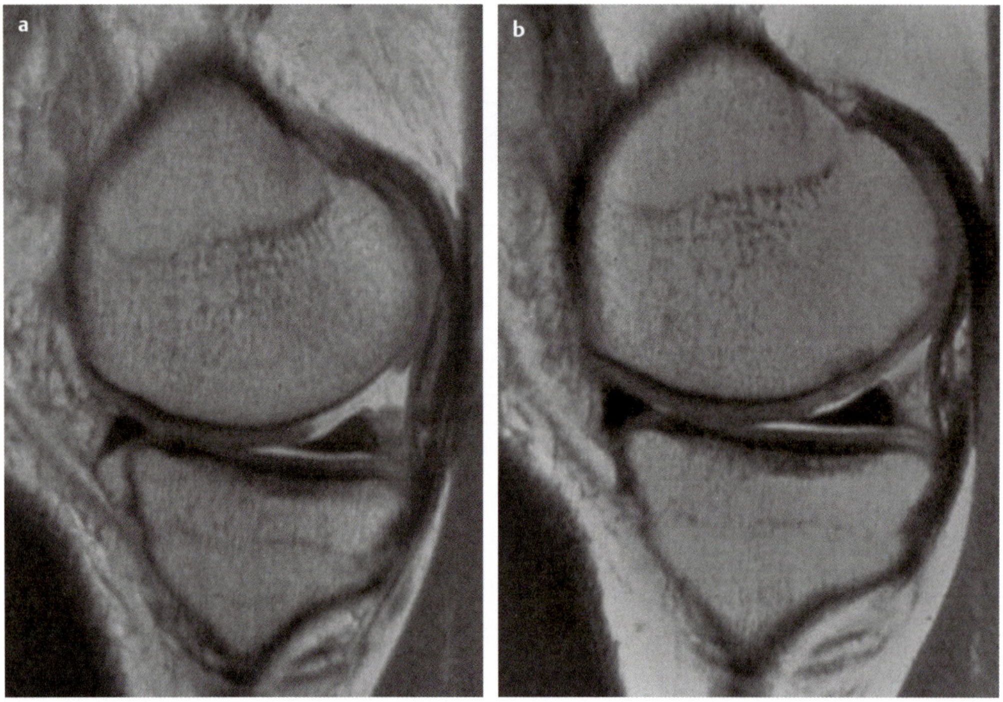

Fig. 18.1a, b Cartilage-specific sagittal magnetic resonance image (MRI) of a full-thickness articular cartilage defect of the central weight-bearing surface of the medial femoral condyle before (**a**) and 24 months after (**b**) autologous chondrocyte transplantation demonstrating almost complete fill, full peripheral and basal integration, nearly normal repair tissue signal intensity, and slightly irregular but intact lamina and subchondral bone.

has further demonstrated that its individual variables are correlated with clinically relevant outcome parameters, such as the visual analog scale score and the KOOS, for the subgroups of pain, symptoms, ADL, sports, and knee-related QOL. Specifically, defect fill, structure of the repair tissue, and repair tissue signal intensity showed a statistically significant correlation with these clinical scores.[56] Similarly, another study demonstrated correlation of defect fill with the ADL score, the IKDC score, and the SF-36 physical component score. Importantly, all knees with good fill demonstrated improved knee function, whereas poor fill grade was associated with limited improvement and decreasing functional scores after 24 months.[57] Besides the standard repair cartilage fill measurements, a fully automatic 3D MRI technique has been recently developed that provides precise quantification of knee cartilage volume and presents an additional valuable tool for clinical outcome studies.[58]

Functional Magnetic Resonance Imaging Evaluation

MRI can provide information on cartilage morphology, its intrinsic composition, and its structure, including its 3D collagen network, proteoglycan content, and interstitial water content. MRI technologies such as delayed gadolinium-enhanced MRI of cartilage (dGEMRIC), T1rho, T2 mapping can provide additional quantitative and qualitative information about the cartilage repair tissue. Quantitative and qualitative magnetic resonance techniques also allow for differentiated zonal assessment throughout the thickness of the repair tissue. Although no MRI parameter is specific to any tissue component, quantitative parameters may show a direct relation to repair tissue biomechanical properties. The functional MRI techniques can be helpful in comparing repair tissue generated after different surgical techniques and can also allow for longitudinal comparison

Table 18.3 Three-dimensional magnetic resonance observation of cartilage repair tissue (3D-MOCART) scoring

Variables

1. Defect fill (degree of defect repair and filling of the defect in relation to the adjacent cartilage)
   - 0%
   - 0–25%
   - 25–50%
   - 75–100%
   - 100–125%
   - 125–150%
   - 150–200%
   - > 200%

   Localization
   - Whole area of cartilage repair
     - > 50%
     - < 50%
   - Central
   - Peripheral
   - Weight bearing
   - Non–weight bearing
2. Interface (integration with adjacent cartilage to border zone in two planes)

   Sagittal (femur, patella, trochlea, tibia)
   - Complete
   - Demarcating border visible (split-like)
   - Defect visible < 50%
   - Defect visible > 50%

   Coronal (femur, tibia); axial (patella, trochlea)
   - Complete
   - Demarcating border visible (split-like)
   - Defect visible < 50%
   - Defect visible > 50%

   Localization
   - Whole area of cartilage repair
     - > 50%
     - < 50%
   - Central
   - Peripheral
   - Weight bearing
   - Non–weight bearing
3. Bone interface (integration of the transplant to the subchondral bone; integration of a possible periosteal flap)
   - Complete
   - Partial delamination
   - Complete delamination
   - Delamination of periosteal flap

   Localization
   - Weight bearing
   - Non–weight bearing
4. Surface (constitution of the surface of the repair tissue)
   - Surface intact
   - Surface damaged < 50% of depth
   - Surface damaged > 50% of depth
   - Adhesions

   Localization
   - Whole area of cartilage repair
     - < 50%
     - > 50%
   - Central
   - Peripheral
   - Weight bearing
   - Non–weight bearing
5. Structure (constitution of the repair tissue)
   - Homogeneous
   - Inhomogeneous or cleft formation

   Localization
   - Whole area of cartilage repair
     - < 50%
     - > 50%
   - Central
   - Peripheral
   - Weight bearing
   - Non–weight bearing
6. Signal intensity (intensity of magnetic resonance signal in the repair tissue in comparison with the adjacent cartilage)
   - Normal (identical to adjacent cartilage)
   - Nearly normal (slight areas of signal alteration)
   - Abnormal (large areas of signal alteration)

   Localization
   - Central
   - Peripheral
   - Weight bearing
   - Non–weight bearing
7. Subchondral lamina (constitution of the subchondral lamina)
   - Intact
   - Not intact
8. Chondral osteophytes (osteophytes within the cartilage repair area)
   - Absent
   - Osteophytes < 50% of the thickness of the cartilage transplant
   - Osteophytes < 50% of the thickness of the cartilage transplant

   Localization
   Size: _____mm (plane: _____) × _____mm (plane: _____)
   - Central
   - Peripheral
   - Weight bearing
   - Non–weight bearing
   - Relation to other alterations within this score of variable no._____

Table 18.3 Continued

Variables	
10. Subchondral bone (constitution of the subchondral bone) ◆ Intact ◆ Granulation tissue ◆ Cyst ◆ Sclerosis Localization ◆ Whole area of cartilage repair • > 50% • < 50% ◆ Central ◆ Peripheral ◆ Weight bearing ◆ Non–weight bearing	11. Effusion (approximate size of joint effusion visualized in all planes) ◆ Absent ◆ Small ◆ Medium ◆ Large

of tissue quality and quantity during the cartilage repair and maturation process within individual repair techniques. Previous reports outlined the utility of dGEMRIC for noninvasive MRI monitoring of the glycosaminoglycan content in patients after different surgical cartilage repair techniques, such as microfracture, ACI, and MACI.[59] Clinical studies to date have been inconsistent in supporting a correlation between dGEMRIC and functional knee scores such as the KOOS.[60] Experimental studies comparing quantitative T1rho and T2 mapping with histology, including safranin-O/fast-green staining and type II collagen immunohistochemistry demonstrated that T2 and T1rho mapping provide indirect assessment of the biochemistry of tissue repair. T2 mapping has been shown to correlate with collagen orientation, which can highlight differences between immature, unorganized, and more hyaline-like repair tissue. However, quantitative T2 has demonstrated poor correlation with collagen content. Clinical correlation studies confirmed that T2 correlates with functional outcome measures such as the Lysholm score and IKDC knee evaluation form.[59] The limitation of MRI is its inability to directly quantify functional properties of repair cartilage tissue. Although functional cartilage MRI provides much promise as an outcome parameter after cartilage repair, future biomechanical correlation and clinical validation studies are needed to determine which morphological or biochemical MRI variables have the highest predictive value and can serve as the best surrogate instruments for clinical outcome.

◆ Conclusion

Articular cartilage injuries in the knee and other joints are observed with increasing frequency, and systematic evaluation of the current and developing treatment options is critically important to establish an effective evidence-based therapeutic algorithm for these often debilitating injuries. To establish efficacy and comparability between established and novel therapeutic techniques, systematic outcome evaluation is critical. Outcome evaluation can be achieved in several ways using patient-reported general health outcome surveys, knee-specific functional outcome scores, athletic activity scales, and structural outcome evaluation of the repair cartilage tissue by macro- or microscopic evaluation or MRI. An increasing number of functional outcome instruments have been validated specifically for articular cartilage procedures and provide reliable and responsive assessment of knee cartilage repair with good overall effect sizes. Increasingly, structural outcome evaluation is being used for quantitative and qualitative evaluation of the repair cartilage tissue with a trend toward noninvasive, standardized, and high-quality longitudinal MRI techniques. However, these techniques require further validation in the clinical setting before they can reliably be substituted for macroscopic or histological repair tissue assessment. The combination of outcome measures and study designs should be individually adjusted to each study hypothesis and the tested cartilage repair technique. Optimally, randomized controlled or prospective cohort studies should be performed using at least one validated outcome measure for general health, pain and knee function, activity level, and a structural outcome parameter. To summarize, much progress has been made in the development and standardization of outcome evaluation of articular cartilage repair in the knee. This methodological progress facilitates a more rigorous future scientific evaluation of the efficacy of novel technologies in the rapidly developing clinical field of articular cartilage repair.

References

1. Arøen A, Løken S, Heir S, et al. Articular cartilage lesions in 993 consecutive knee arthroscopies. Am J Sports Med 2004;32(1):211–215
2. Obremskey WT, Pappas N, Attallah-Wasif E, Tometta P 3rd, Bhandari M. Level of evidence in orthopaedic journals. J Bone Joint Surg Am 2005; 87(12):2632–2638
3. Jakobsen RB, Engebretsen L, Slauterbeck JR. An analysis of the quality of cartilage repair studies. J Bone Joint Surg Am 2005;87(10):2232–2239
4. Kocher MS, Steadman JR, Briggs KK, Sterett WI, Hawkins RJ. Reliability, validity, and responsiveness of the Lysholm knee scale for various chondral disorders of the knee. J Bone Joint Surg Am 2004;86-A(6):1139–1145
5. Horas U, Schnettler R, Pelinkovic D, Herr G, Aigner T. Osteochondral transplantation versus autogenous chondrocyte transplantation: A prospective comparative clinical study [in German]. Chirurg 2000;71(9):1090–1097
6. Steadman JR, Rodkey WG, Briggs KK, Rodrigo JJ. The microfracture technic in the management of complete cartilage defects in the knee joint [in German]. Orthopade 1999;28(1):26–32
7. Alfredson H, Lorentzon R. Superior results with continuous passive motion compared to active motion after periosteal transplantation. A retrospective study of human patella cartilage defect treatment. Knee Surg Sports Traumatol Arthrosc 1999;7(4):232–238
8. Bouwmeester SJ, Beckers JM, Kuijer R, van der Linden AJ, Bulstra SK. Long-term results of rib perichondrial grafts for repair of cartilage defects in the human knee. Int Orthop 1997;21(5):313–317
9. Hangody L, Kish G, Kárpáti Z, Udvarhelyi I, Szigeti I, Bély M. Mosaicplasty for the treatment of articular cartilage defects: application in clinical practice. Orthopedics 1998;21(7):751–756
10. Minas T. Autologous chondrocyte implantation for focal chondral defects of the knee. Clin Orthop Relat Res 2001;391(Suppl):S349–S361
11. Salai M, Ganel A, Horoszowski H. Fresh osteochondral allografts at the knee joint: good functional results in a follow-up study of more than 15 years. Arch Orthop Trauma Surg 1997;116(6-7):423–425
12. Bahuaud J, Maitrot RC, Bouvet R, et al. Implantation of autologous chondrocytes for cartilagenous lesions in young patients. A study of 24 cases [in French]. Chirurgie 1998;123(6):568–571
13. Erggelet C, Steinwachs MR, Reichelt A. The operative treatment of full thickness cartilage defects in the knee joint with autologous chondrocyte transplantation. Saudi Med J 2000;21(8): 715–721
14. Micheli LJ, Browne JE, Erggelet C, et al. Autologous chondrocyte implantation of the knee: multicenter experience and minimum 3-year follow-up. Clin J Sport Med 2001;11(4):223–228
15. McConnell S, Kolopack P, Davis AM. The Western Ontario and McMaster Universities Osteoarthritis Index (WOMAC): a review of its utility and measurement properties. Arthritis Rheum 2001; 45(5):453–461
16. Roos EM, Roos HP, Lohmander LS, Ekdahl C, Beynnon BD. Knee Injury and Osteoarthritis Outcome Score (KOOS)—development of a self-administered outcome measure. J Orthop Sports Phys Ther 1998;28(2):88–96
17. Sackett DL, Rosenberg WM, Gray JA, Haynes RB, Richardson WS. Evidence based medicine: what it is and what it isn't. BMJ 1996;312(7023):71–72
18. Gikas PD, Morris T, Carrington R, Skinner J, Bentley G, Briggs T. A correlation between the timing of biopsy after autologous chondrocyte implantation and the histological appearance. J Bone Joint Surg Br 2009;91(9):1172–1177
19. Tins BJ, McCall IW, Takahashi T, et al. Autologous chondrocyte implantation in knee joint: MR imaging and histologic features at 1-year follow-up. Radiology 2005;234(2):501–508
20. Hoemann C, Kandel R, Roberts S, et al. International Cartilage Repair Society (ICRS) recommended guidelines for histological endpoints for cartilage repair studies in animal models and clinical trials. Cartilage 2011;2:153–172
21. US Food and Drug Administration. Guidance for industry. Preparation of IDEs and INDs for products intended to repair or replace knee cartilage. http://www.fda.gov/downloads/biologicsbloodvaccines/guidancecomplianceregulatory. Accessed December 2011
22. Greco NJ, Anderson AF, Mann BJ, et al. Responsiveness of the International Knee Documentation Committee Subjective Knee Form in comparison to the Western Ontario and McMaster Universities Osteoarthritis Index, modified Cincinnati Knee Rating System, and Short Form 36 in patients with focal articular cartilage defects. Am J Sports Med 2010;38(5):891–902
23. Hambly K, Griva K. IKDC or KOOS? Which measures symptoms and disabilities most important to postoperative articular cartilage repair patients? Am J Sports Med 2008;36(9):1695–1704

24. Bekkers JE, de Windt TS, Raijmakers NJ, Dhert WJ, Saris DB. Validation of the Knee Injury and Osteoarthritis Outcome Score (KOOS) for the treatment of focal cartilage lesions. Osteoarthritis Cartilage 2009;17(11):1434–1439
25. Nunnally JC. Psychometric Theory. 2nd ed. New York, NY: McGraw-Hill; 1978
26. Irrgang JJ, Anderson AF, Boland AL, et al. Development and validation of the international knee documentation committee subjective knee form. Am J Sports Med 2001;29(5):600–613
27. Garratt AM, Brealey S, Gillespie WJ; DAMASK Trial Team. Patient-assessed health instruments for the knee: a structured review. Rheumatology (Oxford) 2004;43(11):1414–1423
28. Anderson AF, Irrgang JJ, Kocher MS, Mann BJ, Harrast JJ; International Knee Documentation Committee. The International Knee Documentation Committee Subjective Knee Evaluation Form: normative data. Am J Sports Med 2006;34(1): 128–135
29. Engelhart L, Nelson L, Lewis S, et al. Validation of the knee injury and osteoarthritis outcome score subscales for patients with articular cartilage lesions of the knee. Am J Sports Med 2012; 40(10):2264–2272
30. Smith HJ, Richardson JB, Tennant A. Modification and validation of the Lysholm Knee Scale to assess articular cartilage damage. Osteoarthritis Cartilage 2009;17(1):53–58
31. Tegner Y, Lysholm J. Rating systems in the evaluation of knee ligament injuries. Clin Orthop Relat Res 1985;198(198):43–49
32. Briggs KK, Lysholm J, Tegner Y, Rodkey WG, Kocher MS, Steadman JR. The reliability, validity, and responsiveness of the Lysholm score and Tegner activity scale for anterior cruciate ligament injuries of the knee: 25 years later. Am J Sports Med 2009;37(5):890–897
33. Krych AJ, Harnly HW, Rodeo SA, Williams RJ III. Activity levels are higher after osteochondral autograft transfer mosaicplasty than after microfracture for articular cartilage defects of the knee: a retrospective comparative study. J Bone Joint Surg Am 2012;94(11):971–978
34. Bekkers JE, de Windt ThS, Brittberg M, Saris DBF. Cartilage repair in football (Soccer) athletes: what evidence leads to which treatment? A critical review of the literature. Cartilage 2012;3(4): 43S–49S
35. Flanigan DC, Harris JD, Trinh TQ, Siston RA, Brophy RH. Prevalence of chondral defects in athletes' knees: a systematic review. Med Sci Sports Exerc 2010;42(10):1795–1801
36. Mithöfer K, Peterson L, Mandelbaum BR, Minas T. Articular cartilage repair in soccer players with autologous chondrocyte transplantation: functional outcome and return to competition. Am J Sports Med 2005;33(11):1639–1646
37. Kreuz PC, Steinwachs M, Erggelet C, et al. Importance of sports in cartilage regeneration after autologous chondrocyte implantation: a prospective study with a 3-year follow-up. Am J Sports Med 2007;35(8):1261–1268
38. Marcacci M, Kon E, Delcogliano M, Filardo G, Busacca M, Zaffagnini S. Arthroscopic autologous osteochondral grafting for cartilage defects of the knee: prospective study results at a minimum 7-year follow-up. Am J Sports Med 2007; 35(12):2014–2021
39. Steadman JR, Miller BS, Karas SG, Schlegel TF, Briggs KK, Hawkins RJ. The microfracture technique in the treatment of full-thickness chondral lesions of the knee in National Football League players. J Knee Surg 2003;16(2):83–86
40. Gudas R, Kalesinskas RJ, Kimtys V, et al. A prospective randomized clinical study of mosaic osteochondral autologous transplantation versus microfracture for the treatment of osteochondral defects in the knee joint in young athletes. Arthroscopy 2005;21(9):1066–1075
41. Krych AJ, Robertson CM, Williams RJ III; Cartilage Study Group. Return to athletic activity after osteochondral allograft transplantation in the knee. Am J Sports Med 2012;40(5):1053–1059
42. Mithoefer K, Hambly K, Della Villa S, Silvers H, Mandelbaum BR. Return to sports participation after articular cartilage repair in the knee: scientific evidence. Am J Sports Med 2009;37(Suppl 1): 167S–176S
43. Bedi A, Feeley BT, Williams RJ III. Management of articular cartilage defects of the knee. J Bone Joint Surg Am 2010;92(4):994–1009
44. Strauss E, Barker J, Kercher J, Cole BJ, Mithoefer K.. Augmentation strategies following the microfracture technique for repair of focal chondral defects. Cartilage 2010;1:145–152
45. Mithoefer K, Williams RJ III, Warren RF, Wickiewicz TL, Marx RG. High-impact athletics after knee articular cartilage repair: a prospective evaluation of the microfracture technique. Am J Sports Med 2006;34(9):1413–1418
46. Brophy RH, Zeltser D, Wright RW, Flanigan D. Anterior cruciate ligament reconstruction and concomitant articular cartilage injury: incidence and treatment. Arthroscopy 2010;26(1):112–120
47. Joseph C, Pathak SS, Aravinda M, Rajan D. Is ACL reconstruction only for athletes? A study of the incidence of meniscal and cartilage injuries in an ACL-deficient athlete and non-athlete population: an Indian experience. Int Orthop 2008;32(1): 57–61
48. Brittberg M, Winalski CS. Evaluation of cartilage injuries and repair. J Bone Joint Surg Am 2003; 85-A(Suppl 2):58–69
49. Saris DBM, Vanlauwe J, Victor J, et al; TIG/ACT/01/2000&EXT Study Group. Treatment of symptomatic cartilage defects of the knee: characterized chondrocyte implantation results in better clinical outcome at 36 months in a randomized trial compared to microfracture. Am J Sports Med 2009;37(Suppl 1):10S–19S
50. van den Borne MP, Raijmakers NJ, Vanlauwe J, et al; International Cartilage Repair Society. International Cartilage Repair Society (ICRS) and Oswestry macroscopic cartilage evaluation scores validated for use in Autologous Chondrocyte Implantation (ACI) and microfracture. Osteoarthritis Cartilage 2007;15(12):1397–1402
51. Nehrer S, Spector M, Minas T. Histologic analysis of tissue after failed cartilage repair procedures. Clin Orthop Relat Res 1999;365(365):149–162
52. Knutsen G, Engebretsen L, Ludvigsen TC, et al. Autologous chondrocyte implantation compared with microfracture in the knee. A randomized trial. J Bone Joint Surg Am 2004;86-A(3):455–464

53. Lane JM, Brighton CT, Ottens HR, Lipton M. Joint resurfacing in the rabbit using an autologous osteochondral graft. J Bone Joint Surg Am 1977; 59(2):218–222

54. Fickert S, Gerwien P, Helmert B, et al. One-year clinical and radiological results of a prospective, investigator-initiated trial examining a novel, purely autologous 3-dimensional autologous chondrocyte transplantation product in the knee. Cartilage 2012;3:27–42

55. Marlovits S, Singer P, Zeller P, Mandl I, Haller J, Trattnig S. Magnetic resonance observation of cartilage repair tissue (MOCART) for the evaluation of autologous chondrocyte transplantation: determination of interobserver variability and correlation to clinical outcome after 2 years. Eur J Radiol 2006;57(1):16–23

56. Moradi B, Schönit E, Nierhoff C, et al. First-generation autologous chondrocyte implantation in patients with cartilage defects of the knee: 7 to 14 years' clinical and magnetic resonance imaging follow-up evaluation. Arthroscopy 2012;28(12): 1851–1861

57. Mithoefer K, Williams RJ III, Warren RF, et al. The microfracture technique for the treatment of articular cartilage lesions in the knee. A prospective cohort study. J Bone Joint Surg Am 2005; 87(9):1911–1920

58. Dodin P, Pelletier JP, Martel-Pelletier J, Abram F. Automatic Human Knee Cartilage Segmentation from 3D Magnetic Resonance Images. IEEE Trans Biomed Eng 2010;6:15–22

59. Trattning S, Winalski C, Marlovits S, et al. Magnetic resonance imaging of cartilage repair: a review. Cartilage 2011;2:5–26

60. Vasiliadis HS, Danielson B, Ljungberg M, McKeon B, Lindahl A, Peterson L. Autologous chondrocyte implantation in cartilage lesions of the knee: long-term evaluation with magnetic resonance imaging and delayed gadolinium-enhanced magnetic resonance imaging technique. Am J Sports Med 2010;38(5):943–949

Index

Note: Page numbers followed by *f* and *t* indicate figures and tables, respectively.

B

P

Q

R

S

T

Index